SPACE STATIONS AND SPACE PLATFORMS— Concepts, Design, Infrastructure, and Uses

Edited by
Ivan Bekey
NASA Headquarters
Washington, D.C.

Daniel Herman
NASA Headquarters
Washington, D.C.

Volume 99
PROGRESS IN
ASTRONAUTICS AND AERONAUTICS

Martin Summerfield, Series Editor-in-Chief
Princeton Combustion Research Laboratories, Inc.
Monmouth Junction, New Jersey

Published by the American Institute of Aeronautics and Astronautics, Inc.
1633 Broadway, New York, NY 10019

American Institute of Aeronautics and Astronautics, Inc.
New York, New York

Library of Congress Cataloging in Publication Data
Main entry under title:

Space stations and space platforms—concepts, design, infrastructure, and uses.

(Progress in astronautics and aeronautics; v. 99)
Includes index.
1. Space stations. I. Bekey, Ivan. II. Herman, Daniel, 1927– .
III. American Institute of Aeronautics and Astronautics. IV. Series.
TL507.P75 vol. 99 629.1s[629.44'2] 85-19972
[TL797]
ISBN 0-930403-01-0

Table of Contents

Foreword

It is the nature of man to strive to improve the quality of his life, to learn more about himself and the universe in which he lives, and to accept the challenge of the unknown. It has only been during the 20th century, however, that dreamers and scientists alike have seriously contemplated the possibility that the struggle to learn, work, and observe might well surpass the confines of the surface and atmosphere of the Earth and move into the space beyond.

Although Everett Hale wrote, in 1869, of a satellite made of brick on which men circled the Earth, he gave a very limited explanation of either how or why it should be done. Soon after the turn of the century, however, Tsiolkovsky wrote of such a spacecraft and postulated engineering solutions for power, life support systems, and overall structure. In 1918, Robert Goddard visualized a spacecraft capable of carrying civilization to distant stars. In 1923, Herman Oberth suggested that manned spacecraft could be useful for communications and refueling stations, and apparently for the first time, gave a name to such spacecraft: space stations. Other writings speculated on design concepts and potential uses for space stations, but it was not until 1952, when Wernher von Braun wrote his article entitled "Crossing the Last Frontier," that the concept of a manned space station captured public attention.

The story of space activities which has followed is well documented. The first satellite; the first manned spacecraft; the first landing on the Moon; the first planetary mission; the first landing on Mars; the Soviets' first launch of a space station—Salyut 1—on April 19, 1971; the American launch of its first space station—Skylab—on May 14, 1973; the first commercial communications satellite; the first in-orbit satellite repair; the first commercial experiments in microgravity: all have been notable steps in man's push toward more effective use of space to advance his knowledge, improve his well-being, and satisfy his drive to explore the unknown.

Today, the United States is developing a true Space Station, a "permanent presence in space." Such a manned spacecraft will serve as a scientific laboratory, a maintenance and repair depot for other spacecraft, a construction base from which to build and assemble large systems beyond the capabilities of individual launch vehicles, and a site from which it will be possible to launch those large systems into higher orbits or beyond. This Space Station, and the unmanned space platforms being developed to operate with it and be serviced from it, will constitute the space facilities with which we can expand our knowledge of man's utility in space and of the systems which will enhance those attributes.

It is no more possible today to predict with certainty the specific directions in which our quest in space will take us during the next 25 years than it would have been 25 years ago. However, it does seem clear that as we continue to learn we will continue to form new questions, each success will bring new opportunities for even greater accomplishments, and each discovery will heighten the human urge to explore the unknown.

The Space Station and its associated platforms will serve as the stepping stone from which this program will move forward. They will serve the scientist, the engineer, the explorer, and the industrialist. They will serve this Nation and the nations of the world.

This book outlines the history of the quest for a permanent habitat in space; describes our present thinking of the relationship between the Space Station, space platforms, and the overall space program; and outlines a number of resultant possibilities about the future of the space program. Therefore, this book is but one chapter in the quest for the effective use of man's last frontier. It is our intention, in putting forth this book, to outline the design concepts that prevail at this time as a means of stimulating innovative thinking about space stations and their utilization on the part of scientists, engineers, and most important for the long future ahead of us, the students of today who will soon replace us.

Philip E. Culbertson
Associate Administrator
Office of Space Station
NASA Headquarters

Chapter I. Introduction: The Space Infrastructure

Introduction: The Space Infrastructure

Ivan Bekey*
NASA Headquarters, Washington, D.C.

Introduction

This book focuses on the Space Station and its associated platforms, which are the central and most visible space facilities in our thrust toward establishing a permanent presence in space. Facilities cannot be operated in isolation, however, and must have transportation for access. This is particularly true for space facilities. In order to attain the goal of permanent presence in space, the space station and its low altitude platforms must be developed and operated with transportation for routine access to and from the Earth; local transportation between the Space Station, its platforms, and other low altitude satellites; and long-range transportation between low and geostationary orbits, and escape orbits. In addition, geostationary facilities and habitation elements must also be developed.

It is in such an assembly of space facilities and transportation elements that will be vested the capabilities to service the Earth, establish bases on the Moon or Planets, and eventually move beyond.

Historical Perspective

The space age was ushered in by hurling ever more sophisticated expendable spacecraft into space on top of expendable launch vehicles. Marvels of engineering, these satellites were designed and fabricated to function without flaw and unattended for years, accessible only by occasional feeble radio signals. The rocket boosters balancing these spacecraft on their fiery tails grew in dependability and

*Director, Advanced Programs, Office of Space Flight.

capability but were unable to return the expensive spacecraft if trouble developed.

While this trend continued, a parallel development was occurring-manned spaceflight. Beginning by contrast with crude and simple containers, the manned spacecraft were designed with redundant systems which could be brought into play by the astronaut, enormously increasing the flexibility and capability of the man-machine combination. Further, at least the "front end" was now recoverable, and with the man, brought back vital observations and data. But perhaps most importantly, some failures could be fixed on the spot which would otherwise have been aborts, and failed components turned to astounding success stories. The culmination of this movement were, of course, the Apollo program Moon landings and excursions--an engineering tour-de-force carrying a large science program on its broad coattails.

However spectacular each of these achievements was, in the post-Apollo period it was recognized that each mission was a massive and costly undertaking, and that a more routine access and a permanent infrastructure had to be developed if space was to be more than front-page heroics. That we had to learn to exist for prolonged periods in space, learn to build spacecraft to be maintained, and learn to access space and return inexpensively. An experimental space station was orbited--the Skylab, and plans were laid for the two cornerstones of permanent presence in space: a permanent space station and an airplane-like Shuttle transportation system to provide routine access to it by crews and cargo. These were recognized as required to vanquish the factors always limiting space achievements: limited staytime and affordable routine access.

But the bulk of scientific and commercial spacecraft were and still are designed as expendable vehicles. Nonetheless, there are winds of rapid change, as the three R's of space -- retrieval, repair, and refueling of unmanned satellites have just recently been demonstrated vividly by the Shuttle program. Even though the subject satellites were not designed to accommodate such services. And thus the vision of a more permanent infrastructure began to take form, with the team of Space Shuttle/Space Station at its core.

The Present

The Space Shuttle is now essentially operational and the Space Station is an approved project destined to fly within a decade. While most satellites are still designed for long life unattended on-orbit, there is a notable trend, starting with the Hubble Space Telescope, to begin to design for space maintainability. While Earth-to-low orbit transportation is now reusable, all transportation to go higher is still expendable. The cost per pound of payload orbited is not greatly lower that it was ten years ago, but the complexity achievable in space operations is already an order of magnitude greater than before.

A vision of a new way of operating in space is emerging, with a number of highly desirable attributes. These include the ability to assemble larger and more complicated spacecraft from two or more launches; the ability to checkout a spacecraft in space prior to its being committed to transfer to a higher orbit or to operate by itself in space; the ability to modify and to repair or even complete a design in space that has either been improper or incomplete; the ability to repair satellites once they have failed; the ability to refuel satellites whose expendables have been expended in either a planned or unplanned fashion; the ability to reconfigure assets on orbit to change their utility and make their capability match a changing set of requirements. To these must be added: the ability to fuel large vehicles from a permanent space base and carry out launch operations there; the ability to carry propellants to orbit "on the margin" for such operations; the ability to conduct man--intensive and extravehicular-activity-intensive operations in space that require high power and long duration, whether they be commercial adventures, highly complex scientific missions or unexpected repairs; the ability to transfer payloads to geostationary or other high orbits at very low thrust in order to avoid deforming flimsy large deployed structures; and the ability to extend man's reach physically and by remote control far beyond his immediate surroundings.

All of these attributes describe a new way of operating in space. This will require not only the Space Station and the Shuttle, but due to the need for a new degree of mobility, access, staytime, and support at various altitudes, will require various other elements of the space infrastructure.

The Future

The space infrastructure envisioned to fill the needs discussed above by about the year 2000 is illustrated in Fig. 1. The center area of the figure is the centerpiece being based upon the Space Station and elements permanently in low-inclination orbits; the left side shows Shuttle-based low-inclination launching and space operations, which in this time period will be in a transitional phase as we learn to operate with the permanent elements. The right side of the figure shows Shuttle-based operations for near-polar

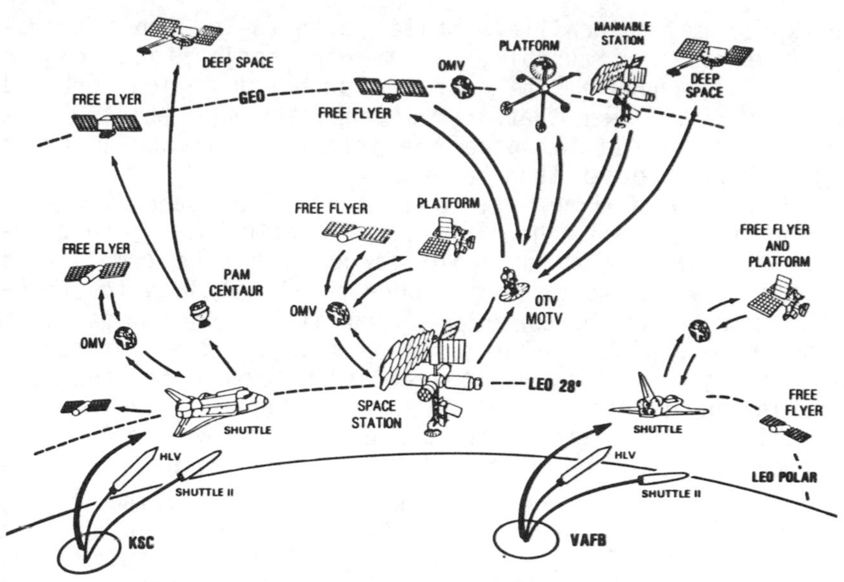

Fig. 1 Elements of space infrastructure.

orbits operating from Vandenberg, since there is no space station currently planned for such inclinations, though a platform will be there. This situation may change, since evolution of the Space Station could take the form of replication in polar orbit, as well as the increase in capability in low orbit that is currently envisioned.

The primary launch vehicle will still be the Space Shuttle. In the future, this may well be augmented by unmanned cargo vehicles if the need for carrying propellants or cargo to the space station warrant it, and a more optimized people carrier may emerge as a second generation Shuttle. Not shown are expendable launch vehicles, which if they will exist in this time frame would be operated entirely by commercial concerns and would probably carry small commercial payloads such as communication or microgravity processing satellites. It is more than likely, however, that the economies and capabilities afforded by the anticipated permanent elements will drive these from the market by the year 2000.

One of the main features of the infrastructure is the local transportation that will be required between the Space Station and its outlying cooperating elements. The Space Station will be far less mobile than the Shuttle, which has limited mobility itself, and thus will not be able to maneuver at all to access free-flying satellites and platforms. Since these are envisioned fundamentally as aggregations of payloads on free-flying spacecraft so designed that service can be provided by the space station or Shuttle, a local transportation vehicle known as the Orbital Maneuvering Vehicle (OMV) will be needed to maneuver between the Space Station, the Orbiter, the platforms, and the free-flyers. This OMV is envisioned as a sort of "harbor tug," which would provide not only deployment and retrieval of such satellites but also the capture and disposal of orbital debris that might otherwise provide a long-term hazard. In addition, when fitted with proper manipulation aids and tankage, could perform some services remotely on platforms or spacecraft without bringing them back to the Space Station or Shuttle. It could thus provide an option for repair and replenishment of satellites which are too hazardous to approach by the Shuttle or to bring to the Space Station.

The unmanned platforms themselves will be vital to the infrastructure. They will house instruments and experiments that by their nature must be isolated from the vibration and contamination levels expected on a manned space station, but which could benefit from manned attendance and servicing. They also provide economy of scale by providing shared resources to the payloads.

Vital elements of the infrastructure will be located in geostationary orbit and eventually accessed primarily from the Space Station, though from the Shuttle initially. These elements will begin as expendable high value satellites, grow into spacecraft that are designed to be serviced or replenished in geostationary orbit, and evolve into geostationary platforms each of which aggregates many communications and observation payloads on a common support bus. These platforms would certainly be serviceable and maintainable on orbit, and have exchangeable front ends to allow their upgrading in orbit. In order to permit the servicing, upgrading and exchange components in geostationary orbit unmanned, remotely controlled, and manned systems will be needed to bring the benefits of attended operations to GEO. Therefore, the geostationary elements of the infrastructure will have to include temporary housing for crews sent on repair missions for short periods of time.

Geostationary assets would be accessed from the Space Station via a reusable two-way long-range transportation system known as the Orbit Transfer Vehicle (OTV). This vehicle is envisioned as based on the Space Station and refueled, repaired, refurbished, and mated with payloads there. It would have the capacity to carry payloads for delivery to geostationary orbit and return to the Space Station; or to bring other payloads back. It would also have the ability to carry crews on round trips to geostationary orbit for servicing missions. Such an OTV will be a key element of the infrastructure. It will probably be modular and thus assemblies of the OTV would also have the capability for launching payloads into trajectories for Solar System exploration and eventually for manned missions to the Moon and the planets.

The functions of all of these elements will be complementary, with the sum of the parts comprising an integrated infrastructure capable of supporting current goals and growing to support larger missions in the future. The Space Station certainly has the key role, as do the platforms, but their functions would not be achievable in isolation without the other facilities and transportation elements. Thus while the remaining chapters of this book detail the development of the concept, configuration, and utility of the Space Station and its associated platforms, they must be viewed within the framework of the larger infrastructure of which they are, of necessity, key but not sole members.

Chapter II. Early Space Station and Platform Planning

Skylab

John H. Disher*
Bethesda, Maryland

Introduction

Skylab was the United States' (and the world's) first manned space station. It was conceived as a large multidisciplinary science and engineering laboratory in space, with extensive participation by its three-man crew in carrying out the planned experiment program, acting as subjects for the medical experiments, and performing certain preplanned maintenance tasks. As it turned out, Skylab successfully accomplished or exceeded all the planned objectives, but, more importantly, through totally unforeseen events, gave new direction and emphasis to the human role in space. This chapter will summarize these Skylab findings as they relate to future space stations.[1]

Skylab Description

First, a few particulars regarding Skylab (Fig. 1): The main module or element of Skylab was called the "workshop" and provided a large habitable volume some 7 m in diameter and 20 m long. Auxiliary modules housed certain of the support systems and experiment hardware. These were called the telescope module, docking adapter, and airlock, respectively. This assembly was launched by a two-stage Saturn V rocket. The three-man crews were launched and returned to Earth separately in Apollo Command Modules that docked with the workshop, which had previously been placed in orbit. Three separate manned missions were flown, for periods of 28, 59, and 84 days, respectively-giving a to-

---

*Consultant.

tal of 513 man days in orbit during the program. The program was completed on Feb. 8, 1974-some eleven years ago. The results from Skylab were, of course, timely and useful in the design of the Space Shuttle and Spacelab. However, the findings of Skylab are of even greater importance to future space station design.

In Fig. 2, Skylab is shown to the same scale as Salyut and Spacelab, the other related spacecraft that have flown. The pressurized and inhabitable volumes of each are shown. Skylab was substantially larger than the other spacecraft shown, and that larger size afforded room for the extensive experimental and medical monitoring equipment carried, as well as for free-flying zero-gravity experiments within the safety of a pressurized environment (Fig. 3.). The specific characteristics of the several spacecraft are tabulated in Fig. 4.

Skylab Subsystems Design

The configuration of the Skylab subsystems resulted from an interesting mixture of influences, combining a strong desire to keep development costs low by using existing available equipment and technology whenever it would do the job, together with a strong motiviation to make the Skylab missions as productive as possible in ev-

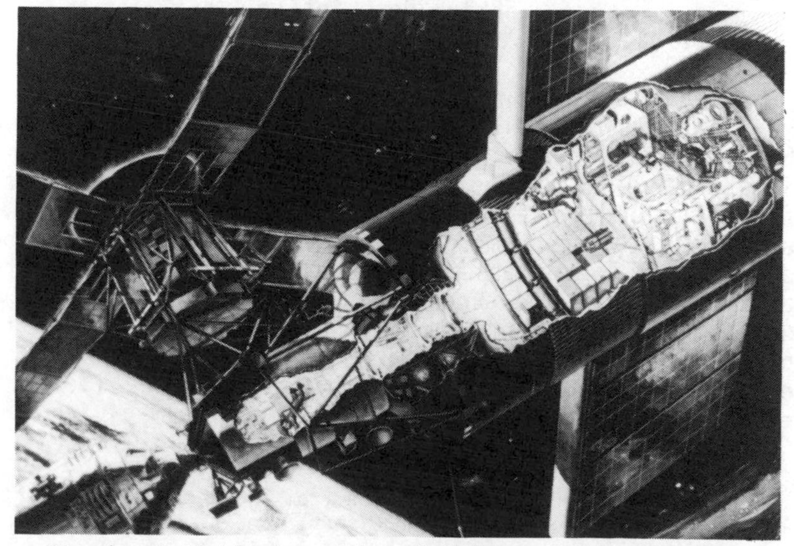

Fig. 1 Artist's impression of Skylab assembly in orbit.

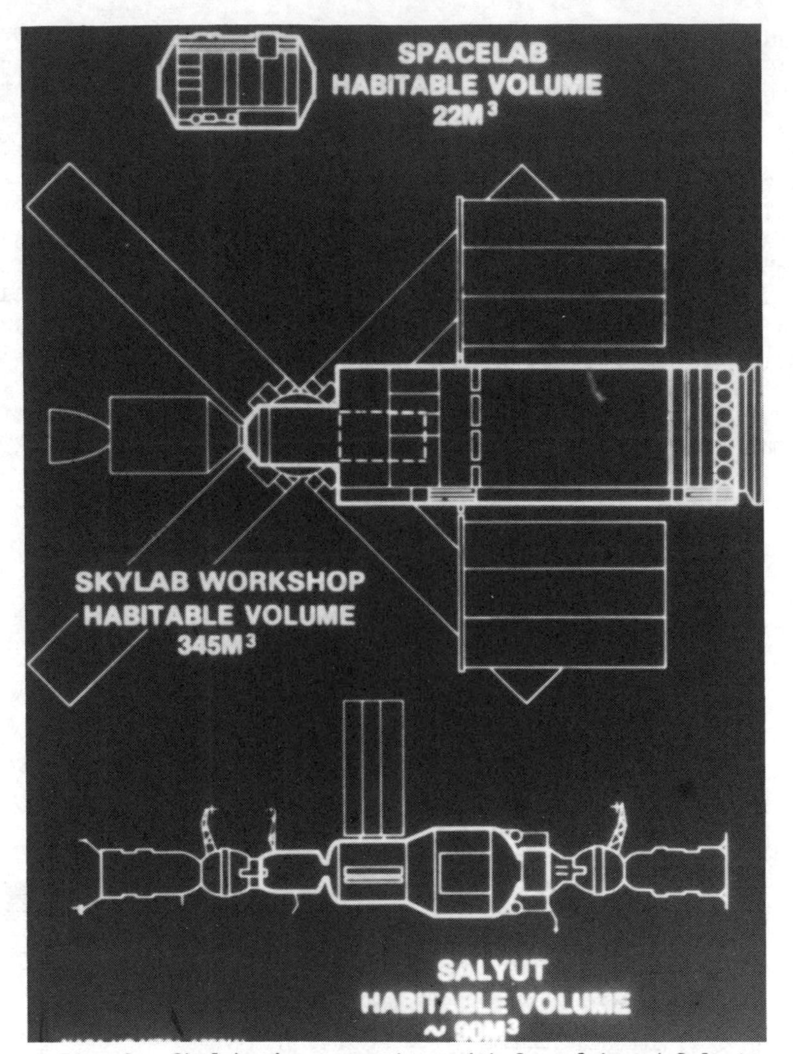

Fig. 2 Skylab size comparison with Spacelab and Salyut

ery area. These factors tended to discourage the introduction of new device technology, while encouraging ingenuity in devising novel operating systems based on the use of existing technology.

At the same time, a number of devices were used that has not seen prior operational use in manned spacecraft. In 1965, a proposal to fly a large control moment gyro (CMG) as a technology experiment in Skylab was withdrawn because it was felt that its development has brought it to readiness for operational use. A year later, CMGs were baselined as the primary operational system for Skylab attitude control. Similarly, the use of molecular sieves for CO_2 removal entered Skylab as an experiment in support of the Air Force Manned Orbital Laboratory (MOL) Program and ended up as the mainline system. On the other hand, it was decided not to attempt to extend cryogenic storage lifetimes, but to accept the weight penalty associated with high-pressure storage of gases for the Skylab atmosphere.

The airlock module (AM) design was based on extensive use of space components available after the completion of the Gemini program, and thus the Skylab had a Gemini hatch for astronaut egress from the airlock module.

General characteristics of the Skylab modules were summarized at the beginning of this chapter. The following

Fig. 3 Astronaut Lousma "free-flying" inside Skylab "workshop".

	SKYLAB	SHUTTLE-SPACELAB	SALYUT
WEIGHT, KG	90,000	4,500	19,000
PRESS. VOLUME, M^3	295-350	8-22	~90
ELEC POWER, KW	25	3-5	3
CREW	3	4	2-3

Fig. 4 Some characteristics of Skylab, Spacelab and Salyut.

sections describe some of the design and operating features of the major subsystems.

Life Support System

Skylab's life support system was an open-cycle system in that none of the expendables were reclaimed for reuse. All consumables, such as oxygen, water, and food, were brought up on the first mission, and no provisions for resupply were made. Skylab's atmosphere is summarized in Table 1. The oxygen was stored in six 45-in.-diam 90-in.-long cylindrical tanks at 3000 psi; the tanks were mounted external to the airlock on the fixed airlock shroud (FAS) and used fiberglass wrapped over a thin steel liner to minimize the hazard in the event of a tank failure. The nitrogen was stored in six 24-in.-diam titanium tanks at 3000 psi, mounted on the external airlock module trusses. The tanks were designed with a 4500-psi capability and could be loaded to support a longer mission.

The gases were regulated from the depleting supply pressure to the required pressure and flow rates. Nitrogen was introduced when the total pressure dropped below 4.8 psi, and oxygen was introduced when the O_2 partial pressure fell below 3.7 psi.

Table 1 Skylab atmosphere and consumables

Atmospheric Composition:	
Total pressure (±0.2)	5 psia
Partial pressure	
O_2 (74% by wt)	3.7 psia
N_2 (26% by wt)	1.3 psia
Humidity	8 mm H_g min
CO_2	5.5 mm H_g max
Air Velocity, ft/min	40
"Comfort Box" temperature, F	
Air	55-90
Wall	93-56
Consumables, lb	
Water	6000
Food	1470
Oxygen	4930
Nitrogen	1320

The life support system also supported extravehicular activity (EVA) operation and provided pressurant for various experiments and the water tanks through suitable valves, regulators, and quick disconnects.

Carbon dioxide and odor removal were accomplished by means of a regenerative system that passed the atmospheric gases through a molecular sieve system. This type of system avoided the weight and volume penalties of a lithium hydroxide system and minimized direct crew involvement in system operation.

Two molecular sieves were located in the structural transition section of the airlock module. One of the molecular sieves was on-line continuously during manned occupancy, while the second provided redundancy. Each molecular sieve contained two sorbent canisters and a charcoal canister. Water and carbon dioxide were removed by the sorbent canisters, odor by the charcoal canister. During normal operation, each sorbent canister operated on a 15-min half-cycle to alternately absorb CO_2 and H_2O from the 15-lb/h gas flow and desorb to vacuum. The canisters were of the adiabatic desorption type containing Linde-type 5A zeolite for CO_2 absorption and Linde 13X zeolite for water absorption.

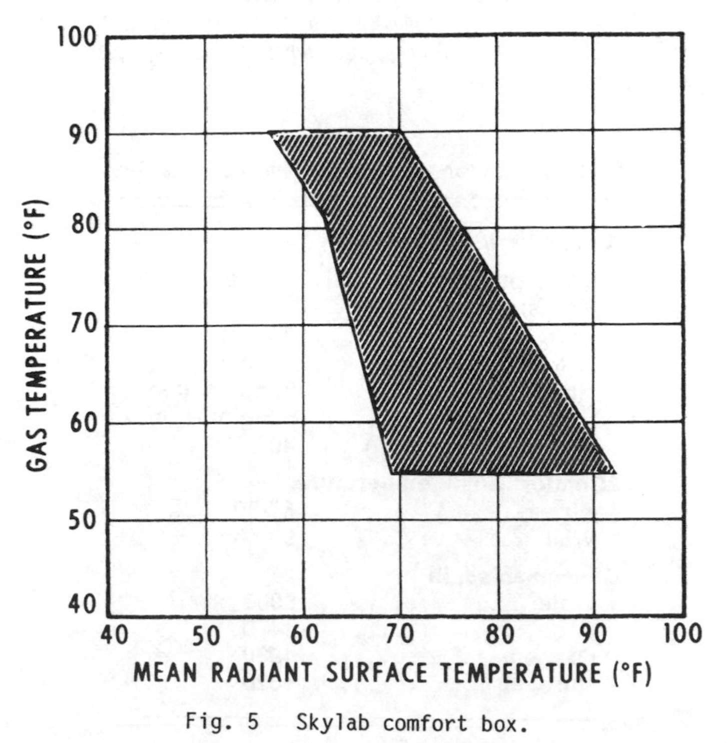

Fig. 5 Skylab comfort box.

Skylab's water system provided water for drinking, food reconstitution, and personal hygiene. The water was stored internally near the top of the workshop in ten cylindrical tanks with a total usable capacity of 6000 lb. Metal expulsion bellows were pressurized by 120 psi nitrogen. An iodine additive was provided to inhibit bacterial growth.

Almost 1500 lb of food was carried in Skylab in individual metal containers, each pressurized to 5 psi. Considerable attention was directed toward food variety. The categories of food included beverages (reconstituted with water), rehydratable foods, thermostabilized foods, frozen foods, and wafer foods. An additional design constraint imposed on the food containers was the capability to withstand the launch pressurization of 26 psi within the Orbital workshop (OWS).

Thermal and Environmental Control System

Skylab's atmospheric control system could be classified as a semiactive system, designed to provide a satisfactory environment through a varying solar exposure and mission profile. In a typical mission, the variations in sun angle relative to the orbital plane (0-73) changed the sunlit part of the orbit from 50% to 100%, with consequent large variations in thermal input. In addition, the experiment and crew activities within Skylab caused significant variations in the amount of internal heat generated. A "comfort box" (Fig. 5), which relates crew comfort to varying combinations of wall temperature and gas temperature, was used as the principal design criterion in the Skylab thermal control system.

Thermally, Skylab resembled a huge thermos bottle. By the use of external paint patterns, a meteoroid shield, which also was planned to serve as a radiation barrier, and various insulation techniques, the internal variations due to solar inputs were to be minimized. Nevertheless, for the coldest parts of the mission, heater power was to provide for crew comfort, and, during the hottest parts of the mission, active heat rejection was to be accomplished by gas interchange between the workshop and the airlock through a bank of heat exchangers in the airlock. The entire system was thermostatically controlled, with manual override available.

Humidity control was provided by means of condensing heat exchangers within the airlock module that processed the gas prior to passage through the molecular sieves. The amount of moisture was controlled to provide a 46 F minimum dew point. The total water contained within the atmospheric gases was generated by the crew's metabolic activities, and no provision was necessary for initial or supplemental humidification.

Most of the active components for systems operations were located within the airlock module and were coldplate conditioned by a dual coolant loop that rejected heat by means of a bifilar radiator. The radiator cylindrically enclosed the multiple docking adaptor (MDA) and the AM's structural transition section and also served as a meteoroid shield. As designed, this radiator was capable of rejecting 16,000 btu/h. The airlock's coolant loops also supplied coolant to the condensing heat exchangers, the astronautical telescope module (ATM) and control and display (C&D) panels, Earth resources experiment packages (EREPs), EVA suits, and provided for workshop cabin heat rejection under high heat conditions. Many of the components utilized in the airlock's coolant loops were available as unused spares at the end of the Gemini program.

Electrical Power System

Skylab had the largest solar electrical power system yet put in space, consisting of two paralleled, but quasi-independent, solar array/chargeable battery systems; one was located on the AM/OWS and the other on the ATM. Each solar array was approximately 1200 ft^2 in area, with the total solar array area comparable to the area of a tennis court. The system characteristics are summarized in Table 2.

The OWS solar array system consisted of two wings of solar panels deployed from beam fairings hinged from the forward skirt of the Saturn IVB stage. The hinges that supported the beam fairings were skewed so that the beam fairings were deployed parallel to the Y axis. Each beam fairing was 4 ft wide by 37 ft long and was attached to the stage by six explosive fittings during launch. Each wing assembly consisted of three wing sections, each of which comprised ten foldable solar panels, a dummy solar cell panel, a truss-type panel, and two parallel stabilizer beams. All the panels were hinged together and were folded accordian-style on one another into the beam fairing. Each panel was 120 in. by 27 in. and was connected to the stabilizer beams with a swivel fitting. Each of the parallel stabilizer beams incorporated spring locks between each segment that locked when the wing section was fully deployed.

The ATM solar array consisted of four wings, each deployed by scissors mechanisms. The wings were attached to the Sun end of the rack and spanned approximately 100 ft.

The power system consisted of eight power conditioning groups (PCGs). Each PCG contained a 33 AH nickel cadmium secondary battery, a battery charger, and voltage regulator. There were two regulated power distribution buses in the AM, each powered by four PCGs. Each regulated bus distributed power to load buses in the AM, OWS, and MDA and was connected to an intermodule power transfer bus. A protected bus tie between the two regulated buses served to balance the power output from each half of the AM system.

The ATM system was similar in principle to the AM/OWS system. The ATM array supplied 18 charger battery regulator modules (CBRMs). The battery was a 20 AH nickel cadmi-

Table 2 Skylab electrical power system characteristics

SKYLAB ELECTRICAL POWER SYSTEM FUNCTION CHARACTERISTICS

GENERAL	
Total power	7530 watts
Bus voltage	25 to 30.5 volts dc
Bus noise	less than 1 volt p-p 20 Hz to 20 kHz
Bus transients	1s thn 50 volts 1s thn 10 microsec
Transfer between systems	2500 watts
Transfer to CSM	2000 watts
Two wire, multiple bus distribution Single point ground	
WORKSHOP SYSTEM	
Power	3814 watts
Array area	1200 ft²
Eight charger/battery units 30% depth of discharge	
ATM SYSTEM	
Power	3716 watts
Array area	1200 ft²
18 charger/battery units 30% depth of discharge	

um type. The output of each CBRM was paralleled through isolation diodes to redundant buses in the ATM power transfer distributor. These buses supplied individual ATM load buses and were connected to the intermodule power transfer buses in the AM. The interconnection between the AM and ATM power system provided the capability to transfer power in either direction.

After activation of the Skylab, the command and service modules (CSM) fuel cells were shut down and the main command module (CM) buses were connected to the intermodule transfer buses in the AM through a plug-in umbilical. Actual power transfer to the CSM was initiated by activation of power relays in the AM. The minimum continuous total power output capability of the two paralleled solar array/battery systems was specified as 7530 W for the solar inertial attitude. This figure applied after eight months in orbit and with the minimum possible orbital light-time to dark-time ratio. At this load, all batteries returned to full charge each orbit and would not discharge more than 30% of their rated capacity. The expected maximum orbit-average total load was less than 7000 W, while the mission average was expected to be less than 6000 W.

Attitude and Pointing Control

The attitude and pointing control system (APCS) provided three-axis attitude stabilization and maneuvering

Table 3 Skylab attitude control

CLUSTER ATTITUDE

Accuracy

Solar inertial (with solar reference)		Local vertical, deg: Earth resources	Local vertical, deg: rendezvous
Pitch	±6 Arc Min	±2	±6
Yaw	±6 Arc Min	±2	±6
Roll	±10 Arc Min	±2	±6

DEADBAND (TACS ONLY)

Attitude, deg ....±0.5
Rate, deg/sec ....±0.05

ATM POINTING

	Accuracy	Stability
Pitch	±2.5 Arc Sec	±2.5 Arc Sec (for 15 min)
Yaw	±2.5 Arc Sec	±2.5 Arc Sec (for 15 min)
Roll	±10 Arc Min	±7 Arc Min (for 15 min)

capability of the cluster throughout the mission, and pointing control of the Skylab experiment packages during experimentation periods. Functional characteristics are shown in Table 3.

Three attitude control subsystems interrelated through the ATM digital computer, with manned control via the ATM control and display unit in the MDA: 1) control moment gyro (CMG) control, the 2) thruster attitude control system (TACS), and 3) experiment pointing and control (EPC).

The combination of the TACS/CMG control subsystem constituted a "nested" control system. In most orbital operations, attitude control was accomplished by the CMG control. The TACS actuated only when CMG momentum buildup reached 95% of its capacity or when the rate and attitude deadbands were exceeded.

After orbit insertion, the thruster attitude control system oriented the attitude of the vehicle with the Z-axis aligned to within ±5 degrees of the center of the sun and the x-axis approximately in the orbit plane, i.e., the solar inertial attitude. The ATM solar panels were then deployed and CMG wheel spinup initiated.

Six control modes were addressable by the control and display console switches for APCS operation:

1) The solar inertial mode maintained the vehicle's minimum moment of the X-axis parallel to the Sun line. During orbit nighttime, this mode was used to perform gravity gradient momentum dump maneuvers for desaturating the CMGs.

2) The experiment pointing mode was used only during solar experimentation. The CMG/TACS nested control system stabilized the vehicle with the ATM experiment package coarse pointed at the Sun, and the EPC subsystem fine-pointed the experiment package.

3) The Z-local vertical (LV) mode was used during the rendezvous phase of the mission or when Earth pointing for experimentation periods was required. Normal vehicle control was effected under CMG/TACS nested configuration.

4) The attitude hold (CMG) mode allowed the vehicle to be maneuvered to any inertial-oriented attitude and held. The nested control system was used to control the vehicle.

5) The attitude hold (TACS) mode was used to maneuver the vehicle to any inertial-oriented attitude and held using the TACS only.

6) The standby mode allowed CSM control of the cluster attitude.

Table 4 Skylab external communications

Frequency, MHz	Power	Modulation	Bit rate	TRNAS RCV	Antenna Type	Antenna Location	Use
230.4	2w	FM/PCM	—	T	Whip		Launch Tm
230.4	10w	FM/PCM	51.2 kbps	T		Extended	Orbital Tm
235.0	10w	FM/PCM	112.64 or 126.72 depending on source selected	T	Discone	from Airlock	Voice, Tm Down data
246.3	10w	FM/PCM		T			
450.0	n.a.	FM	200 bps 20 characters per second	R	Whip	AM	Ground-command teleprinter
296.8	7.6w	AM 3.95KHz	n.a.	T	Helix	AM	CSM ranging
259.7	n.a.	247Hz 31.6KHz TONES		R			
231.9	10w	FM/PCM	72kbps	T		ATM	ATM telemetry,
237.0	10w	FM/PCM	72kbps	T		solar array	real time or recorded
450.0	n.a.	FM	200bps	R		ATM solar array	ATM ground command
243.0		ICW		T		CM	Recovery beacon
259.7	7.6w	AM	n.a.	T			
259.7	n.a.	AM	n.a.	R	Scimitar	SM	Voice ranging to Skylab
296.8	7.0w	AM	n.a.	T			
296.8	n.a.	AM	n.a.	R			
2106.4	250mn 2.8w	PM/PCM	200bps	R			Voice, range code to Gnd,
2287.5	11.2w	PM/PCM	51.2kbps	T	OMNI	CM	up data down
2272.5	selectable	FM	2MHz	T			telemetry TV, TM

This system was sized to overcome disturbance torques arising from gravity gradient effects, crew motion within the vehicle, unbalanced impact by the CSM during docking, and unbalanced torques from spacecraft leakage and venting. Nominal operation could continue with any one of the three CMGs inoperative, and the cold gas supply for the TACS included 500 lb budgeted for TACS-only operation for troubleshooting the CMG control system.

Communications

The external communications links for Skylab are summarized in Table 4. The communications system included the following capabilities.

1) An audio distribution hardline network with 13 communications panels located throughout the Skylab workshop (SWS) to provide voice communications conference capability to crewmen located anywhere orbital assembly (OA) or when performing EVA. Each crewman was equipped with a personal communications system (PCS) for biomedical data and for voice communications.

2) An instrumentation and telemetry system including a measuring subsystem, redundant pulse code modulation programmers, three tape recorders, and separate autonomous data systems for a) experiments M509 (astronaut maneuvering unit) and T020 (foot-controlled maneuvering unit), b)

experiment T013 (crew-vehicle disturbance), and c) the Earth resources experiment package (EREP).

3) An instrumentation and telemetry system for the ATM including a measuring subsystem, redundant pulse code modulation/digital data acquisition system (PCM/DDAS) assembles, and the auxiliary storage and playback (ASAP) assembly.

4) A television system including a video distribution network, for routing video signals from any of the five television cameras of the ATM or from portable television cameras operating from any of several locations throughout the SWS. A video coax switch in the MDA enabled the selection of the video signal to be routed to the CSM for transmission.

Significance of Skylab

Without question, the most significant findings from Skylab came about because of the major mechanical failure that disabled Skylab during launch. About 60 seconds after launch of the Skylab workshop, the meteoroid shield, clasped tightly about the structure, failed and, in turn, damaged the large solar power arrays. On attaining orbit, the workshop was badly overheating without its protective sunshade (a second function of the meteoroid shield) and was operating with only a fraction of its electric power. Early on, it became clear that without emergency repair,

Fig. 6 Skylab with umbrella-type sunshade in place.

Fig. 7 Artist's sketch of astronauts freeing jammed solar array.

the $2 billion program would be lost--an all-out attempt had to be made to save it.

The first manned launch was delayed 10 days while a deployable sunshade and special tools for deploying part of the damaged solar panel were built and tested. Neutral buoyancy testing, under water, proved invaluable in perfecting tools, equipment, and procedures for these emergency zero-gravity repairs in space. By early June, two weeks after their launch, Astronauts Conrad, Kerwin, and Weitz had deployed an umbrellalike sunshade (Fig. 6) and freed the stuck solar array (Fig. 7) to provide satisfactory operation conditions for their laboratory.

Freeing the array, in particular, had shown that physically demanding emergency repair tasks could be carried out by a well-trained crew in space suits under zero-gravity conditions. Based on early experience in the Gemini program, Extravehicular activity had been approached very cautiously. Simple tasks had left earlier crews quite exhausted. Successful experience with the emergency repair of the solar array gave newfound insight and confidence in EVA repair, maintenance and assembly. Before the Skylab program was over, EVA procedures had been used to carry out a number of repairs, including the following:

1) Assembly of a large sunshade over the complete Skylab workshop (Fig. 8). The original parasol-type sunshade was deployed like an umbrella through an airlock

Fig. 8 Assembled Skylab sunshade.

opening in the workshop. There was concern for deterioration of this original sunshade, and, therefore, on the second mission, a larger, more durable shade was assembled in orbit. This task was carried out by Astronaut Garriott and Lousma on Aug. 6, 1973, during a 6-h 3-min EVA.

A 55-ft-long A-frame was made of 5-ft sections of aluminum tubing. In order to assemble this frame, Lousma first crawled, hand over hand, to the base of the telescope module, where the apex of the A-frame was to be attached and where he mounted temporary foot restraints. Garriott then transferred a mounting plate and the pole sections to Lousma, who assembled and attached them to the base plate. The outward section of each pole had an eyelet through which a continuous loop of rope was threaded. The folded "tarpaulin," made of fabric treated with a silicone rubberbased paint, was then pulled out along the poles in a manner similar to raising a flag and secured. This A-frame assembly task and "tarpaulin" deployment to this day composes the largest space structure ever assembled in space; an assembly approximately 25 m long and 7 m wide.

2) Replacement of malfunctioning control gyroscopes located outside the telescope module.

3) Release of a sticking battery relay.

4) Replenishment of a leaking cooling system.

5) Repair of Earth resources microwave antenna.

6) Replacement of a faulty sun-monitoring TV camera.

7) Replacement of the thermal channel of a multispectral Earth resource scanner.

8) Opening and closing of bulky instrument doors.

In addition, a host of smaller repairs were made to keep individual equipment operable.

With these demonstrations of man's ability, it became clear that very large, complex, and expensive systems in space, just as here on Earth, would require, or benefit strongly from, man's presence to deal with emergencies and unanticipated problems. It is now generally accepted that the economic viability of certain large multibillion dollar enterprises in space will require the capability for manned intervention or periodic attention. This is not to say that automation and robots will not be required in space-rather to say that their use will be greatly enhanced by the human capability for on-site decision making, troubleshooting, and repair.

Skylab Recommendations

In planning for the capability for manned intervention in space operations, the following specific recommendations were made by Skylab Program personnel concerning in-flight maintenance:

Selection of Spares

Spares selection should include repair parts for critical items whose design permits in-flight bench repair, as well as replaceable assembles. Skylab has proven that the crew, when provided the proper tools, procedures, and parts, is capable of performing bench repair of failed assemblies beyond prior expectations. Although there were initially no repair parts aboard, these were provided on subsequent revisits and used successfully.

A good example is the teardown of tape recorders by the crew of SL-3 and the subsequent furnishing of repair parts and repair by the SL-4 crew. This reduced the volume requirements for resupply by providing a few repair parts instead of an entire new assembly. This philosophy could reduce the number of primary spares required onboard initially, if the capability to repair the failed items is provided.

Other examples of detail repair on Skylab were the repair of the teleprinter and replacement of the printed circuit boards in the videotape recorder.

The flight backup and test units on a limited-production program should be considered as spares within reasonable refurbishment effort, launch delay, and reprocurement time considerations.

Other major findings of Skylab, relevant to future space stations, are discussed in the following sections.

Criteria for Design

Initial design concepts should include in-flight maintenance provisions, with the necessary design features to facilitate failure detection, isolation, corrective action, and verification of repair. Provisions should be made for tools, spares, maintenance equipment, and space for maintenance work.

Accessibility to equipment attaching hardware, electrical connections, and plumbing is imperative, even in areas where maintenance is not planned. All contingencies cannot be anticipated, but corrective maintenance action can be taken if the general design is consistent with this approach.

In much of the unplanned Skylab repair work, it was necessary to remove cover plates held in place by an inordinate number of fasteners, which were not always of the design best suited for operational removal. Allen head screws and hexagon head bolts were much preferred over other types by the crew.

A substantial effort had to be spent in identifying, to and by the crew, components, cables, and tubing to be repaired or replaced. A simple system of identification decals should be used to facilitate identification.

Selection of Tools

Tools initially selected for Skylab were primarily those required for specific tasks. A few contingency tools were included such as a pry bar, a hammer, and the Swiss Army knife, which proved to be valuable assets. Wrenches were provided only for specific applications. The crew activities and evaluation indicate a tool kit should contain all the tools normally found in a tool collection for comprehensive home usage, as well as the special tools required for special aerospace hardware. Good quality off-the-shelf hand tools are adequate, and no special features are required for use in space. An improved tool caddy for carrying tools from place to place should be developed for easy location of the needed tool after arriving at the work station. Transparent material would be desirable. The caddy should also hold small parts in an accessible manner as the work is done, since containing and locating these items was a problem in zero gravity.

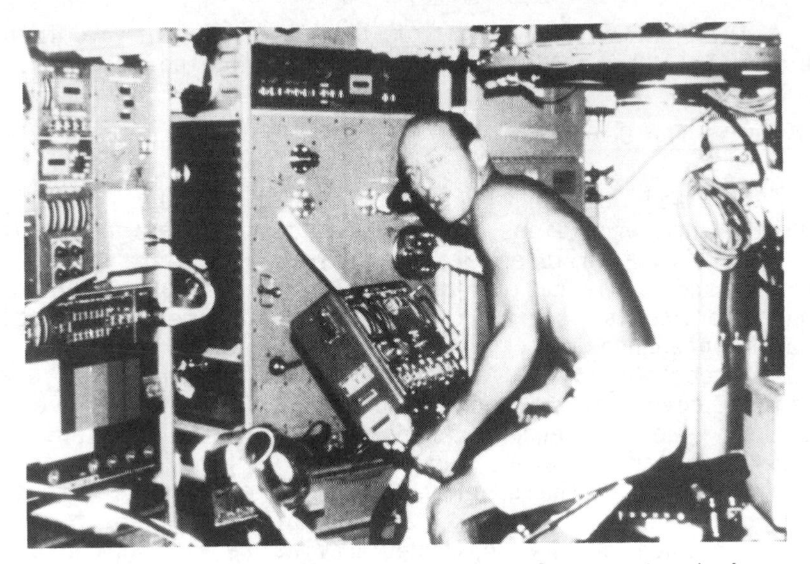

Fig. 9 Astronaut Conrad operating bicycle ergometer during Skylab mission.

Physiological Findings

The Skylab medical program was very thorough, with exhaustive pre- and postflight evaluations,[2] as well as extensive in-flight evaluations with special equipment such as the rotating chair, lower body negative pressure device, and bicycle ergometer (Fig. 9). A medical doctor, Dr. Joseph Kerwin, also flew one of the missions. As a result of the evaluation of the progressively longer 28-, 50-, and 84-day missions, no duration constraints were found. It was concluded that 1 to 1-1/2 hours of daily exercise were required by each crewman to maintain reasonable physical condition of the muscles. None of the crew sustained physical deterioration from the mission, and no discernable differences were detected for the crew of the 84-day mission as compared with the crew of the 28-day mission.

Inasmuch as approximate 90-day crew rotation cycles have been considered desirable for other reasons, such as logistics, it was concluded that Skylab experience verifies the acceptability of planned space station crew duty cycles. In addition, there were no trend data to suggest that longer crew duty cycles at 0-g would not be acceptable. Findings from longer-duration USSR flights since Skylab also indicate the acceptability for longer durations so long as adequate exercise regimens are followed.

Fig. 10 Skylab internal architecture.

Fig. 11 Skylab sleep restraint system.

Habitability Considerations

A discussion of several habitability findings follows.

Visual Orientation

The architecture of the Skylab workshop was gravity oriented (Fig. 10). This orientation permitted ease of ground testing and crew training. In flight, this convention provided the crew with a familiar coordinate system permitting easy orientation, location recognition, and equipment identification. The majority of crew members favored this architectural arrangement but did not feel it to be a hard requirement or constraint.

Sleep Arrangement

Sleep restraints (Fig. 11) should provide substantial hold-down of the body against a firm back. Ventilation flow should be head to foot rather than "up the nose." Skylab crews were extremely sensitive to auditory disturbance while attempting to sleep, and strong attention to auditory isolation should be given in future station design.

Hygiene

The portable spray shower (Fig. 12) provided in Skylab was favorably regarded. However, because of the time

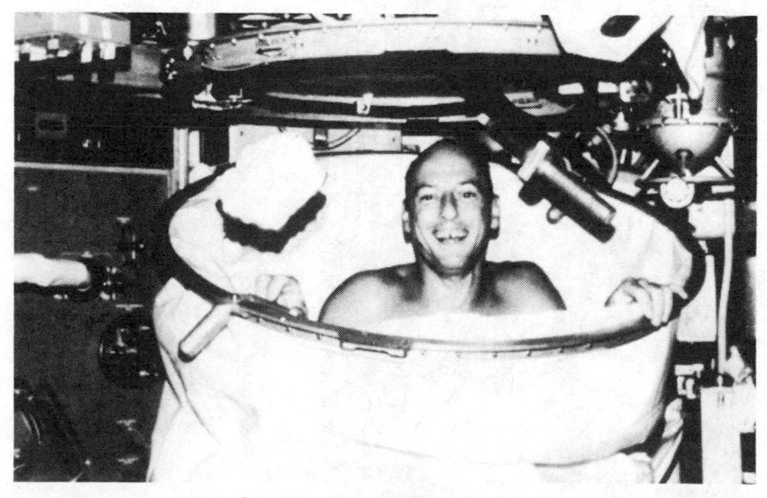

Fig. 12 Skylab portable shower.

required to set it up and clean up after use, it was not used universally. A more convenient shower would have been much appreciated, but a washcloth sponge bath was considered adequate. This approach is being used in Shuttle-Spacelab.

Waste Management

The airflow entrapment system for collecting feces and urine (Fig. 13) worked well for Skylab, and this concept is recommended for future long-duration spacecraft (if the added volume and weight can be tolerated-compared with simpler designs). For fecal collection, a higher airflow rate than used on Skylab would be desirable. The seat should be made of a softer material, and the outer diameter increased to provide a better airflow seal. The lap belt and handholds were absolutely required. The urine collection system should provide for a volume of at least 4000 ml/man/day. The urine separator should not be as noisy as that on Skylab. The cuff system for collecting urine was satisfactory as a contingency mode. The urine collector should be refrigerated or stored in a sealed condition to prevent odor buildup. The waste management compartment should be located sufficiently far from the sleep compartment to minimize noise disturbance to sleeping crew members.

Trash should be separated into biologically active and inactive material. Daily disposal of active material is necessary, whereas less frequent disposal of inactive material is satisfactory. Stowage of collected trash "external" to the habitable volume of the spacecraft is highly desirable. Food containers make up the bulk of the trash and should be designed to consume minimum volume when expended. A compacter would be a desirable feature. Backups and contingency plans are necessary.

Virtually all loose debris (solid or liquid) in the workshop migrated to the air mixing chamber screens in the dome. This phenomenon should be exploited in future designs by strategically locating the environment return air vents and planning to collect loose items there. They should be easily cleanable or should have replaceable filters.

Clothing

Two-piece garments (such as pants and shirts, as opposed to coveralls) proved convenient in Skylab for the same reasons as on Earth (Fig. 14). Two-piece garments are

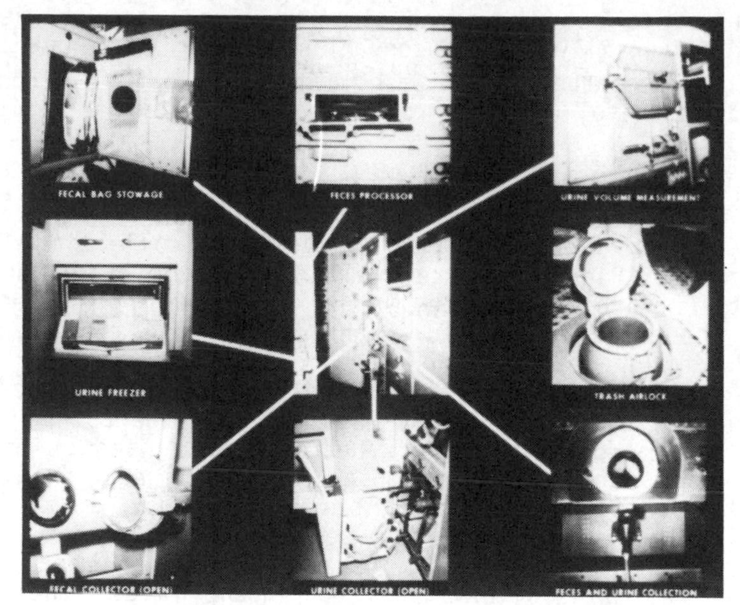

Fig. 13 Skylab waste management system.

Fig. 14 Skylab crew outer garments.

less sensitive to fit, are more adjustable to clothing requirements and are more convenient to personal hygiene procedures. Spacecraft interiors are very clean, and therefore clothing gets soiled principally from the wearer's body. More underwear and socks and fewer outer garments should be provided. Elastic or knitted cuffs to prevent sleeves and pant legs from riding up are not required and were sometimes an inconvenience. Head gear as a protective device is not required. Location and design of pockets warrants engineering study and a realization that the types and locations found practical in aircraft flying suits and in street wear may not be so practical in space station garments. Skylab crews felt that outer garments could be worn one or two weeks without needing change, but underwear and socks should be changed daily.

Food

The food system for Skylab is shown in Fig. 15 and provided for a variety of quite palatable food. Food service for long-duration flights can be optimized if standard menus are planned and provided so that all crewmen are eating the same basic meal at the same time (in contrast with individual selection as was provided on Skylab). Individual preferences can be provided in beverage, snack, spices, and dessert selection from an onboard pantry. Particularly for long-term flight, it is recommended that a pantry-type food storage rather than a meal-sequence system be implemented. In this type system, all

Fig. 15 Skylab food system.

identical foods are stored in the same location, and a meal is prepared by selecting the desired foods from the storage area in much the same manner as a home pantry. This storage arrangement provides flexibility in implementing real-time desires and planning as may be required to support changes to mission duration, timelines, crew health and so forth.

Contamination Control As A System

As a result of the Skylab program, it is evident that contamination control should be integrated into the design criteria on a level comparable to major functional systems. It should be considered from the initial stage through mission support. Future missions should consider a contamination control system that integrates the degradation effects resulting from interactions between all contributory systems.

Systems affected by contaminatiion on Skylab were thermal, power, attitude pointing control, environmental control, crew safety, and all experiments or critical and operational surfaces such as windows and antennas.

Contamination Control Working Group

A contamination control working group (CCWG) was used for integrating contamination design requirements and determining systems interactions and contamination effects on all systems, as well as managing technical contamination studies.

Contamination was recognized early in the Skylab program as a potential critical problem for the experiment optical systems and for other external Skylab surfaces. For this reason, the CCWG was created to coordinate the technical efforts of various groups studying the contamination problem.

Under the guidance of the CCWG, extensive ground testing of Skylab systems, including the waste management system, was performed to predict contamination levels. Design and operational procedure changes were recommended by the CCWG to limit contamination. Rigorous analyses were performed in conjunction with testing to model performance of the contamination producing systems and to predict contamination levels. Flight experience confirmed that this multidiscipline approach is successful and required for complex space vehicles.

Contamination Modeling

Surface deposition contamination and induced cloud brightness levels were found to be predictable to within 30%. Total contamination of the vehicle was predicted fairly accurately with the use of periodic updates of mission critical parameters and as-flown conditions.

Premission predictions of surface deposition and induced cloud brightness around the Skylab vehicle were adequate but required frequent updating as mission changes, anomalies, and contingency operations had major impacts on the contaminant environment.

The line-of-sight model for surface deposition contamination was shown to predict contamination levels within 10%+0 20%. Modeling of the induced cloud brightness around the Skylab was found to be dependent on parameters identified during the Skylab Contamination Ground Test Program and which were mission dependent.

Contamination Monitors

Instrumentation to measure contamination deposition and cloud brightness was found to be invaluable in assessing and predicting experiment degradation, contamination levels on critical surfaces, and for updating contamination prediction models. Mass deposition monitors, low-pressure sensors, residual gas analyzers, and cloud brightness monitors are recommended for future consideration.

Quartz crystal microbalances (QCMs) were successfully used on the exterior of Skylab to monitor mass deposition rates. Cloud brightness monitors measured the brightness of the induced atmosphere around the vehicle. The accuracy of the prediction model was improved by using these measurements as specific reference points.

Material Source Properties

Uniform materials testing criteria were used to determine parameters required for accurate modeling and assessment.

Success of contamination modeling and subsequent countermeasure is dependent upon extensive materials testing for source rates over the range of temperatures, times of exposure, and ambient environment interactions anticipated. Resultant deposition capability must be determined for major sources as a function of temperature variations of source and sink and the contaminant effects on signal attenuation.

Overboard Venting

The Skylab contamination ground test program demonstrated the need for testing of vent systems to determine operating parameters and to evaluate vent nozzle designs.

Experience indicated that desired particle size distribution, direction, and velocity can be created by proper nozzle design and flow rates for a given liquid. Advantage can be taken of sublimation characteristics and ambient atmosphere drag effect to minimize contamination potential. Given these parameters, modeling can determine proper timelines and vent sequences. Alternative methods to venting overboard for sources unacceptable in a vent mode should be established.

Experiment Exposure

Proper timelines for experiment exposure or operation in relation to engine firings, outgassing levels and overboard venting are necessary to ensure low contamination levels. The clearing time of particles and molecular interactions with the ambient atmosphere should be considered.

Modeling of vents and engine firings successfully indicated periods during which an experiment or critical surface required protection on Skylab. Particles were observed when the predicted clearing time was not observed.

Waste Tank

Testing and flight observations have shown that discharging waste liquid into a screened waste tank, exposed to vacuum, can be successful in allowing only vapor to escape.

The Skylab waste tank concept was instrumental in keeping the brightness of the induced atmosphere within the limits established for normal operation by ground testing.

Significant Contamination Sources

Of all the sources of contamination on a manned vehicle, materials outgassing and engine firings are major problems because of the continuous, longlasting nature of outgassing and the necessity for engine firings at times which may not be propitious for contamination control. Other major venting can be adequately designed, controlled, or timelined to minimize cloud brightness or deposition potential.

Deposition rates on Skylab mass detectors confirmed that outgassing sources and engine firings were the major sources. Other vented or leaked material did not deposit significantly at the temperatures of the Skylab exterior.

Contamination Control During Ground Handling

Proper contamination control of experiment and vehicle components requires uniform specifications encompassing the susceptibility of critical surfaces. Documentation and monitoring by a single organization should exist from production to launch for continuity of control and a comprehensive central record.

In general, Skylab prelaunch cleanliness was well controlled, and no adverse effects resulted from prelaunch contamination.

Onboard Cleaning and Storage of Optics

Skylab optical cleaning kits for accessible optics consisted of a mild detergent solution, distilled water, lint-free cotton, lens tissues, and air syringe and were successful in removing contamination from certain Skylab surfaces. However, these techniques will not remove many contaminations such as deposits outgassed from external sources. Storage of optics in both gaseous nitrogen (GN) and vacuum was satisfactory.

The capability to clean accessible optics and the development of techniques to clean remote optics are highly desirable. New techniques for contaminant detection and cleaning include Auger spectroscopy, binary scattering, metastable beams, ion sputtering, and activated plasmas.

Skylab Subsystems Operational Experience

Subsystem experience from Skylab is generally applicable to space station design, with the degree of applica-

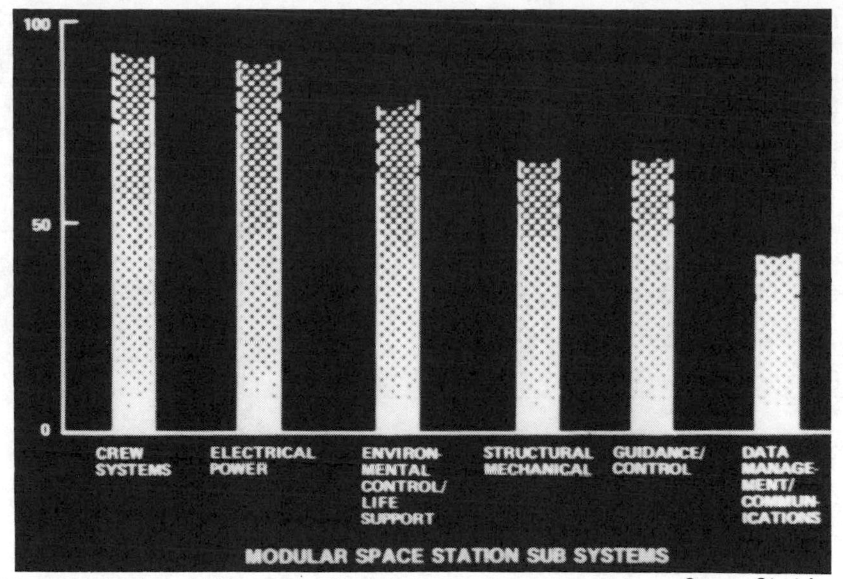

Fig. 16 Applicability of Skylab technology to future Space Stations.

bility estimated, as shown in Fig. 16. As indicated, crew systems data were most directly applicable, and those findings have been discussed here in some detail. Other subsystem experience, such as that with solar/battery electrical power and control moment gyroscopes for attitude control, has not been discussed herein, but is well documented in Skylab system evaluation reports.

Concluding Remarks

Eleven years ago, perhaps ahead of its time, the first US manned space station was launched. The first weeks of its life were trying indeed. But from that trying period and the subsequent months of operation, we learned many things we might have otherwise waited years or decades to discover. Such hardwon information should not be overlooked by today's space station designers.

References

[1]"Skylab, Our First Space Station," NASA SP-400, 1977

[2]"Biomedical Results from Skylab," NASA SP-377, 1977

The Soviet Salyut Space Station Program

Marcia S. Smith*
Congressional Research Service, Washington, D.C.

Introduction

It is 1971. America has welcomed home Alan Shepard, Stuart Roosa, and Edgar Mitchell from a successful landing on the Moon, the third US crew to accomplish such a feat. The Moon race ended two years ago--and America won, hands down. The Soviets, having lost the race, turned their immediate attention to Earth orbit, and in 1970 extended the duration of manned spaceflight to 18 days with the Soyuz 9 mission. The results were discouraging, though, for the crew returned home in poor physical condition. As they say, the fall is fine, it's the landing that can kill you. Weightlessness, too, is fine, it's readapting to Earth's gravity that is difficult.

Undeterred, the Soviets forge ahead and, on April 19, launch the world's first space station, Salyut 1. (Actually, the Soviets consider the Soyuz 4/5 docking in 1969 to have been an experimental space station, but since the crew had to move from one spacecraft to the other via EVA rather than through a hatch, this assertion is not generally accepted in the West.) It is the first step in an evolutionary program that by the end of 1983 will see the launch of eight Soviet space stations (two of which are failures) and 33 crews to occupy them (six of which do not dock successfully). As they gain experience, the Soviets will demonstrate in-orbit refueling, manned space construction, and perform impressive amounts of work in materials processing, biological research (both on the crews and with plants and other specimens), and Earth resources and astronomical observations. They will also gain practical experience in the everyday trials and tribulations of

---

*Specialist in Aerospace and Telecommunications Systems, Science Policy Research Division. The views expressed in this report do not necessarily represent those of the Congressional Research Service, any member or committee of Congress, or staff thereof.

creating a permanent manned presence in space: docking failures, repairs, emergency landings under adverse weather conditions, and loneliness for the crews living in space away from family and friends for as long as seven months. Tragically, they also are faced with the death of another crew, the three men of Soyuz 11. (The first cosmonaut killed during a spaceflight was Vladimir Komarov, who died on April 24, 1967, upon impact with the Earth when the parachute lines on his Soyuz 1 spacecraft tangled during descent.)

This is the history of the Salyut Space Station as it had unfolded by the end of 1983.

First-Generation Salyut Stations

There were a total of six first-generation Salyuts launched from 1971 to 1976, two of which (Salyut 2 and Kosmos 557) failed before they were occupied. (Kosmos 557 failed so early in its mission that the Soviets didn't even call it Salyut.)

Station Characteristics

Salyut 1 weighed about 18.6 tons, was 13.5 m long, with a maximum diameter of 4.15 m, and had a habitable volume of 90 m^3. With the Soyuz ferry craft attached, the weight increased to 25 tons, the length to 21 m (23 m if protruding antennas are included), and the habitable volume to 100 m^3.

The characteristics of Salyut did not change very much for the first-generation design (in fact, the major modification for the second-generation stations was the addition of a second operational docking port; the stations' dimensions have remained the same). Alterations to the first-generation stations included the number and placement of solar panels (Salyut 1 had four stationary panels, two at each end, while all the other Salyuts have had three individually rotatable panels in a T formation), the use of return modules on two of the stations (probably for returning exposed film), advances in on board systems such as navigation, and interior decorating improvements for psychological support. Salyut 3 and 5 had showers for the crew, while Salyut 4 did not (no mention of a shower was made on Salyut 1).

The Salyuts were launched unmanned by the D-1 (Proton) launch vehicle, with crews sent to the stations later using the Soyuz spacecraft. The Soyuz would dock at the forward end of the space station, and the crew would enter through a hatch into the transfer area with its several windows for observing Earth and space. (There are reports in the West that with Salyut 3 and 5, the military Salyuts, the docking port was located at the aft end instead.) The main part of the space station contained areas for living and working, and finally there was an unpressurized area that contained the fuel tanks and orientation systems.

Military Versus Civilian Space Station Distinction

During the early 1970's, Western observers of the Soviet program concluded that the Soviets had two parallel space station programs: one optimized for civilian research, the other for military reconnaissance. The distinction was drawn with the launch of Salyut 3, the second successful space station, in 1974. Salyut 3 was placed in a lower orbit than Salyut 1, and the crews sent to the station were all military instead of mixed military/civilian. When the crew entered the station, they switched to military telemetry, and, after the station had been unoccupied for several weeks, a small pod was ejected and recovered on Earth (a traditional technique for recovering photographic reconnaissance film). When Salyut 4 was launched, it had characteristics similar to Salyut 1. Salyut 5 was like Salyut 3. With hindsight, the conclusion could be made that had they been successful, Salyut 2 would have been the first military space station, and Kosmos 557 would have been civilian, thus giving an alternating civilian/military pattern to launches. Since the Soviets do not admit to using space for military purposes, they obviously have never admitted to such a dual program, and the distinction disappeared with the introduction of the second-generation space stations, which apparently are used for both.

Salyut 1 and the Soyuz 11 Tragedy

The first crew sent to the world's first space station, Soyuz 10, docked, but for reasons the Soviets have never made clear, could not enter the station and had to return home. The second crew, Soyuz 11, had better luck and successfully completed a three-week mission aboard Salyut 1, the longest space mission at that time.

The three-man crew (Georgiy Dobrovolskiy, Viktor Patsayev, and Vladislav Volkov) tested the station's design, tried out various methods of orienting and navigating the station, performed medical experiments, astronomical and atmospheric observations, and tried to grow plants in a hydroponic farm.

On June 29, 1971, the men died during their return to Earth. The Soyuz spacecraft has two compartments, which separate during reentry, with one part burning up in the atmosphere and the other returning the crew to the surface. In this instance, the crew had improperly sealed the hatch between the two sections, so when they separated, the spacecraft atmosphere vented into space. The Soviets had long since dispensed with the practice of wearing spacesuits, because of confidence in their hardware and the fact that three spacesuited cosmonauts could not be accommodated in the small quarters of Soyuz. The men died of pulmonary embolisms.

No further crews were sent to Salyut 1, and it was deorbited into the Pacific (the practice with all Soviet space stations) on Oct. 11, 1971. The Soviets returned to two-man crews so that they could wear spacesuits during launch and reentry, a practice followed to this day. Another three-man crew was not flown until 1980, after the Soyuz and the spacesuits were modified so all could be accommodated.

Two Space Station Failures

Twenty-two months after the Soyuz 11 tragedy, the Soviets were ready to try again. Salyut 2 was launched on April 3, 1973, and announcements were made that crews were preparing for launch. Something went wrong, however, and on April 14 it was reported that the space station had undergone a "catastrophic malfunction" that ripped off the solar panels and other externally mounted equipment, leaving it tumbling in orbit without telemetry return. Either an explosion or a misfiring thruster was blamed, although the most widely held theory was that the upper stage of the launch vehicle had exploded, and its debris damaged the station.

Not discouraged, on May 11, 1973, another space station was launched. This one failed so early in its mission, that rather than naming it Salyut, it was designated Kosmos 557. (Kosmos is a generic category in the Soviet space program that is used for a variety of scientific, applications, experimental, and military programs and is often used for spacecraft failures.)

Salyut 3: Back in Business Again

On June 24, 1974, the Soviets finally succeeded in putting another space station in orbit, Salyut 3. As noted earlier, this is categorized in the West as the first military space station because of its lower orbit, military crews, military telemetry, and use of a return capsule.

The first crew sent to the space station, Soyuz 14, docked successfully and conducted a 16-day mission. This was the first operational flight of the ferry craft version of Soyuz, which had the solar panels removed to increase the maneuverability of the spacecraft and allow more weight for equipment. In the ferry mode, it theoretically would not need an independent flight capability, since using batteries, it had a lifetime of 2.5 days, and docking was normally accomplished after one day.

Reports from TASS stated that the crew was performing "Earth observations," further lending credence to the supposition that their main task was reconnaissance. They also made atmospheric observations and performed medical experiments.

Five weeks after Soyuz 14 returned, Soyuz 15 was launched, but docking was not achieved, apparently because a failure of a new, automatic docking system. This flight demonstrated the problem of applying theory to practice, since without the solar panels, the ship did not have enough power to remain in orbit while the problem was solved, and the crew had to return home before the batteries were depleted.

On Sept. 23, 1974, two months after the Soyuz 14 crew left and one month after the Soyuz 15 crew failed to dock, a capsule was ejected from the space station and recovered. The Soviets have never mentioned what was in the capsule, but it is assumed in the West to have been photographic film taken by a camera on the space station while the station was in an unmanned mode.

Salyut 4: Two Successes and the First Manned Launch Abort

Salyut 4, another civilian Salyut, was launched on Dec. 26, 1974, and the first crew, Soyuz 17, docked successfully on Jan. 12, 1975. The two men remained in space for 30 days, performing the now routine Earth resources and atmospheric observations, but from a higher orbit than the Salyut 3 crew. In addition, the space station carried the Orbital Solar Telescope (OST) for extensive studies of specific areas of the Sun. Other astrophysical observations were made using two x-ray telescopes. Observations were made of X-1 Cygni, which was then suspected of being the ever-elusive black hole (although it is now classifed as a neutron star). There were continued attempts to grow plants, a prerequisite for a closed-cycle space station that does not require resupply from Earth, and, although certain vegetables were grown and eaten by the crew, they did not have any success in getting plants to go from seed to flower to seed again. Systems tests were conducted of a new navigation system for the space station, and, for the first time, a teletype was used for sending messages up to the crew so they did not have to be present when the message arrived.

The next launch to the space station gave the Soviets another in their long list of space firsts: the first time a manned launch was aborted before reaching orbit. The third stage of the A-2 launch vehicle failed, and the crew landed 1600 km downrange and only 320 km north of the Chinese border. The launch occurred on April 5, 1975, and, as expected, was not given a Soyuz name. Instead, the Soviets traditionally refer to it as the "April 5 Anomaly," although it is commonly called Soyuz 18A in the West. The main significance of the failure was that it occurred only three months before the scheduled launch of the Apollo-Soyuz Test Project (ASTP), and there was some concern about the Soviets' ability to accomplish their part of the mission. The Soviet explanation was that it was an old launch vehicle that had been sitting in storage for some time, and a new one would be used for ASTP, so there was no cause for concern.

The next flight was designated Soyuz 18 when it was launched on May 24, 1975, and it carried two men to Salyut for a 63-day mission, during which time the Apollo-Soyuz mission took place. The fact that the Soviets would have two crews in orbit at the same time worried certain members of Congress, but NASA was convinced that the Soviets had the capability to provide command, control, and tracking services for two flights at one time (after all, in 1969 they had had three spacecraft--Soyuz 6, 7, and 8--in orbit at once). The ASTP mission proceeded as scheduled.

The Soyuz 18 crew continued the experiments begun by their colleagues on Soyuz 17. One significant procedure that was tested on this flight was spraying a new reflective coating on the mirror of the solar telescope, which had become degraded. Additional experiments with spraying coatings were conducted on Salyut 6 and 7.

Although the crew did not set a new world duration record, they did more than double the length of any previous Soviet manned mission.

Soyuz 18 was the final crew to occupy Salyut 4, but another Soyuz docked with the space station in Nov. 1975. Soyuz 20 carried biological specimens (turtles, fruit flies, and various plants and seeds), but its primary mission was testing the Soyuz systems in a powered down condition for 90 days to ascertain whether or not they could be successfully reactivated. The ship undocked and was recovered on Feb. 15, 1975.

No further spacecraft docked with Salyut 4, and it was deorbited on Feb. 2, 1977.

Salyut 5: Two Successes and Another Docking Problem

By now, the Soviets had made significant advances in their space station program. The Salyut 4 program, despite the April 5 Anomaly, had been a great success, and the crews had had time to perform a wide variety of experiments during their longer-duration missions.

The next space station, however, followed the pattern of Salyut 4 and apparently was optimized for reconnaissance observations. While the crews did not continue the broad array of experiments begun on Salyut 4, they did perform a number of other experiments which heralded the future of Soviet space station activities, particularly several connected with materials processing in space.

Two crews successfully docked with the space station (Soyuz 21 and 24), but a third (Soyuz 23) was unable to dock because of a failure of the automatic docking system, similar to that encountered with Soyuz 15. The Soyuz 23 crew landed under emergency conditions at night in a partially frozen lake, 2 km from shore. Recovery rafts could not reach the spaceship because of ice, so helicopters were used to tow it to shore. This was the first "water" recovery in the Soviet manned program!

The Soyuz 21 crew was launched on July 6, 1976, and returned on Aug. 24, making a rare night landing. There were rumors, which remain to this day, that an acrid odor had developed on the space station, necessitating an earlier-than-expected return.

As noted above, the next mission, Soyuz 23, failed to dock. Soyuz 24 docked with the space station on Feb. 8, 1977, but remained for only two weeks. Western observers speculated that they returned so quickly because the military reconnaissance photographs they had taken were needed on the ground, and, in fact, the return capsule was ejected only one day later, unlike Salyut 3, where it was ejected a full month after the Soyuz 15 crew would have docked.

Experiments were conducted on this flight that had later application on Salyut 6 and 7. In one case, the air was changed in the space station. Although it was speculated this was done because of the aforementioned odors on the Soyuz 21 mission, the Soviets stated that it was a test related to future missions (compressed air has been used on Salyut 6 and 7 to replace air that escapes when airlocks are opened for trash disposal, experiments, or EVAs). A 1983 article in Krasnaya Zvezda[1] described the Salyut 5 operation as having been quite complex and requiring the development of torqueless nozzles to prevent the station from losing its orientation while the air was vented into space. Thus, there continues to be no substantial evidence of any problem on Salyut 5 that caused an early return for the Soyuz 21 crew.

Among the materials processing experiments were use of the Kristall furnace for crystal growth experiments; the Sphere experiment for studying the melting and hardening of molten metals in weightlessness; and the Reaktisya experiment for soldering stainless steel with a manganese-nickel solder. (The Soviets had conducted the first "materials processing" experiments aboard Soyuz 6 in 1969, when they tested various forms of welding.) Another device, Potok, was used to evaluate the possiblity of building capillary pumps in space for liquids without using electricity.

End of the First Era

By the end of February 1977, the Soviets had gained considerable experience with space stations, both positive (six crews had occupied four different space stations) and negative (two space stations were total failures, two crews were unable to dock, another crew docked but was unable to enter the station, a launch was aborted before reaching orbit, and three men had died). The crews had spent a considerable amount of time performing medical and biological experiments, and Earth resources, astrophysical, and atmospheric observations, and had started materials processing investigations. The crews themselves had adjusted to remaining in space for as long as 63 days, performing as much as 2.5 hours of exercise a day to ensure that they would be able to readapt to Earth's conditions. By and large, the program was successful, and the Soviets were ready for the next step.

Second-Generation Salyut Stations

On Sept. 29, 1977, the Soviet Union took an important stride towards achieving a permanent manned presence in orbit with the launch of a second-generation space station, Salyut 6. Although the station still had the same essential dimensions as its predecessors, it had a second operational docking port. From the time Salyut 6 was launched until the end of 1983, all manned space missions that have been launched by the Soviets have been designed to dock with either Salyut 6 or its successor, Salyut 7. Since it was during the time of Salyut 6 that the United States entered a hiatus in manned spaceflight , waiting for the Space Shuttle to be launched, the Soviets steadily moved ahead in accumulating person hours (and therefore experience) in space, although they still have not rotated crews on the station to achieve the long sought after "permanancy."

The second-generation stations, and associated vehicles, are stepping stones to an even more ambitious space station era that will involve the modular assembly of large orbiting complexes.

Salyut 6: Host to 16 Space Crews

Designed to last 1.5 years, Salyut 6 hosted crews from 1978 to 1981, followed by a year of unmanned operations with Kosmos 1267, and was finally deorbited on July 29, 1982, after almost five years of operational life. This long lifetime could be attributed to frequent repairs by knowledgeable crews, and the ability to refuel the space station in orbit.

The most significant change in the space station was the addition of a second docking port, which enabled the Soviets to launch additional crews to "visit" the main crew on the station, and unmanned resupply flights using the unmanned Progress vehicle (a modified Soyuz) to deliver experiments, food, water, air, personal items, and, importantly, fuel for transfer into the space station's fuel tanks. (In some cases, the Progress engines themselves were fired to change the station's orbit.) Progress can carry up to 2300 kg of cargo to the space station and a total of 12 Progress missions were flown to Salyut 6. The addition of the second docking port required substantial redesign of the engine area that had formally occupied that space.

A total of 16 crews occupied Salyut 6. (In order: Soyuz 26, 27, 28, 29, 30, 31, 32, 35, 36, T-2, 37, 38, T-3, T-4, 39, and 40.) Two others tried to do so, but the very first mission (Soyuz 25) could not dock, and another (Soyuz 33) suffered an engine malfunction. Soyuz 34 and T-1 were unmanned.

Of the crews that did dock, five were long-duration crews, the longest of which increased flight duration to 185 days. Most of the others were "visiting" crews that stayed for approximately eight days and brought new experiments and took back the results of those which had been performed. (Progress can only deliver cargo to the space station; it cannot bring anything back, since it disintegrates during reentry.) Nine of the visiting missions involved cosmonauts from the Interkosmos countries (Bulgaria, Cuba, Czechoslovakia, East Germany, Hungary, Mongolia, Poland, Romania, and Vietnam). Although seen primarily as a public relations part of the space program, each country provided at least one of its own scientific experiments. (The Bulgarian cosmonaut was aboard Soyuz 33, which could not rendezvous with the space station.)

Another function of the visiting crews was to rotate spacecraft so that a fresh Soyuz would be available for the long-duration crews. Based on the Soyuz 20 test, it was assumed that the design life for Soyuz in a powered down condition was 90 days. Thus, for example, the Soyuz 26 crew, which stayed in orbit for 96 days, would need a new Soyuz to bring them home. The Soyuz 27 visiting crew therefore returned in the Soyuz 26 spacecraft, leaving the newer Soyuz 27 to bring their colleagues home at the end of the mission. This procedure became routine during the Salyut 6 days and necessitated a number of do-si-do maneuvers in space. The refueling lines are located only at the aft docking port, so if a Progress docked to refuel the space station, the aft port had to be used. In some cases, however, the visiting crew would have docked at the aft end, but returned in the Soyuz parked at the forward end. The long-duration crew would then have to suit up, enter the Soyuz and undock from the space station, wait for ground controllers to rotate the station 180 deg so the forward docking port faced them, and redock. This complicated procedure was repeated many times during Salyut 6's lifetime.

In 1979, the Soviets introduced an upgraded version of the Soyuz spacecraft, designated Soyuz T, with upgraded avionics, a better fuel system, and solar panels (which had been removed for the ferry version of Soyuz). The latter were reinstalled both so that Soyuz can fly independently for longer periods of time and so that the electricity they provide can be tied into the space station's power system. The interior has also been redesigned so that, along with new spacesuits, three cosmonauts can be accommodated and, in November 1980, the Soyuz T-3 crew became the first three-man Soviet crew since Soyuz 11. The original Soyuz is no longer used for the manned program.

In 1981, the Soviets docked an unmanned module, Kosmos 1267, with Salyut 6, and announced that it was a test of a modular space station, although a Western magazine claimed that it was a battle station in orbit armed with meter-long interceptors.[2] The Soviets occasionally reported on tests conducted with Kosmos 1267, which they probably would not have done if it were a military vehicle, and the entire assembly was deorbited in July 1982 with no firings of the alleged interceptors. On the basis of available evidence, it seems that Kosmos 1267 was exactly what the Soviets said it was--a test of a modular space station.

Salyut 7: The Beginning of Modular Space Stations

At the end of 1983, the Salyut 7 space station was in orbit awaiting a new crew. The space station had been launched on April 19, 1982, the eleventh anniversary of the launch of Salyut 1. Three crews occupied the station during 1982: the Soyuz T-5 crew, which set a new duration record of 211 days, and the two visiting crews, Soyuz T-6, which included the first French "spationaut," and Soyuz T-7, which took the second woman into space, Svetlana Savitskaya. (The Soviets had launched the first woman into space, Valentina Tereshkova, in 1963 on Vostok 6.) Four Progress resupply missions were flown.

In 1983, the first operation conducted with Salyut 7 was the docking of Kosmos 1443, an unmanned spacecraft which the Soviets described as an operational version of Kosmos 1267. As the weeks progressed, the Soviets described it as much more than a modular space station extension, though. Rather, Kosmos 1443 was a multipurpose vehicle that could transport 2.5 times as much cargo to the space station as Progress and return up to 500 kg to Earth, serve as a space tug, and become an extension to the space station, increasing the habitable volume by 50 m^3. The spacecraft was as large as Salyut, weighed 20 tons, and had its own solar panels (Fig. 1).

The first crew sent to the station in 1983, Soyuz T-8, could not dock because the rendezvous radar did not deploy correctly, and the crew could not judge their closing speed accurately enough for a "seat-of-the-pants" docking. On June 27, 1983, the Soyuz T-9 crew was launched and became the first occupants of the new "space complex" (Fig. 2a).

Kosmos 1443 performed its space tug role several times during the mission, raising and lowering the orbit of the complex. It also served its cargo role, delivering supplies to the space station and returning material to Earth. Surprisingly, though, it never really acted as a space station extension. After the crew unloaded the cargo and emplaced the materials to be returned to Earth, the entire spacecraft was undocked, not just the descent module. The descent portion returned 350 kg of material to Earth on Aug. 14; the main body burned up during reentry on Sept. 19. Whether something went wrong with either Kosmos 1443 or the space station itself that caused a change in mission profile, or if Kosmos 1443 never was intended to serve as a space station extension, is unknown. The fact that it is still called Kosmos, and has not been given a name such as Progress or Soyuz, suggests that it may still be experimental despite what the Soviets have said (it may be that the Russian and English meanings of "operational" and "experimental" are not the same).

The Soyuz T-9 crew remained on Salyut 7 for 149 days, longer than originally planned. In fact, there is substantial evidence that this was to have been the first time that crews were to be rotated, rather than leaving the station unoccupied for a period of time before launching a new crew. On Sept. 26, 1983, however, the Soviets added another of those space firsts that they probably would rather not have--the first use of emergency abort procedures on the pad. Seconds before the Soyuz T-10 crew would have lifted off, a fire erupted in the launch vehicle. Using the escape tower, the two-man crew was carried away in the Soyuz to safety, landing 3 km away from the pad. Interestingly, the two men (Titov and Strekalov) were two of the three crewmembers aboard Soyuz T-8, which could not dock with Salyut 7 earlier in the year.

The Soviets conceded that this event had occurred at an international space conference in Budapest, Hungary, two weeks later. In December 1983, they referred to Titov and Strekalov as the crew that "would have relieved" the orbiting crew, lending credence to the theory that this would have been the first crew rotation.

With their replacements still on Earth, the T-9 crew's mission was extended and on Nov. 1 and 3, they conducted two EVAs to install additional solar panels to the existing arrays. This was the first time a Soviet crew conducted

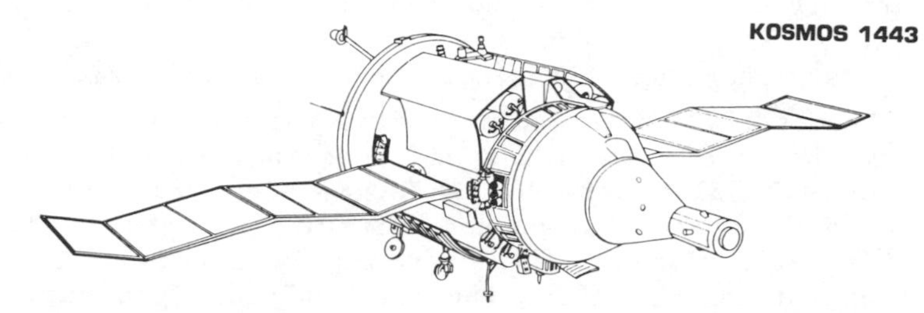

Fig. 1. Drawing based on an illustration that appeared in the Soviet newspaper Pravda, July 3, 1983.

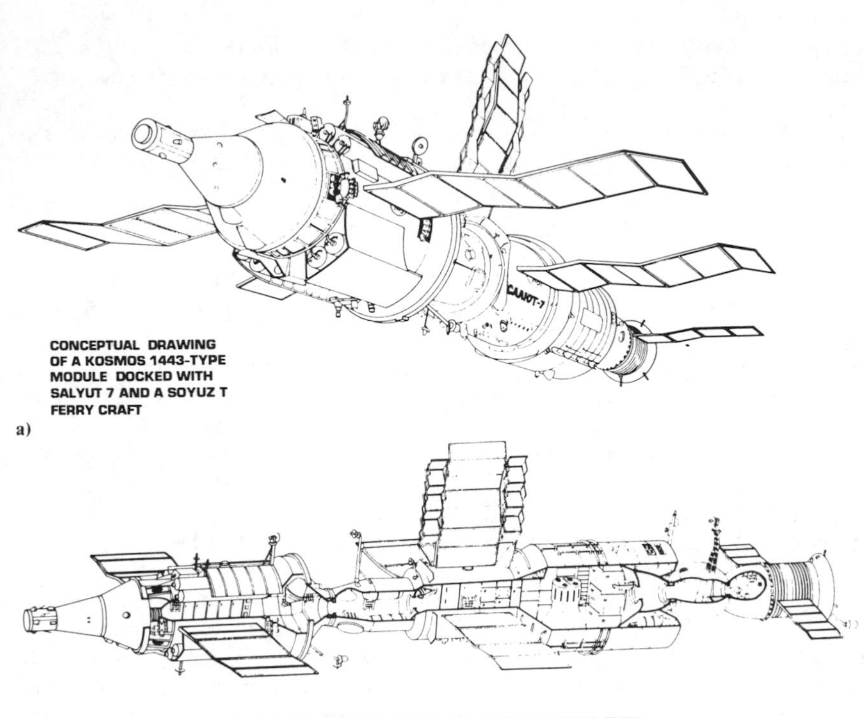

Fig. 2. Drawings based on illustrations that appeared in the Soviet magazine Soviet Union, No. 2, 1984. Although the drawings were labeled in the Soviet magazine as being Kosmos 1443-Salyut 7-Soyuz T-9, in fact, the Kosmos 1443 module had undocked months before the crew deployed the two additional solar panels that appear on the vertical solar array. Hence, this has been labeled a "conceptual" drawing of what the complex would have looked like had Kosmos 1443 still been docked through the end of the Soyuz T-9 mission.

two EVAs on the same mission. The solar arrays had been delivered to Salyut on Kosmos 1443, so the fact that they waited until November implied that they had been waiting for the T-10 crew, and the Soviets later stated that the original plan had called for all four men to install the arrays together.

In any case, the T-9 crew successfully attached the two panels to the existing arrays, which according to the Soviets increased the power by 50%.[3] Pictures of the installation which appeared in _Pravda_ showed them to be only about one-third the size of the main arrays, however, and, since there were only two additional arrays and three main arrays, it is unclear how that adds up to a 50% increase, unless the main arrays were degraded in capability.

In addition to the cargo delivered by Kosmos 1443, two Progress missions brought supplies to the T-9 crew.

A Tour of the Interior of Salyut 6 and 7

Salyut 6 and 7 are virtually identical, and the interior detail is much the same as with previous Soviet space stations (Fig. 2b).

Arriving through the transfer compartment at the forward end of the station, the cosmonaut enters an area with several windows where items such as pressure suits are stored. Next is the work compartment, with walls that are removable and covered with soft, bright-colored cloth. Following the recommendations of previous space station crews, additional sound insulation was installed to lessen the noise of onboard equipment.

Almost immediately in front of the access hatch is the station's central control post. Behind that is a table for working and eating and, nearby, a water tank (some water is delivered via Progress, while other water is regenerated from water vapor in the air produced by breathing and perspiration). Cupboards for storing food are located to the left and right of the working compartment.

Past the table is the scientific equipment area for experiments such as the BST-1m submillimeter telescope and the MKF-6M Earth resources camera. The shower is located to the right of the equipment compartment on the "ceiling" and close to that are two airlocks through which garbage (including human waste from the toilet) is disposed. (These "honeybuckets" are deployed into space through the airlock by springs and decay naturally after about one month.) Two of the materials processing devices used on Salyut 6 (Splav and Isparitel) were alternately emplaced in one of these airlocks. Behind the scientific equipment area is the toilet, with a soft door with a zipper fastener. Past that is the hatch into the intermediate chamber, which leads to the aft docking hatch.

There are 20 portholes in Salyut for visual and photograph observations. Problems have developed with the optical quality of the portholes, as dust collects on them, and, in a few cases, they have been struck by meteorites.

Atmospheric pressure on Salyut is maintained at 700-960 mm Hg, with a partial oxygen pressure of 160-240 mm Hg and carbon dioxide, no more than 7-9 mm Hg. Temperature is maintained at about 20-22°C.

Refueling with Progress

As noted earlier, the Progress spacecraft is used to refuel the space station, and the refueling lines are located at the aft docking port. When Progress docks, the refueling lines are automatically coupled. There are two refueling units: one for the propellant (hydrazine) and one for the oxidizer. Salyut has six tanks (three for propellant, three for oxidizer), while Progress has four (two for each). (The third set of tanks on Salyut is held in reserve.)

The Salyut tanks have an accordion-type device in the middle. When the tanks are full, the accordion is flat; as they empty, nitrogen is forced in the accordion to force fuel into the engines. Thus, the first task in refueling is to remove the nitrogen from the fuel tanks and pump it back into its storage tank. Once this is accomplished, the propellant is pumped in, then the oxidizer. Compressed nitrogen at 8 atm is used to force the fuel from Progress into Salyut. The Soviets have performed fuel transfers with the stations both in an occupied and unoccupied mode.

Salyut 6 and 7 Experiments

The vast array of experiments conducted by the Salyut 6 and 7 crews are far too numerous to discuss here, so only a few examples are provided. (Details on the experiments and all other aspects of the Soviet manned space program can be found in Ref. 4.)

Medical/Psychological. A wide range of biomedical experiments were conducted to determine the causes and cures of space motion sickness (from which the cosmonauts suffer at least as much as the US astronauts), and the reaction of humans to long-duration exposures to weightlessness (the longest Soviet mission is 211 days compared with 84 days for the Americans). The latter includes both physiological (such as hearing, taste, speech, and coordination) and psychological changes.

The crews are required to perform 2.5 hours of exercise every day and have returned to Earth in generally good condition. Areas continuing to be of concern to Soviet physicians are space motion sickness, cardiovascular deconditioning, blood cell mass loss, and bone mineral loss. Nevertheless, the Soviets appear optimistic about even longer duration missions.

Psychologically, the Soviets continued to place considerable emphasis on maintaining contact with family and friends through letters and two-way television conversations. The first two-way television was installed on Salyut 6 in 1979 for the Soyuz 32 crew, who reportedly received the first images of Cheburashka, a popular hero of children's cartoons, with "joyful exclamations." Personal items can be requested by the crews for transport via Progress (Soyuz 29's flight engineer, Ivanchenkov, had his guitar delivered, for example). Insight into the reaction of the crews to their lonely vigil in space was provided by the release of portions of Valentin Lebedev's diary.[5] He was a member of the 211-day mission, and the diary excerpts detailed his loneliness, problems with insomnia despite a tiring workload, and the art of establishing a working relationship with his fellow crewmembers (including those on visiting missions).

Biological. On Salyut 7, the Soviets finally succeeded in growing a plant from seed, through its entire lifecycle, back to seed again, using an Arabidopsis seed. Although the crews routinely grow vegetables for their own consumption (onions, cucumbers, radishes, parsley, etc.), this particular goal had eluded them since Salyut 1.

The effects of weightlessness on bacteria were studied on both Salyut 6 and 7 with the joint Soviet/French Cytos experiment. Experiments were also conducted with tadpoles to see if there would be differences in adapting to weightlessness if they were born on Earth or on the space station (those born on the space station had no difficulties, while those brought from Earth swam in a disorderly manner for more than two weeks), as well as the traditional heredity experiments with fruit flies (drosophila).

Earth Resources. Both space stations carried the MKF-6M camera built by East Germany and tested on the Soyuz 22 mission (which did not dock with a space station). The camera records in six bands: 0.46-0.50 microns, 0.52-0.56 microns, 0.58-0.62 microns, 0.64-0.68 microns, 0.70-0.74 microns, and 0.78-0.86 microns. Each film cassette weighs 13 kg and takes 1200 frames of film. One image shows an area 255x155 km with a resolution of about 20 m. Other cameras are used to augment the MKF-6M studies, including the KATE-140 topographic camera, which, like the MKF-6M, is attached to the space station, and several hand-held cameras. Earth resources observations using the cameras and visual observations seem to consume about half of the crew's experiment time. From 1977-1980 alone, 9500 photographs were taken with the MKF-6M and another 4500 with the other cameras. The number of photographs increased with Salyut 7, and the long-duration crew in 1982 took more than 20,000 MKF-6M and KATE-140 photographs. As noted earlier, one problem has been getting the exposed film cassettes back to Earth, so on Salyut 7 a device was used to video-record and relay back to Earth information on observed areas to avoid that delay (and the Kosmos 1443 descent module could also return some film).

Atmospheric Studies. On Salyut 6, the BST-1m submillimeter telescope was used for atmospheric observations in the infrared, ultraviolet, and submillimeter wavelengths. It used cryogenically cooled sensors that had to be calibrated each time it was used and, added to the fact that it required a large amount of electrical power (1.3-1.5 kW), resulted in it not being used as often as other instruments. Data from this instrument revealed anomolously strong emissions in the submillimeter band in areas of thunderstorm formation.

Other equipment carried for atmospheric studies were the Yelena gamma ray device, and the Duga electrophotometer (made by Bulgaria).

Astronomical Observations. The only extensive astronomical observations conducted on Salyut 6 involved the KRT-10 radio telescope, which was delivered to the station via a Progress mission, assembled by the crew, and deployed out the aft docking port. The telescope had a 10-m-diam dish with four horns in the 12-cm band and a spiral irradiator in the 72-cm band.

Operations were conducted with the KRT-10 for about three weeks (several days were needed just for calibration), after which the crew attempted to jettison it to free the docking port. At that point, the radio telescope hooked on a part of the space station, requiring the crew to go out on EVA and literally kick it loose. (There is some evidence that the radio antenna caught on the space station during deployment, not when it was being jettisoned, and the antenna never fully opened. The Soviets have not admitted this.) Both celestial and Earth observations were made with the KRT-10.

The Salyut 7 crews have performed a few astronomical observations as well, primarily using two French-built devices that are both for studying the universe and the Earth's atmosphere. The first, Pirimig, is a highly sensitive camera for studies of the upper atmosphere, interplanetary medium, and galaxies in the visible and near-infrared bands; the other is PCN, for nighttime studies of weak light sources such as the luminosity of interstellar dust and luminescent clouds and lightning. The space station is also equipped with the SKR-OM x-ray spectrometer for observations of interstellar space.

Materials Processing and Other Technical Experiments. The Salyut 6 crews worked with two materials processing furnaces (Splav for metal alloys and Kristall for crystal growth), and an experiment called Isparitel was used to study the depositing of protective film layers. The Salyut 7 crews seem to be focusing on electrophoresis experiments using the Tavria unit (little mention has been made of Splav or Kristall), studies of the impact of space conditions on thin film coating (the Electrophotograph experiments), and studies of heat transfer in multiphase media (using the Pion device).

On Salyut 6, over 30 materials were processed in Splav and or Kristall, for a total of 300 different samples. Among the materials that were subjected to repeated experiments were cadmium-mercury-telluride, cadmium sulphide, gallium arsenide, indium antimonide, and indium arsenide. The original Kristall unit ceased functioning early in the mission, and a replacement was delivered by one of the Progress flights. The Splav furnace was placed in an airlock when in use so the heat would be radiated into space, but Kristall was designed so that its exterior temperature would not rise more than 50 K above ambient temperature.

The Isparitel experiment was designed to study the process of evaporation and condensation of different materials in space. Coatings were sprayed onto metal, glass, or plastic plates by vaporizing the coating material (silver, gold, or alloys containing aluminum, copper, and silver) with two electron guns. The vapor would condense on the plate over a period of 1 second to 10 minutes, depending on the thickness desired. Results were expected to be applied to the construction of future space stations. Approximately 186 samples were produced with this device, and the results were reported to be promising.

On Salyut 7, the T-9 crew performed 19 experiments in the electrophotograph series for studying the effect of space flight conditions on thin film coatings that were exposed to the conditions of space flight in an airlock. Degradation of materials in space continues to be a subject of concern to the Soviets.

The Soviets experimented with producing interferon on both Salyut 6 and 7 and with urokinase on Salyut 7. They have found that they can purify substances between 10 to 15 times better in space than on Earth, and the productivity in space is hundreds of times higher. A pure protein substance was produced from the membranes of an influenza virus by the Soyuz T-9 crew in 1983, which the Soviets heralded as "a domestic standard with which we can now compare all the vaccines that are produced."[6]

∗ **TOTAL LAUNCHES 1957-1982**		
	U.S.	796 successes 106 failures
	U.S.S.R.	1538 successes 187? failures
∗ **TOTAL LAUNCHES 1983**		
	U.S.	22
	U.S.S.R.	98
∗ **TOTAL MANNED LAUNCHES SINCE 1961**		
	U.S.	42 (3 to a space station, all successful)
	U.S.S.R.	56 (32 to space stations, 25 successful) +1 launch failure
∗ **CUMULATIVE HOURS IN SPACE**		
	U.S.	29,153:06 (longest flight – 84 days)
	U.S.S.R.	70,407:46* (longest flight – 211 days)

*Does not include Soyuz T-10/T-11 mission.

Fig. 3. US/USSR comparisons. Data current through May 15, 1984.

Another series of experiments performed by the Soyuz T-9 crew used the "Pion" device for studying heat and mass transfer to, and the physics of, multiphase media in weightlessness. The apparatus could record changes in the density and temperature of substances as the container in which they were held was heated, and the process was recorded on film and videotape using a holographic device.

Military Experiments. As noted earlier, the Soviets do not admit to using space for military purposes, so it is difficult to ascertain whether or not military experiments are conducted on Salyut. Part of the problem is defining what a "military" experiment is. Earth resources observations may well prove useful to the national security establishment in the Soviet Union, as it would in the United States. Similarly, certain materials processed on Salyut might be useful for military as well as civilian devices (e.g., cadmium-mercury-telluride is used for infrared sensors, which might be applied in medicine as well as in military reconnaissance). Using Salyut as a target for laser ranging experiments is another gray area that could have both civil and military applications.

The Western media have reported some incidents of "military experiments" on Salyut, such as observations of the bioluminescence of plankton, which might reveal the path of a submarine, and pointing and tracking tests similar to those planned for the US Talon Gold experiment, which is scheduled to be flown on the Space Shuttle in

1985, and could be related to development of a space laser system.

So many experiments could have both military and civilian applications that it is impossible to state that the Soviets do not conduct any military experiments aboard Salyut, but, with silence on the part of both the Soviets and the US Government, no definitive conclusions can be drawn. On average, it is probably safe to say that the percentage of time spent by Salyut crews on "military" tasks is approximately the same as that which will be spent by US Shuttle crews on "military" tasks through the end of this decade.

Conclusion

Throughout the history of its manned space program, the Soviet Union has followed a slow but steady course towards attaining a permanent manned space presence. They have not leaped forward with technological advances the way the United States has with the Space Shuttle, but instead have followed a consistent goal using a conservative design philosophy. As a result, they have a significant lead over the United States in experience in manned space operations, although the United States holds the lead in technology (Fig. 3).

Although the Soviets may not publish very much information about what they plan to do this year or next, the long-term future is frequently discussed. In November 1983, a Pravda article discussed what the Soviets view as the future for space stations: a complex of large facilities in orbits ranging from 200-4000 km,[7] serviced by freight and passenger transport spacecraft. Included in the list of separate installations are research laboratories, housing modules, powerful energy installations, a refueling station, repair workshops, and construction sites for producing and installing standardized components. Among the benefits of this "orbital complex" cited in the article would be the ability for continuous monitoring of the state of the atmosphere and crops, detecting forest fires, and producing mineral resources surveys; tracking ships and aircraft; providing stable television reception and permanent radio and television communications; batch production of materials unattainable on Earth; and serving as a base for "ambitious space projects like nighttime illumination in regions of the far North using reflected sunlight."

Clearly they have big plans. In the nearer-term future, it would not be unexpected to see more frequent launches of the Kosmos 1443-type spacecraft, perhaps outfitted as industrial plants in space that could operate autonomously for a time and then dock with a Salyut for change out of materials by a space station crew.

References

[1]Krasnaya Zvezda, Sept. 17, 1983, p. 3.

[2]Aviation Week and Space Technology, Oct. 26, 1981, p. 15, and Nov. 30, 1981, p. 17.

[3]Pravda, Nov. 4, 1983, p. 3.

[4]Smith, M.S., Hellman, A., and Dodge, C.H., "Soviet Space Programs 1976-1980 (With Supplementary Data Through 1983), Part 2, Manned Space Programs and the Space Life Sciences," prepared for the Senate Committee on Commerce, Science, and Transportation, US Government Printing Office, Washington, D.C., 1984.

[5]Gubarev, V., "His Heart Remains on Earth," Pravda, August 15, 1983, p. 7.

[6]Leningradskaya Pravda, Nov. 29, 1983, p. 4.

[7]Paton, B. and Semenov, Y., "For Orbits of the Future," Pravda, Nov. 28, 1983, p. 7.

Space Telescope: The Proto-Space Platform

Herbert Gursky*
E.O. Hulburt Center for Space Research,
Naval Research Laboratory, Washington, D.C.

Introduction

It seems difficult for writers to avoid superlatives when discussing the Space Telescope. What is this program that is variously described as the most important advance in astronomy since Galileo first used a telescope for astronomical purposes and the scientific mission that will provide answers to the major astronomical questions? What is it that makes it a space platform? Why has it been singled out of all of NASA's upcoming science missions to merit inclusion as a separate section in this book?

The truth, of course, is that it is neither a new beginning nor a culmination. As a science mission, it comprises a 2.4-m-diam normal incidence mirror that will yield approximately 0.1 arc sec angular resolution in the visible and ultraviolet wavelength ranges. Instruments in the focal plane will provide spectroscopic and photometric data and imagery on a great variety of astronomical objects. Later sections of this chapter will explain in more detail what these statements mean.

As a space project it incorporates a great deal of the understanding that has developed in the 35 years or so that astronomy has been conducted from space. It is also a very expensive project; however, it is by no means the most expensive science project that NASA has carried out. The Space Telescope has been the subject of a number of substantial reviews recently. International Astronomical

---

*Superintendent, Space Science Division.

Union Colloquium No. 54 (Ref. 1) dealt with scientific research on the Space Telescope. A current description of the Space Telescope as an astronomical observatory was the subject of a special session of the IAU 18th General Assembly.[2] Also, C. R. O'Dell, the first project scientist, described the Space Telescope in Ref. 3.

Astronomy deals with the study of objects seen in the sky well beyond the confines of the Earth. What constitutes "objects" and "confines" has changed with time. The Sun, planets, comets, the fixed stars, and "new" stars (i.e., the sudden appearance of a star) all were known as long back as we have records of astronomy. From the earliest times, there were speculations on what these objects were. Somehow astronomy has always stirred a deep human emotion; indeed, it cannot be an accident that the Bible begins with a statement regarding the origin of the universe.

The history of astronomy has been highly episodic--great advances have occurred in short times and have frequently been related to periods of broad cultural or technological changes. The 20th Century ushered in such a period. At the beginning of this century, scientific and technical advances took place that had a profound effect on astronomy. The discovery of the nature of electromagnetic radiation and of the atom and its nucleus and the introduction of quantum mechanics and of special and general relativity laid the basis for almost all the understanding that we have today of astronomical phenomena. At the same time, a burgeoning industrialism made possible the construction of optical telescopes of a previously unimagined size and quality along with the technical means to study the focused radiation. Figure 1 illustrates an example of a result obtained around 1900 from a newly commissioned telescope that led to the discovery of white dwarfs.

These advances would have been enough for future generations to make note of this period of time as constituting a major scientific revolution. However, humans were also learning to lift themselves above the surface of the Earth, first in balloons, subsequently with airplanes, and finally with rockets. It was natural that these carriers were used for scientific investigations, and it was almost predictable that important and unexpected astronomical discoveries were made. Thus, within the first two decades of the 20th Century, scientists in balloons discovered "ultrahigh-energy radiation" or cosmic radiation, the true nature of which was not realized for another two decades.

These early balloon experiments were the true forerunners of space astronomy as we now know it. In the first place, carrying instruments to balloon altitudes allowed overcoming limitations on measurements imposed by the Earth's atmosphere, principally its absorption. Secondly, astronomical information was obtained from other than the narrow range of visible light wavelengths between about 0.3 and 0.6 μm. Finally, the direct descendants of these early balloon scientists were among the first to use sounding rockets and orbiting satellites for astronomical research.

Following World War II, the technical means existed to carry scientific instruments well above the sensible atmosphere and the scientific understanding existed to provide a firm basis for at least four kinds of astronomical investigations. First, as noted, the cosmic ray scientists knew that even the slightest residue of atmosphere had a deleterious effect on the incoming radiation. Second, it was theorized that the Sun was a strong source of ultraviolet and x-radiation based on solar related effects observed in the Earth's ionosphere. Third, from a general understanding of radiation processes, it was known that many stars and other objects had to be copious sources

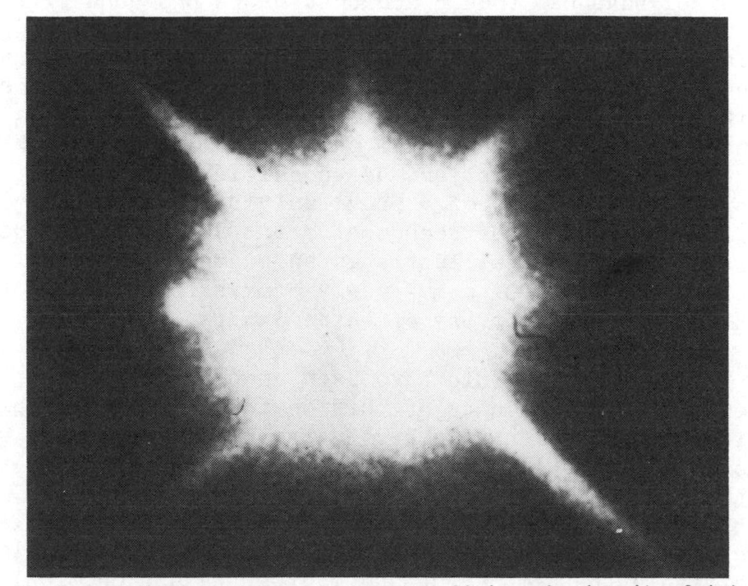

Fig. 1 Image of the bright nearby star Sirius showing its faint, hot companion Sirius B. This photograph was made possible through the introduction of high-quality optics with minimum scattering. The nature of Sirius B was not understood until the discovery of the quantum mechanical principle of degeneracy.

of ultraviolet radiation. Finally, the discovery of radio emission from astronomical objects during the 1930's sensitized astronomers to the possibility that astronomical objects might be found that radiate in every portion of the electromagnetic spectrum. Great discoveries were made during investigations conducted along each one of these lines. Thus NASA, even at its birth, had a broad constituency of astronomers jostling for room aboard whatever carriers could be provided.

During the past 25 years, there has emerged the basic "mission" concept, namely, an orbiting satellite along with scientific instruments, generally dedicated to a specific branch of astronomy. It has frequently been the case that a series of satellites was configured, where each satellite was more or less the same, but carrying a different set of scientific instruments.

One of NASA's first space astronomy missions was the Orbiting Astronomical Observatory (OAO), four of which were launched, two successfully. This mission was principally concerned with the observation of the ultraviolet emission from stars. At about the same time, flights of the Orbiting Solar Observatories (OSO) were begun, eight of which were eventually flown. These were principally dedicated to studies of the Sun, but carried secondary experiments dedicated to other disciplines as well.

Since these earliest missions, the United States has developed and orbited the Small Astronomy Satellite (SAS) and the High-Energy Astronomy Observatory (HEAO) series, each of which comprised three satellites covering the disciplines of x-ray, gamma rays, and cosmic rays. In addition, single missions have been configured--Skylab, International Ultraviolet Explorer (IUE), the Solar Maximum Mission (SMM), and the Infrared Astronomical Satellite (IRAS)--in the disciplines of solar physics, ultraviolet astronomy, and infrared astronomy. Finally, two major missions are now in the construction phase and scheduled for launch in the late 1980's--the Gamma Ray Observatory and the Space Telescope.

Origin of the Space Telescope

Although it is clear from the above discussion that US scientists and engineers have been responsible for a progression of astronomical missions, the essential idea for the Space Telescope emerged well before NASA's existence (see Ref. 4, for a detailed history of the idea for a space telescope). The key point is that from the very earliest of the nationally based studies established to advise NASA on priorities for the space sciences, the idea of a large, high-quality telescope in space has received serious consideration. To quote one of these, "We conclude that a space telescope of very large diameter, with a resolution corresponding to an aperture of at least 120 inches, detecting radiation between 800 A and 1 mm, and requiring the capability of man in space, is becoming technically feasible, and will be uniquely important to the solution of the central astronomical problems of our era."[5]

NASA formally began studies of the Space Telescope in 1971 (then called the Large Space Telescope). The announcement of opportunity soliciting proposals from the scientific community was issued in 1977. Based on this solicitation, four groups were given responsibility for focal plane instruments and a number of individuals were named to other responsibilities relating to the scientific objectives of the mission. The primary spacecraft contractors were selected at about the same time and the Space Telescope Science Institute was established in 1981. Initially, launch was scheduled for 1985, but various technical problems have resulted in delays totaling a year or more.

Science with the Space Telescope

As noted in the introduction, astronomy deals with the study of objects at great distances from us, and, until recently, all investigations were conducted from the surface of the Earth. The Earth's atmosphere represents a considerable barrier to such observations, both by degrading the coherence of incoming radiation (by scattering and refraction) and by diminishing its intensity (by absorption and scattering). Also, the atmosphere is itself a source of radiation. The Space Telescope (ST) is not subject to these constraints. The limitation on the image quality will be determined by the diffraction limit of the primary mirror and by other hardware constraints to be about 0.1 arc sec. On the ground, it is possible to obtain 1 arc sec images, but not consistently. Also, the short wavelength response will also only be limited by the mirrors, in this case, their reflectivity. With ST, mirror coatings are being applied that will allow good performance to about 0.12 μm rather than the 0.3-0.35 μm normally encountered on the Earth's surface. There is no natural long wavelength limit; however, longward of a few microns, ST loses many of its advantages over ground-based observatories. These performance advantages translate into the following observational advantages for astronomy:

1) Much better spatial resolution in extended objects. Our ability to discern features and resolve stars will be improved by a factor of 10 compared with the best ground-based photographs of objects such as planetary nebulae, supernova remnants, globular clusters, external galaxies, and quasars. At least some nearby stars will resolve into disks, others will show up as binaries; objects now believed to be pointlike will show structure.

2) Improved signal-to-noise for studying point images and spectra. For faint objects, currently the noise (background) is determined by the sky-background. The ST will see at least a factor of 2 less background at visual wavelengths because of the absence of the atmosphere, but more importantly, an additional factor 100 less background for point sources because of the higher angular resolution. For spectral data, which are one-dimensional, the additional reduction may only be a factor of 10.

3) Improved signal/noise for studying short-time variability. Atmospheric conditions vary on all time scales, imposing limits, which are more or less independent of the brightness of an object, on how the variability can be studied. Above the atmosphere, however, such effects will be absent, allowing measurements to approach limits imposed by count statistics and other intrinsic constraints, independent of brightness. Earth occultation will make it difficult to observe phenomena that vary on time scales of the order of 1 hour.

4) It will be possible to obtain near-simultaneous photometric and spectral data over a very broad spectral range from about 0.1 to 1 μm.

Another aspect of operating in space compared with carrying out ground-based optical astronomy is a matter of operating efficiency. Ground-based telescopes are notoriously inefficient. Between the Sun, the Moon, the weather, atmospheric conditions, city lights, meteoric dust, etc., ground-based optical observatories achieve a net operating efficiency of less than 10%. The equivalent number for ST will probably be 40%-50%, with the principal limitation being Earth occultation. Overall, recent NASA astronomical missions such as IUE and Einstein have achieved a very high degree of operational efficiency, and this situation should continue to be possible for ST.

Since it is still the case that much of astronomy depends on information obtained in the ST wavelength range, these advantages mean that ST represents virtually a new beginning for astronomers. Almost all existing optical data could be improved upon, and it will be possible to obtain entirely new kinds of information. Indeed, the principal limitation to how broadly ST will be applied will, in fact, be access to observing times.

But astronomy is more than an encyclopedic collection of facts; there are major unresolved problems, the solutions to which may require the kind of data provided by the Space Telescope. Examples of these are cited below.

Planetary Science

Although most of the detailed, specific information we have on planets will come from probes, it is the case that synoptic imagery taken from ST will provide unparalleled information on a variety of atmospheric phenomena known to occur. In the case of Venus and Mars, ST will provide images with special resolution of 20 and 35 km, respectively, at the time of closest approach, which is comparable with terrestrial pictures currently being provided by National Oceanic and Atmospheric Administration's (NOAA) geosynchronous weather satellites.

Globular Clusters

These groupings of stars are the oldest condensed objects which we can conveniently study, and their nature has an important bearing on the origin of stars and galaxies generally. In fact, there may be a minor embarrassment brewing in that the ages of globular clusters as determined by their stellar content may be greater than the age of the universe as determined by the Hubble constant (H_0) and the deceleration parameter (q_0). Fortunately, observations with ST can be used to improve all the relevant parameters. Globular clusters--large numbers of stars (point sources) with severe crowding in their central regions--are ideal to study with ST. A particular point of interest is that the very inner regions of globular clusters may contain black holes, formed either during the initial collapse of the gas cloud that formed the cluster or subsequently by the coalescence of the innermost stars. Depending on its mass, such a black hole will be easy to detect based on the motion of nearby stars.

Galaxian Distances

This parameter is a crucial starting point for a variety of investigations, most notably the determination of the Hubble constant. Currently, one of the limits in our ability to do so is the necessity to resolve stars in galaxies and measure their brightness, an area in which ST will excel.

Evolution of Galaxies

Galaxies are dynamically stable agglomerations of stars and must represent one of the earliest and primary condensations from the primordial gas that constituted the universe 10-15 billion years ago. Thus, from a starting point as a tenuous gas made up of hydrogen and helium, we now have a vast richness of stars, clusters, supernovae remnants, etc., and chemical elements up to Uranium. We know more or less how this came about, but because of the finite travel time of light, we actually can go back into the past and see the process evolve. Presently, our principal limitation is spatial resolution--with ST we can go back ten times further than we can from the surface of the Earth with the same image quality. We can even reach back into the era when quasars were plentiful. Thus, we can see the universe at much earlier times than we now can and try to unravel the evolutionary processes without having to sort through conditions in our own and in nearby galaxies.

To summarize, the importance of ST is that it provides so much higher-quality data than has been available in the past that it provides almost a new beginning in the area of optical and ultraviolet astronomy. Furthermore, this region of the electromagnetic spectrum provides a variety of information only marginally available in other wavelength ranges. These data bear on certain of the fundamental issues of astronomy--the origin and evolution of the universe and the origin and evolution of the stars and the chemical elements, for example.

The Technology of the Space Telescope

The current layout of the Space Telescope is shown in Fig. 2. The design is dominated by the following factors:

1) The volume (4.6 m in diameter and 14 m in length) and weight (17,000 kg) available on the Shuttle.

2) The primary mirror diameter of 2.4 m (f:24 Cassegrain configuration) requires overall precision of tens of microns. The mirror surface itself must be precisely held to fractions of a micron.

3) Since many of the detectors operate in an integrating mode, the telescope must be pointed stably with a precision of greater than 0.1 arc sec during exposures in order to preserve the image quality.

4) The focal plane must be able to accommodate a variety of instruments; in fact, there are six different instruments incorporated into the ST--the fine guidance sensors, the wide field camera, the faint object camera, the faint object spectrograph, the high-resolution spectrograph, and the high-speed photometer.

5) Stray light from the Earth, Sun, and Moon could represent a significant limitation on the achievable scientific results.

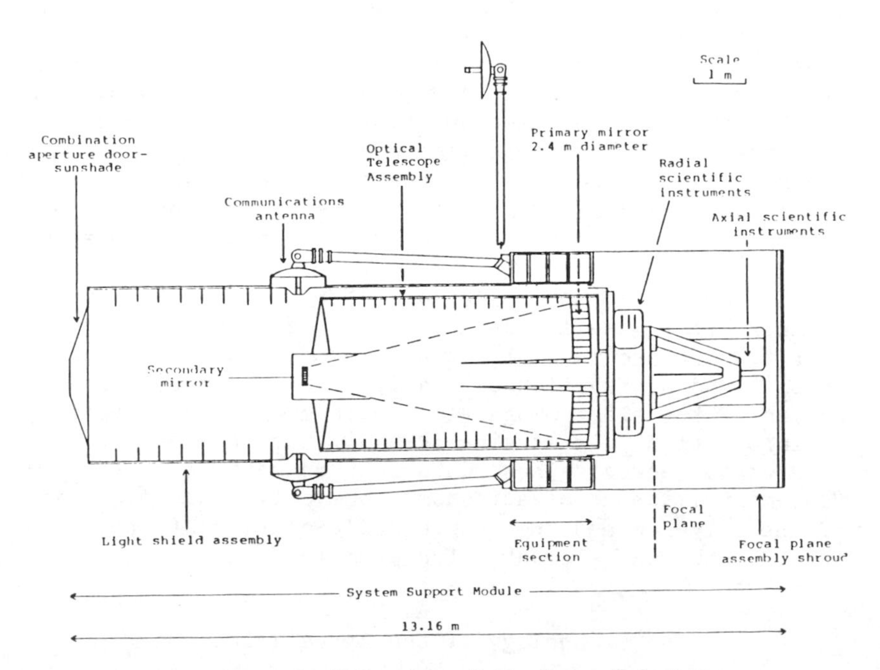

Fig. 2 Conceptual drawing of the Space Telescope.

These factors have been the principal drivers on the technology associated with ST. For example, the very high precision has required the development of a graphite-epoxy truss to assure mechanical and thermal integrity of the optics. The primary mirror itself incorporates activators to control the figure.

The complexity and the power of the Space Telescope is well illustrated in Fig. 3, which shows how the focal plane of the telescope is utilized. Note first that light is utilized from a 15-arc min radius image area in the focal plane. The maximum resolution, 0.1 arc sec is preserved to within about 5 arc min from the center of the field. At 10 arc min, the resolution will degrade to 0.3 arc sec, and over the entire 15 arc min the resolution is still substantially better than what is achievable on the Earth.

The focal plane is divided into four regions. The innermost 3 arc min by 3 arc min section is currently reserved for the wide field camera, which can operate continuously and obtain images while observations are being conducted with other instruments. Next, between about a 4- and 9-arc min radius, are four axial sections reserved for other instruments. The outermost ring of three sections is

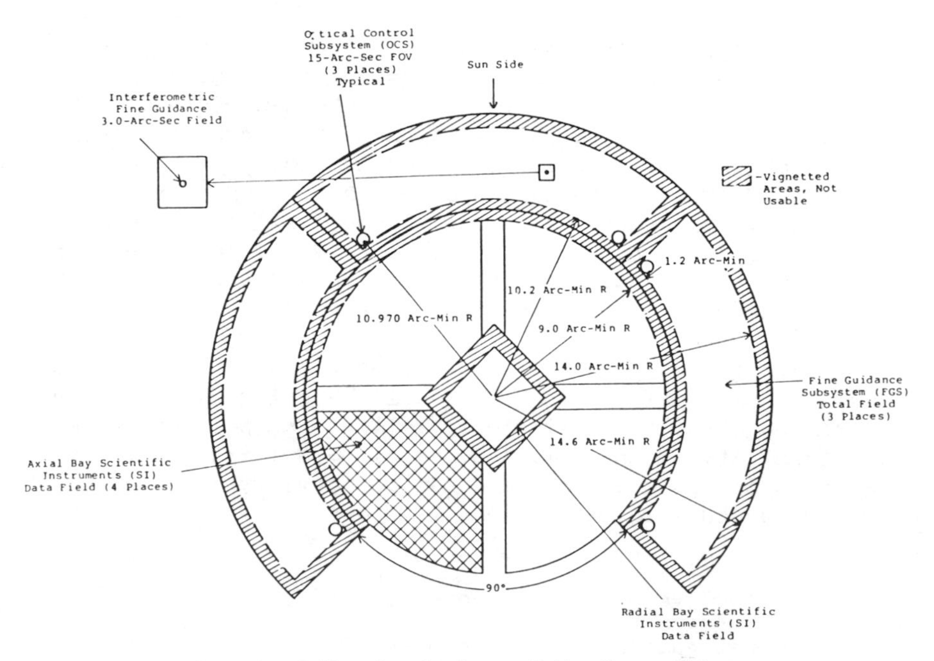

Fig. 3 Layout of the focal plane of the Space Telescope.

reserved for use by star trackers. Thus, there are eight distinct sections that can accommodate measuring instruments, which allows for multiple observations during a given observing run.

Space Telescope is currently being flown with six distinct instruments, each with a number of operating modes. As noted, the wide field camera occupies the central radial position and provides continuous imagery. The axial positions are occupied by a faint object camera, two spectrometers, and a photometer. These instruments are designed to make detailed measurements on single objects. Finally an astrometric capability, the fine guidance sensor, is incorporated into each of the three outermost segments. These instruments are among the most complex built for astronomy, ground or space. For example, the faint object camera offers the following options.

1) The f:96 relay: 34 filters, 5 attenuators, 3 polarizers, 2 objectives prisms, and apodizing mask, and occulting fingers.

2) The f:48 relay: 12 filters, 2 objective prisms, 1 fixed grating, and 1 focal plane slit.

There are also factors relating to the mission that dictate the design; most notable is the fact that astronomical observations are being conducted meaning that the ST must be inertially pointed for extended periods of time. Given the variety of observations to be conducted, there must be great flexibility in how the spacecraft can be operated and how the data are retrieved from orbit. Normally, observations can be planned well in advance of their implementation; however, there is a requirement to respond to transient phenomena of high interest; e.g., the appearance of a new comet or a supernova or nova outburst.

The ground operations are dominated by the fact that a large number of investigators will be using ST. That, plus the fact that ST intrinsically allows for a large degree of flexibility, means that the overall complexity of mission operations is much greater than is normally associated with an unmanned science mission.

The Space Telescope Science Institute

In recognition of the breadth of science that can be accommodated by ST, the complexity of the operations, and the necessity to involve the broadest possible group of scientists, NASA has made a major change in how it intends to interface with the scientific community by establishing an independent research institute, the Space Telescope Science Institute (STScI), with that responsibility. Chosen on the basis of competitive proposals, STScI is operated by the Association of Universities for Research in Astronomy Inc. (AURA) on behalf of NASA and is located on the campus of The Johns Hopkins University in Baltimore, Md.

STScI staff will also be located at Goddard Space Flight Center (GSFC), which retains its traditional responsibility for capturing telemetered data, command and control of the spacecraft, and monitoring the health of the spacecraft.

The responsibilities of the STScI includes processing and archiving the scientific data, selecting and funding observers, evaluating the scientific peformance, and conducting research through its internal staff. As an independent institution, STScI will have oversight by an independent visiting committee and provide peer review for observing proposals by a time allocation committee, very much in the style of other national scientific facilities. Thus, NASA recognizes that Space Telescope has a role in national affairs transcending its position as a NASA-operated facility.

Space Telescope as a Space Platform

In contrast with other space astronomy missions, it is hard to find a discussion of the Space Telescope without a linkage to manned operations. For example, Lyman Spitzer commented in 1974 in his contribution to Large Space Telescope--A New Tool for Science that "the development of the Space Shuttle...offers the prospect of regular visits by trained astronauts, who could maintain and update the instruments."[4] James Downey, in the same document, identifies the following guidelines for maintenance of the ST:

> 1. Return of LST [Large Space Telescope] to the ground is primary mode to accomplish major servicing.
>
> 2. The capability for on-orbit servicing shall be considered.
>
> 3. The first opportunity for LST ground return shall be assumed to occur after 2½ years in operation.

4. The scientific instruments shall be arranged to permit on-orbit replacement of any instrument or appropriate sub-assemblies.

5. Relatively easy access, both on ground and in orbit, to all LST components of modules is a design goal.

More generally, scientists have approached the question of man-supported operations more gingerly. For example, the following comments appear in the 1969 position paper of the Astronomy Missions Board[6]:

"On the one hand, it appears clear that an entirely unmanned astronomy program is possible.... On the other hand, it seems entirely possible that the range of astronomical research in space could be widened to a decisive degree if manned operations in space were applied to space astronomy."

NASA has approached the subject equally cautiously. The Skylab mission clearly demonstrated how astronauts could contribute directly to the success of a major astronomical mission. More currently, NASA had made important commitments to manned operations in its Shuttle program by specifically incorporating instrument operations into the activities of the Shuttle personnel. These range from simple turn-on/turn-off to the launch and retrieval of small packages that need to operate independently from the Shuttle. More significantly, NASA is committed to the concept of payload specialists, individuals with a technical background, selected and trained with a specific mission and set of experiments in mind. For example, at my own institution, the Naval Research Laboratory (NRL), two NRL staff members, Dr. John-David Bartoe and Dr. Dianne Prinz, are in training for the Spacelab II mission to assist with the operation of the High Resolution Telescope and Spectrograph (HRTS) instrument, a solar spectrometer developed by NRL for which they are co-investigators. Not only has NASA provided for such individuals on all its major Shuttle missions, but, not unexpectedly, the positions have become highly competitive.

Regarding the Space Telescope itself, maintainability in space is being built into the spacecraft. The entire payload is designed to be brought back into the Shuttle bay, from which it was originally launched, and requires folding up solar cells, antennas, and various other appendages that are deployed during operations. Back in the Shuttle, the payload can be refurbished and carried to a higher altitude. The ST altitude is quite critical; too high and it will encounter an increasing radiation flux which produces an undesirable background of events in its focal plane detectors and can even interfere with the operations of its computers; two low, and the effects of aerodynamic drag interfere with the ability to point accurately. Eventually, the payload will drop in altitude to where orbital operations are not possible. Thus, this reboost capability is essential for long life.

In principle, if the ST is brought back into the bay, it could also be brought back to Earth. As attractive as this might seem, it may not be practical to do so. First, the telescope would be out of operation for many years during a ground refurbishing cycle. Second, a site and crew would need to be prepared to handle the payload. The expense of such operations might, in fact, rival the cost of a new mission.

It is the case that the major subsystems and certainly the focal plane instruments are being designed to be replaced in orbit. Thus, the principal aspect of the maintainability is likely to be replacement of focal plane instruments as newer or more powerful instruments become available and the replacement of subsystems that have failed. With this capability and the reboost capability, the ST is assured an almost indefinite lifetime as a space observatory.

Thus, with Space Telescope, space operations comprise much more than simply retrieval of data by telemetry and changes of configuration by command. A true, but limited, ability to maintain in space exists that will greatly expand the capability of this remarkable instrument. It will be an important step along the way of developing a routine presence in space for society.

References

[1]Longuir, M. S., and Warner, J. W. (Eds.), Scientific Research with the Space Telescope, IAU Colloquium No. 54, US Government Printing Office, Washington, DC, NASA CP-211, 1979.

[2]Hall, D. N. B. (Ed.), The Space Telescope Observatory, IAU 18th General Assembly, NASA CP-244, 1982.

[3]O'Dell, C. R., "The Space Telescope," in Telescopes for the 1980s (G. Burbidge and A. Hewitt, Eds.), Ann Reviews, Palo Alto, Calif., 1981.

[4]Spitzer, L. Jr., "A History of Large Space Telescope," in Large Space Telescope, a New Tool for Science, AIAA 12th Aerospace Sciences Meeting, Washington, DC, 1974.

[5]Space Research: Directions for the Future, Part 2, Astronomy and Physics, National Academy of Sciences, Pub. No. 1403, 1966.

[6]A Long-Range Program in Space Astronomy, Position Paper of the Astronomy Missions Board, NASA SP-213, 1969, p. 271.

The European Reusable Space Platforms SPAS and EURECA

Dietrich E. Koelle*
Messerschmitt-Boelkow-Blohm-ERNO, Ottobrunn, Federal Republic of Germany

Abstract

The paper deals with the new space systems, "Reusable Platforms," which have been made feasible by the U.S. Space Shuttle. The first of this kind was SPAS-01, an experimental platform developed as industrial venture by MBB, and flown twice in 1983 and 1984. The first operatinal reusable platform for micro-g-applications, called Eureca, is now under development for ESA as a European program. Both platforms are described with their technical and operational features.

Reusable Platforms

Reusable space platforms as a new type of orbital system were conceived in Europe as a logical application of the new reusable Shuttle/Orbiter transportation system. In 1976 the idea of a simple Shuttle-dedicated satellite platform that would not only be launched by the Shuttle but also retrieved for multiple reuse was conceived by the advanced project team at Messerschmitt-Bölkow-Blohm GmbH in West Germany. The feasibility, cost advantages, and a number of potential applications and users were identified by internal studies in 1977, and the new concept was presented at the 1978 Goddard Memorial Symposium in Washington, D.C.[1] and subsequently discussed in the press.[2]

---

*Director, Advanced Space Systems and Technology, Space Division.

In the early study phases, it became evident that mission-dedicated space platforms were required in contrast to multipurpose platforms designed for a large number of different experiments. Three basic types of reusable platforms were identified: 1) platforms for astronomical observations (Astro-Platform), flying different telescopes in sequence; 2) platforms for remote sensing and global monitoring (Earth-oriented, high data rate); and 3) platforms for space processing experiments and operations (Sun-oriented for power reasons). These three basic platform missions require different orbits, different orientation in space, and different subsystems.

The inherent advantage of small, dedicated, reusable platforms is their versatility and inclination as well as the relatively low investment required for construction. Reusable platforms represent not only a stepping stone towards permanent space platforms but will probably also be a versatile supplement for special missions to future permanent platforms in space.

The SPAS-01 Demonstration Platform

When seeking support for the new concept of a reusable platform, the MBB team soon recognized that years would be required for the introduction of such a new tool into the official space program planning. Being fully convinced of the new approach, especially with respect to future commercial space activities, MBB decided in 1979 to go ahead with its own experimental demonstration project called Shuttle Pallet Satellite No. 1 (SPAS-01). The German Ministry for Research and Technology (BMFT) supported this industrial venture by allocating a number of scientific and technological experiments under development in West Germany to be flown on SPAS-01 for a fixed fee. This covered about half of the total cost but would not have allowed for a free-flyer mission. Therefore, it was a fortunate development that NASA was then seeking a device for exercising rendezvous and retrieval operations for future Shuttle payloads. There was no other alternative available than the SPAS-01 extended to a free-flyer system. The negotiated special launch agreement resulted in a cost contribution of some 20% by NASA. The rest was MBB Company funding, except for a small charge from European Space Agency (ESA) for flying three small experiments (1%), a real bargain for this organization.

The SPAS configuration and structural construction was completely different from conventional spacecraft designed for

Fig. 1 SPAS-01 with the project team and NASA payload specialists J.M. Fabian and Sally Ride.

expendable launchers with a symmetrical spacecraft adapter. The truss structure bridges the orbiter cargo bay and interfaces with two longeron trunnions and one keel fitting. The structure is completely modular, consisting of carbon-fiber tubes and titanium nodes and carrying all equipment on standardized mounting plates. This provides ideal conditions for assembly and testing (Fig. 1). A large safety and mass margin were inherent features of the design as well as the use of commercial components (like diver pressure bottles for the N_2 gas storage). The total mass of SPAS-01 was 1501 kg, including a 900-kg payload consisting of eight experiments and three NASA cameras.

Originally planned for launch in June 1982, the STS-7 flight finally took place from June 18 to June 24, 1983, and resulted in a fully successful flight and retrieval of SPAS-01 (Fig. 2). This first demonstration of orbital spacecraft operations by the Shuttle paved the way for follow-on activities like the Solar Max spacecraft repair. It also brought back from space a series of Shuttle orbiter pictures as well as the first series of scenes from the Earth surface taken by the MOMS Camera (Modular Opto-electronic Multispectral Scanner), the first fully electronic camera flown in space. In order to demonstrate reusability and to gain additional data and experience, SPAS-01 was flown a second time on the eleventh Shuttle flight (mission 41B), Feb. 3-12, 1984.The payload equipment flown was identical to the first flight, however, a training unit for the planned SMM spacecraft repair was added (Fig. 3).

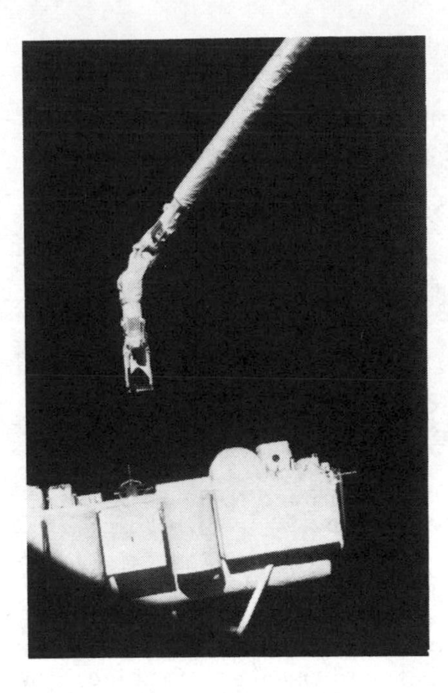

Fig. 2 SPAS-01 in space after release.

The EURECA Platform

Studies performed by ESA in 1981-82 to define a Spacelab follow-on program resulted in the preference for a reusable automatic payload carrier or platform over the other alternatives studied. In addition, a survey of future payload and user requirements confirmed a strong demand for such platforms: 75% of all proposals preferring a platform flight opportunity.

A phase A study was performed in 1982 with two basic candidates: one using the Spacelab Pallet as structure and the other using the SPAS concept. The SPAS-configuration was selected by ESA for a detailed phase B study, performed in 1983 84 by MBB-ERNO in Bremen. The core payload selected concentrates on material processing and biology and leads to the implementation of a dedicated Sun-oriented spacecraft, as already conceived in 1978 (Ref. 1).

Figure 4 illustrates the so-called European Retrievable Carrier (EURECA), with a total mass of more than 4000 kg (8800 lb), 1000 kg being payload mass (Ref. 4). This first dedicated, reusable platform project will be realized in the 1984-88 period as an ESA program. Follow-on launches are foreseen every two years.

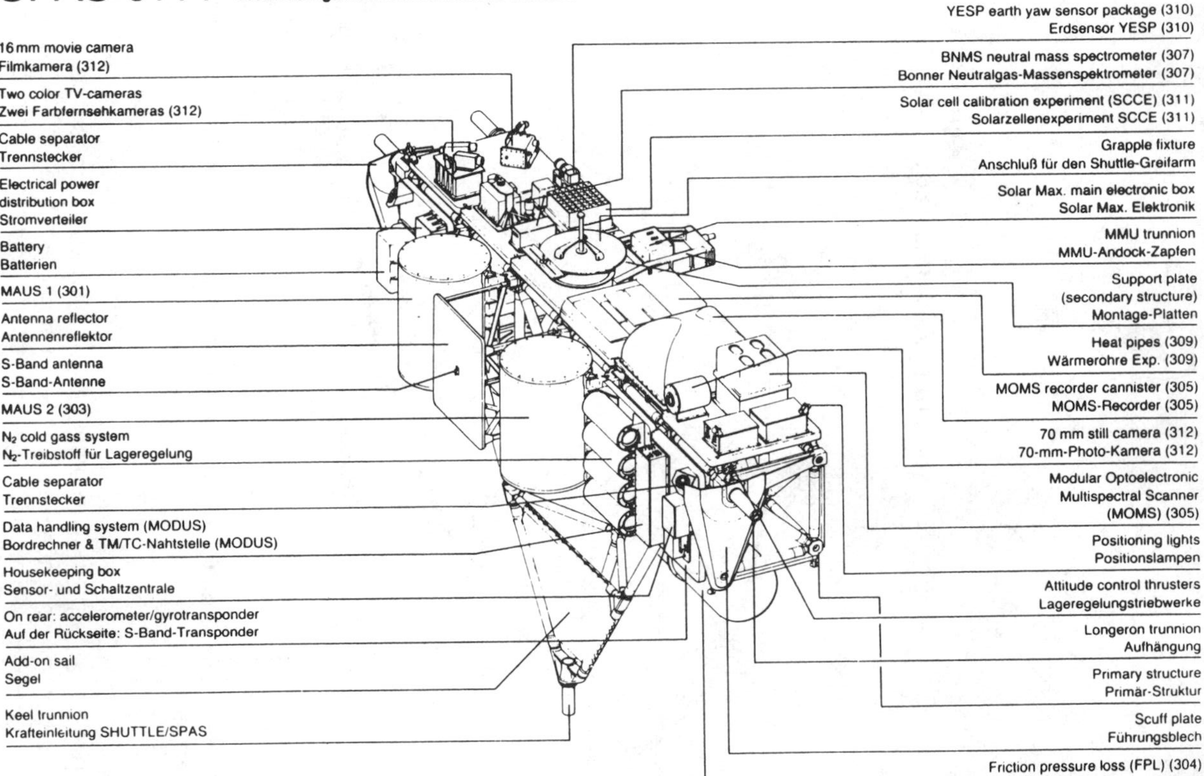

Fig. 3 SPAS-01 A (second flight) overall configuration.

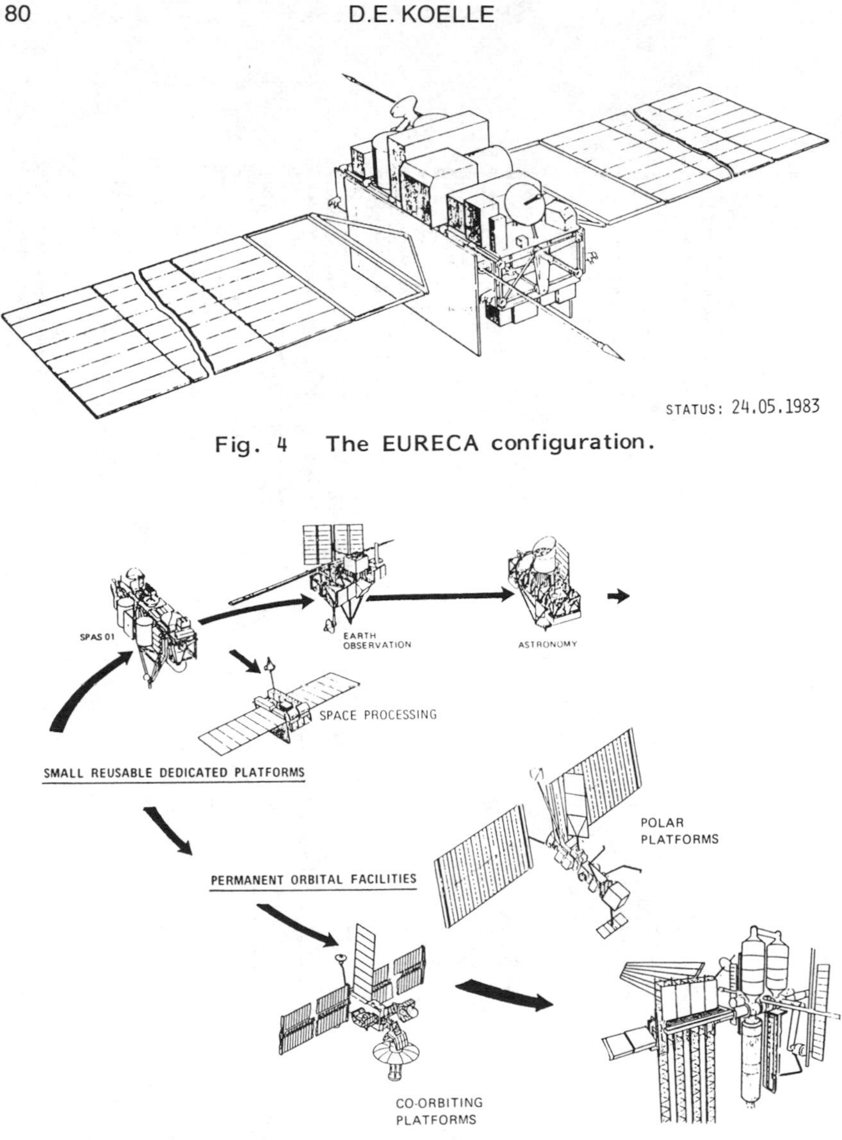

Fig. 4 The EURECA configuration.

Fig. 5 Space platforms scenario with reusable, polar, and coorbiting platforms.

The first mission of EURECA in 1988 will comprise a number of material science facilities (core payload) plus some add-on experiments for biology and solar physics. For the second mission in 1990 a rendezvous and docking exercise is foreseen for EURECA with the SPAS platform as an orbital test of the newly developed European RdV equipment.

EURECA provides major improvements over the first experimental SPAS platform, such as an orbit stay time capability of six to eight months, a large solar array power supply (5.4 kW) and radiator surfaces (11 m^2), twice the experiment payload, and an improved attitude and orbit control system.

EURECA is also equipped with an on-board propulsion system, comprising eight 20 N thrusters, and six propellant tanks with 620 kg Hydrazine. The propulsion system is required to increase the orbit altitude from the Shuttle release altitude of some 300 km to the platform initial mission altitude of 500 km.

It can be shown that the micro-g conditions on this free-flying platform are better by at least one order of magnitude compared with Shuttle-based systems (such as Spacelab). Also, the specific costs are considerably lower, which will be a major criterion for any future commercial activity in this area.

Figure 5 shows a platform scenario as it probably will develop in the future from dedicated reusable platforms, to permanent polar platforms, to co-orbiting platforms in the vicinity of a large manned space station serving as an operations center.

References

[1]Davidts, D. and Koelle, D.E., "The Shuttle-dedicated Pallet Satellite (SPAS) and its Applications", 1978 Goddard Memorial Symposium, Washington, D.C., March 8-10, 1978.

[2]"Germans Develop Multimissiom Satellite," Aviation Week, May 8, 1978.

[3]"Orbiter Deploys, Retrieves SPAS,"Aviation Week, June 27, 1983.

[4]Mory, R.L., Nellessen, W., ESA/ESTEC Noordwijk, and Wienss, W., Hochgartz, K., MBB-ERNO Bremen, "European Retrievable Carrier - an Evolutionary Approach toward a European Space Platform", IAF-Congress Paris/France, 27 Sep. - 2 Oct., 1982, Preprint No. IAF-82-11.

Chapter III. Introduction: Space Station and Platform Roles in Supporting Future Space Endeavors

Introduction: Space Station and Platform Roles in Supporting Future Space Endeavors

Daniel H. Herman*
NASA Headquarters, Washington, D.C.

President Reagan has committed the United States to a program that will ultimately establish the human species as a permanent space occupant. Mankind will finally have the facility to fulfill the vision of Tsiolkovski, who wrote, "The Earth is the cradle of mankind, but man cannot stay in the cradle forever." This facility will be developed by increments in an evolutionary manner, each increment expanding man's capability to function in space. The first increment will be operational in the early 1990's.

Many endeavors have been analyzed and enunciated that necessitate the development of this facility. The phenomenal results derived from space astronomy missions, such as the Infrared Astronomical Satellite (IRAS), have generated a desire to install the equivalent of a Mount Palomar observatory in space. This concept entails that this observatory will be structured so that focal plane instruments can be changed, with the implication that this observatory will be maintained in situ while conducting continuous observations of the universe.

The yield from Earth observation instruments, such as those obtained from the Landsat satellites and the Shuttle Imaging Radar (SIR-A), has demonstrated the potential for a greater understanding of our natal planet that would result from the development of an Earth observing instrument platform deployed in a near-polar Sun geosynchronous orbit. This platform would have the ability to house a set of Earth viewing instruments in various spectral bands that could be commonly boresighted and registered. This capability would have a profound influence in enhancing our knowledge of the atmosphere, the oceans, and the solid Earth.

---

*Director, Engineering Division, Office of Space Station.

A permanently manned Space Station would enable the development of commercial laboratories that, due to the unique environment afforded by the weightlessness of space, could produce materials not obtainable elsewhere. The pharmaceutical materials thus generated will contribute significantly to the quality of life by eliminating many debilitating diseases. Crystals generated in space could fuel our burgeoning information industry. In the future, a Space Station, in low Earth orbit, could serve as a staging base for the deployment of large systems to geosynchronous altitude, the Moon, and, ultimately, to Mars and nearby asteroids. The resources of our solar system will be available for the benefit of mankind.

These multifaceted sets of functions dictate an architectural approach that is widely distributed. Pressurized laboratories with manned intervention are necessary for the conduct of manufacturing processes. A servicing facility is needed for the maintenance and repair of unmanned space assets. An unmanned platform, with manned accessibility, is required to incorporated delicate astronomical instruments. Transportation between the Space Station and these platforms is required. A similar platform for Earth observations is required at a polar inclination. Ultimately, a manned base in polar inclination will support these Earth observation systems. Finally, an assembly capability must be established in space, coupled with a transportation node, to deploy large space systems to geosynchronous altitude and beyond. The very presence of man in space allows for research that will sustain the human species in space for longer and longer periods.

In the early days of space flight, man's basic tolerance of weightlessness was a matter of considerable debate. Many learned scientists feared the worst, and a variety of research programs were quickly started to gain some insight into this new problem, which might seriously curtail the future of manned space flight. Without solid data as to the true physiological effects of weightlessness, artificial gravity was considered as a logical solution to the problem. Fortunately, the rapidly accumulated experience of the Mercury and Gemini programs demonstrated that human physiology was quite tolerant of weightlessness, and subsequent return to Earth revealed no lasting untoward effects.

During the Apollo and, particularly the Skylab missions, more detailed investigations indicated that weightlessness indeed caused significant alterations in human physiology. In some cases, the altered effects were short-lived and self-limiting, as with the space adaptation syndrome (motion sickness). In other cases, simple countermeasures appeared adequate to preserve terrestrial norms, as in the use of lower body negative pressure to reduce postflight orthostatic intolerance. Finally, in a few cases, the effect is likely cumulative, as in the case of bone demineralization and radiation exposure. Although the current countermeasures are not perfect, they do appear to forestall the effects of weightless exposures and preserve a physiological state that is readily reversible to terrestrial norms upon return to Earth.

With the advent of a Space Station, it is necessary to again consider the physiological effects of weightlessness, no longer in an acute sense but now in terms of long-term exposure. Recent Soviet experience on Salyut has indicated that man can remain aloft for at least seven months and perhaps a year. The planned NASA Space Station will initially operate with three-month stay times for crew members. When longer stay times will become necessary is not yet clear. However, the Space Station is an ideal laboratory for developing a more detailed understanding of the physiological and psychological effects of living in weightlessness. The perfection of more effective and efficient countermeasures that do not significantly inhibit crew productivity will also be developed using the Space Station. Finally, the evaluation of new alternatives to our current array of countermeasures, such as the systematic, short exposure to various levels of artificial gravity, will provde the knowledge base that will enable planetary exploration.

Initially, man's use in space will be limited and very precious. His efficiency must be enhanced by making full use of automation technologies. Sophisticated end-effectors must be developed for servicing vehicles. Artificial vision systems must be developed for the repair and maintenance of remote assets. Robotic technologies must be developed to enable the assembly of large structures in space. These systems must be controlled by "expert systems" leading toward a degree of artificial intelligence in space. With these tools, man can function as an intelligent/adaptive supervisor, controller, manager, and explorer.

The advent of man's permanent presence in space will alter, to a major degree, endeavors that have been constrained by a lack of this capability. The support of major astrophysical observatories will be realized. We will have the capability to assemble extremely large apertures in space in various spectral bands that might yield a major insight as to the evolution and future of the universe. We might obtain fundamental knowledge as to the nature of matter and energy.

The Space Station, as a staging base for manned planetary exploration, will enable us to repeat on the Moon the same profound advances in knowledge made on the Antarctic continent during the International Geophysical Year. We can now consider, in a realistic manner, the return of a distributed selected sample of Martian material to the Earth. The analysis of this material will give us a major insight into the evolutionary history of the terrestrial planets. This staging base capability will enable us to use the resources in near-Earth asteroids to sustain our space facilities. Ultimately, we can envision manned exploration of the solar system.

We are on our way to expanding man's vista beyond the Earth. Some of the consequences of this capability are now discussed.

Astronomy and the Space Station

Herbert Gursky*
E.O. Hulburt Center for Space Research,
Naval Research Laboratory, Washington, D.C.

Introduction

Space is the natural site from which to conduct astronomy. From the surface of the Earth, the atmosphere is a murky, opaque medium through which passes only a tiny fraction of the radiation that carries astronomical information. Thus, astronomers have always sought the high ground--carrying instruments to mountaintops and flying them on rockets and balloons. It is natural that astronomers have been among NASA's earliest and best customers.

It is also natural that astronomers are able to take advantage of man's presence in space--as scientists to operate the instruments and understand their responses and as technicians to build facilities and maintain them.

Indeed, all these things have been happening and will continue as long as the opportunities will be provided. Skylab carried a cluster of solar instruments that was operated by astronauts. Astronomical instruments were carried along as part of the Apollo missions and some were operated from the surface of the moon. The Shuttle is heavily used by astronomers, and, in 1984, the Solar Maximum Mission, a traditional, free-flying satellite, was repaired in space by astronauts from the Shuttle.

The advent of the space station should provide the astronomical community with an exciting new potential for extending their observational capability. The Space Science Board, the principal scientific advisory panel in the United States, noted in a report to the NASA Administrator that the present space transportation system,

---

*Superintendent, Space Science Division.

Shuttles and expendable launch vehicles, is adequate for the purposes of space science for the next 15 years or so. The board went on to state, "In the longer term, the Space Science Board sees the possibility that a suitably designed space station could serve as a very useful facility in support of future space science activities. Such a space station could provide means for erecting and fabricating large and novel structures in space, and for servicing, fueling, and retrieval of payloads in orbit."

Astronomy as a Science

Astronomy comprises the study of the universe, beginning with the Sun and extending to the most remote distances from the Earth. The Universe includes a great variety of objects, gases, fields, and radiation. These interact and evolve, and, most important, radiate. It is the study of the radiation that is the work of the astronomers. In addition, astronomers and physicists are concerned with how it all began--the origin of the Universe as we see it around us.

In an astounding century, advances in astronomy have been truly astounding, and much of the new understanding has originated from space-borne instruments. I will use the Sun a fairly common variety of star to illustrate the kinds of investigations conducted by astronomers.

The Sun must be about 6 billion years old. Most of its radiant output emerges as thermal (black-body) radiation from its surface, which peaks at around 0.5 micron in the visible wavelength range. The surface is not uniform. It is pocked with sunspots of somewhat lower temperature and is broken up on a small scale into granules and supergranules. The sunspots are indicative of enhanced magnetic fields that comprise active regions. The granules are apparently convective cells and reflect the upwelling of material from below. Thermal energy from nuclear reactions taking place deep inside the Sun flows to the surface by radiation and by convection. Two very different kinds of experiments provide direct information regarding the interior. One is the study of neutrinos emerging from the nuclear reactions, and the other is the study of surface oscillations (small changes in brightness or velocity) that are driven by the convective flow and may provide information on the size and distribution of the convective cells.

The solar atmosphere is a complex region of high temperatures and tangled magnetic fields. The lowest region, the chromosphere, comprises gas at temperatures between 10^4 and 10^5 K. It is especially bright in the active regions and elsewhere shows the granulation of the underlying surface. Higher up is the corona where gas temperatures range from 10^5 to 10^6 K. At these temperatures, the radiation is principally in the form of ultraviolet radiation and x-rays.

The overwhelming aspect of the Sun's atmosphere is the variety of structures that are seen and the great sense of some underlying order, as is illustrated in Fig. 1. Active regions are the principal focus of the structure. The basic structural element is a loop which appears in all sizes and shapes, either singly or in clusters. Apparently there is an intense magnetic field running parallel to and just below the surface. Sometimes the field will emerge from the surface and form a loop. As an example of the order that exists on the Sun, the north magnetic pole of the major loop structures (where the magnetic flux emerges in the active regions is always east

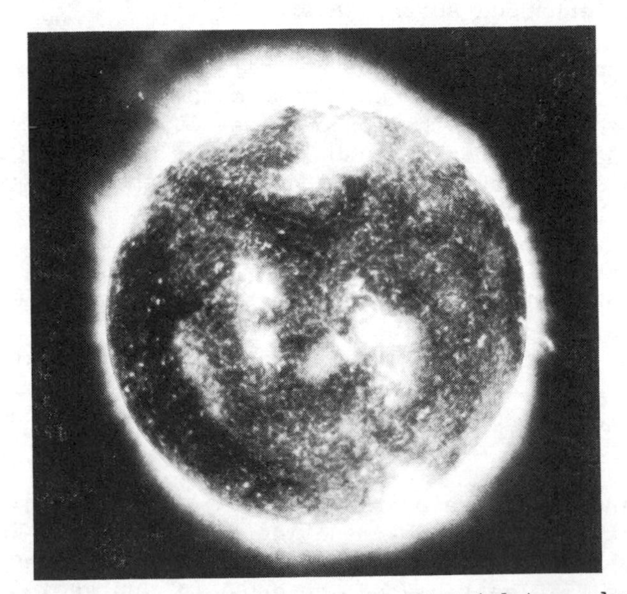

Fig. 1 The Sun as seen in the extreme ultraviolet wavelengths. The chromosphere is seen as the mottling over the whole surface; the corona is seen as the bright, single region and the enhancements at the disk. The solar rotation axis emerges from the Sun near where the disks show minimum emission. Photograph courtesy of the Naval Research Laboratory.

(or west) of the south pole in one hemisphere and reversed in the other hemisphere. In addition to loops are prominences--large regions of cool gas that hang suspended above the surface.

These features exhibit varying degrees of stability and instability. The basic time element is the 11-year solar activity cycle, although some phenomena (such as the east-west loop orientation) require two solar cycles to complete. During the minimum of the cycle, sunspots and active regions essentially disappear from the surface. We know that this activity cycle may disappear entirely for approximately 100 years, perhaps every few hundred years. Individual active regions may persist for many months, but during that time the structures making up the region will change considerably. Occasionally, specific features will become unstable on a very short time scale. The two most dramatic examples are the sudden release of prominences which will propel away from the Sun at velocities of hundreds of kilometers per second, forming coronal mass ejections. The other is the sudden heating, literally explosion, of a loop of gas, forming a solar flare. These events can have dramatic effects on the Earth; for example, enhanced auroral activity.

Coronal mass ejections and solar flares trigger a variety of other phenomena. The mass ejections will create shock waves that propel away at speeds of approximately 1000 km/s. Solar flares may also generate a similar shock wave. These shock waves are believed to be responsible for the acceleration of electrons that emit radio waves and are responsible for radio bursts observed at the Earth. Protons and other nuclei must also be accelerated during mass ejections and solar flares. Gamma-ray emission coincident with flares show clear evidence of discrete lines associated with specific nuclear reactions involving protons and neutrons. Also energetic neutrons and protons are observed in the vicinity of the Earth following flares.

There is also a continuous flow of gas and magnetic field--the solar wind--emanating from the Sun. This wind extends beyond the Earth as far as our measurements have taken us, at least beyond the orbit of Jupiter. The flow is not continuous. A remarkable spatial feature is the sectoring associated with coronal holes. In those regions, the solar magnetic field is more open and perhaps weaker than it is elsewhere on the Sun, allowing the escape of the solar wind at a much higher velocity than otherwise.

Implicit in the above discusson is a high degree of time variability. On a short time scale, solar flares and coronal mass ejections are initiated in seconds, although secondary processes frequently linger for hours. The surface granulation changes in times of the order of minutes. Active regions can change on a day-to-day basis even though, as a distinct region, they persist for months. The solar rotation period of 27 days modulates all these phenomena. The level of solar activity goes up and down with the 11-year cycle, which itself seems to be subject to large changes on time scales of hundreds of years.

Virtually every radiation measuring technique has been used to study the Sun. The electromagnetic emission from the Sun ranges from low-frequency radio to MeV gamma rays. In addition, energetic neutrons, protons, and heavier nuclei are observed. Also, the plasma and magnetic fields comprising the solar wind are studied in situ.

Very little of these emissions can be studied from the surface of Earth. The Sun is the subject of intensive ground-based investigations in the visible light domain (0.3-0.8 microns) and at radio frequencies. Where the atmosphere is reasonably transparent, however, in the visible domain, image quality is limited to approximately 1 arc sec resolution by atmospheric effects, and it is difficult to obtain the continuous stretches of data needed to study time varying phenomena.

The Sun is typical of other astronomical objects in the range and diversity of the phenomena it displays. However, in reaching out to the stars and other objects we suffer an enormous reduction in flux--the light from the nearest stars is attenuated by 10^{11}-10^{12} compared with that from the Sun. Thus, we must employ very large instruments in order to gather an adequate number of photons. Similarly, very high angular precision is a requirement, both to be able to acquire and to remain pointed at particular regions in the sky for extended intervals of time and to achieve as fine a spatial scale as is possible.

In summary, the basic reasons for conducting astronomy from space is that only very limited astronomical information is available on the Earth's surface. The breadth of the phenomena to be studied imply the use of a wide variety of instruments covering every accessible wavelength of radiation.

Current Array of Astronomy Missions

Table 1 lists the major US space astronomy missions, beginning with the Orbiting Astronomical Observatory (OAO) series that have been flown or that have been approved for flight. Not included are other satellite astronomy experiments in the category of single experiments of limited data return. For example, a number of experiments flew as part of the Apollo missions.

There are a number of conclusions that can be drawn from this table.

1) With the exception of the Infrared Astronomy Satellite (IRAS) and the International Ultraviolet Explorer (IUE), these missions are in low-inclination, low-altitude orbits. To some extent, this is a technical constraint reflecting the payload weight penalty for other orbits. There is a real effect relating to the radiation background that is minimized for these orbits. The principal drawback of the low-altitude orbits is the solid angle subtended by the Earth, which can result in a loss of observing time. IRAS was in a polar, sun-synchronous orbit reflecting one of the unique problems of infrared astronomy relating to the thermal background. IUE is in a geosynchronous orbit, which allows for real-time control and for continuous observation periods not limited by the orbital period.

2) A wide variety of disciplines is represented; specifically, observations include infrared, visible, ultraviolet x-ray, gamma-ray, and cosmic ray astronomy and solar. The photon energy range extends from about 0.01 eV (infrared) to about 100 meV (gamma-ray) or 10 orders of magnitude. It is not surprising, therefore, that "space astronomy" breaks up into a number of subdisciplines, each with its specialized technology and its unique requirements. Furthermore, single categories break down into finer units. Gamma-ray astronomy, as an example, is made up of at least three distinct energy ranges comprising 0.1-5 meV, 5-50 meV, and 50-200 meV. In each case, the physics of the production processes must change, and each is associated with radically different observational

Table 1 Major space astronomy mission

Name of Mission	Discipline	Number Flown
Orbiting Astronomical Observatory (OAO)	Ultraviolet astronomy	3
Orbiting Solar Observatory (OSO)	Solar observations	6
Small Astronomy Satellite (SAS)	X-ray, gamma-ray astronomy	3
Apollo Telescope Mount (ATM)	Solar observations	1
High-Energy Astronomy Observatory (HEAO)	X-ray, gamma-ray Cosmic ray astronomy	3
International Ultraviolet Explorer (IUE)	Ultraviolet astronomy	1
Infrared Astronomy Satellite (IRAS)	Infrared astronomy	1
Solar Maximum Mission (SMM)	Solar observations	1
Gamma Ray Observatory (GRO)	Gamma ray astronomy	1987
Hubble Space Telescope (HST)	Visible, ultraviolet astronomy	1986
Cosmic Background Explorer (COBE)	Radio, infrared astronomy	1988
Extreme Ultraviolet Explorer (EUVE)	Extreme ultraviolet astronomy	1989
X-ray Timing Explorer (XTE)	X-ray astronomy	1990

techniques. Also several subdisciplines, (notably, radio astromomy, relativity experiments, and ultra high-energy gamma-rays) are not yet represented in space but could be important entries in the space station era.

3) There has been a maturation in certain of the disciplines involving the evolution into larger, more powerful space facilities. In the case of ultraviolet astronomy, as an example, the sequence goes from the

Orbiting Astronomical Observatory (OAO) to the International Ultraviolet Explorer (IUE) to the Hubble Space Telescope (HST). The HST represents a mature, almost classical astronomical facility comprising a large, high-quality mirror, carrying a number of focal plane instruments and capable of conducting a great variety of observational programs. The same sequence is evident in x-ray astronomy and is likely to occur in infrared astronomy. In gamma-ray astronomy and other areas, single instrument facilities may not emerge; rather, the maturation will take the form of much larger instruments, more sensitive and more refined than their predecessors.

Many of the newer missions are intrinsically capable of very long lives, of which NASA is actively planning to take advantage. Phenomena are being investigated, such as the Sun, with its 11-year cycle, that require a long baseline of measurements. Also, many of the subdisciplines have such broad observational potential that very extended lifetimes can be justified. An example of the impact of the long life has just occurred in the Solar Maximum Mission (SMM) repair, in which the Shuttle was employed to retrieve, repair, and return to orbit the SMM spacecraft.

A final noteworthy aspect of Table 1 is simply its length. Only the planetary scientists can boast such a broad involvement with space missions. It implies, and it is the case, that astromomers are well equipped in terms of well-formulated scientific objectives and direct experience to deal with the problems of configuring missions for space, analyzing the retrieved data, and folding the scientific results into their main body of knowledge.

Future Astronomy Programs

As is the case in other scientific disciplines, recommendations for major facilities emerge from the community as formally expressed through both standing and ad hoc committees of scientists. Final decisions on support for specific space missions rest with NASA. In the case of space astronomy, principal guidance is provided by the Space Science Board, a committee of the National Academy of Sciences, and one of its discipline oriented subcommittees, the Committee on Space Astronomy and Astrophysics. A significant recent advisory activity has been the convening, by the academy, of an astronomy survey committee under the chairmanship of George Field and the publication of its report in 1982. The report covers more than just space activities, and its recommendations, as they relate to space, are listed below:

A. Major New Programs. The Committee believes that four major programs are critically important for the rapid and effective progress of astronomical research in the 1980's.

1. An Advanced X-Ray Astrophysics Facility (AXAF) operated as a permanent national observatory in space, to provide x-ray pictures of the Universe comparable in depth and detail with those of the most advanced optical and radio telescopes. Continuing the remarkable development of x-ray technology applied to astronomy during the 1970's, this facility will combine greatly improved angular and spectral resolution with a sensitivity up to one hundred times greater than that of any previous x-ray mission.

2. A Large Deployable Reflector in space, to carry out spectroscopic and imaging observations in the far-infrared and submillimeter wavelength regions of the spectrum that are inaccessible to study from the ground, thus extending the powerful capabilities of NTT to these longer wavelengths. Such an instrsument, in the 10-m class, will present unprecedented opportunities for studying molecular and atomic processes that accompany the formation of stars and planetary systems.

B. Moderate New Programs. In rough order of priority, these are:

1. An augmentation to the NASA Explorer program, which remains a flexible and highly cost-effective means to pursue important new space-science opportunities covering a wide range of objects and nearly every region of the electromagnetic spectrum.

2. A far-ultraviolet spectrograph in space, to carry out a thorough study of the 900-1200-A region of the spectrum, important for studies of stellar evolution, the interstellar medium, and planetary atmospheres.

3. A space VLB [Very Long Baseline] interfermetry antenna in low-Earth orbit, to extend the powerful VLBI technique into space in parallel with the rapid completion of a ground-based VLB Array, in order to provide more detailed radio maps of complex sources, greater sky coverage, and higher time resolution than the Array can provide alone.

4. An Advanced Solar Observatory in space, to provide observations of our Sun--the nearest star--simultaneously at optical, extreme ultraviolet, gamma-ray, and x-ray wavelengths, to carry out long-term studies of large-scale circulation, internal dynamics, high-energy transient phenomena, and coronal evolution.

5. A series of cosmic-ray experiments in space, to promote the study of solar and stellar activity, the interstellar medium, the origin of the elements, and violent solar and cosmic processes.

6. An astronomical Search for Extraterrestrial Intelligence (SETI), supported at a modest level, undertaken as a long-term effort rather than as a short-term project, and open to the participation of the general scientific community.

C. Programs for Study and Development. Planning and development are often time-consuming, especially for large projects. It is therefore important during the coming decade to begin study and development of programs that appear to have exceptional promise for the 1990's and beyond. Projects and study areas recommended by the Committee in this category include the following, in which the order of listing carries no implication of priority:

1. Future x-ray observatories in space;
2. Instruments for the detection of gravitational waves from astronomical objects;
3. Long-duration spaceflights of infrared telescopes cooled to cryogenic temperatures;
4. A very large telescopes in space for optical, ultraviolet, and near-infrared observations;
5. A program of advanced interferometry in the radio, infrared, and optical spectral regions;
6. Advanced gamma-ray experiments; and
7. Astronomical observatories on the Moon.

The Requirements of Astronomical Missions

The preceding sections have laid out the history of space astromomy in terms of major missions and an outline of its future as seen by the recent Astronomy Survey Committee. Implicit in that discussion is an enormous amount of detail concerning how one conducts astronomy from space, some of which has been alluded to. It is appropriate to abstract that detail here as a guide to how the space station would be viewed by the community.

The dominant hardware problems in astronomy tend to be the requirement to detect large numbers of photons (or particles), the necessity to reduce irrelevant background radiation, and the requirement for inertial stability (that is fixed to a stellar reference system). In one sense the systems drivers on astronomy are easy to state. Throughout their history, astronomers have always striven for "more": more sensitivity, more angular resolution, more spectral resolution, more.... The reason is that astronomers are not like other scientific explorers. We cannot, like the planetary scientists, send our missions to the objects of interest, nor can we, like physicists or biologists, devise intricate laboratory experiments to determine the characteristics of objects of interest or like archeologists, venture into the field to excavate the remains of ancient cultures. We are forced, at least in our present circumstances, to remain in the vicinity of the Earth. To gain more information, we invariably have no choice but to build larger and more refined instruments.

Beyond this, the laws of physics and the properties of matter limit choices and configurations. Again, the limitations can be simply described. The ability of radiation to reflect and refract allows the construction of focusing telescopes with attendant improvements in sensitivity, reduction of background, and the ability to resolve spatial detail. The size of the telescope provides a measure of its light gathering power and, up to a point, its spatial resolution. Diffraction effects limit telescopes to angular resolution in radians of λ/D, where λ is the wavelength of the focused radiation and D is the diameter of largest dimension of the telescope. However, in order to be able to construct a diffraction limited telescope, the reflecting surface must be smoothed and shaped to a degree of precision much smaller than λ. In practice, one is thus limited by the intrinsic graininess of materials, and it is probably not possible to construct diffraction limited telescopes at wavelengths below about 1000 A (UV). At shorter wavelengths, the best telescopes will probably perform at about the 0.1 arc sec range, comparable to that of the Space Telescope. There is no obvious intrinsic limit to angular resolution at longer wavelengths aside from how large a telescope one can build or, more realistically, how widely separated one can place individual telescope elements. Thus, somewhat ironically, the current record for angular resolutions exists with the radio astronomers working at centimeter wavelengths who achieve resolutions of milliarcseconds using arrays of telescopes spread over continental distances. Both radio and optical astronomers describe systems capable of microarcsecond resolution.

Finally, at very short wavelengths (~1 A in the x-ray range), it becomes impossible to construct practical focusing instruments at all, and one resorts to simple area detectors. The design of these devices is dominated by the specific mode of interaction of the photon and the range of the electrons and photons that result from these interactions. Indeed the subdisciplines comprising x-ray and gamma-ray astronomy tend to divide precisely at points where the physics of the interactions change, along the lines outlined in Table 2. The last entry in the table, gamma-rays > 100 meV involving gas Cerenkov telescopes, has not yet had a major space mission.

Aside from such a general discussion, how else can one define "more" in terms of what are likely to be the astronomer's requirements? A straightforward approach is to scale from existing missions, which I have done for some of the subdisciplines in Table 3.

I have taken the writer's priviledge of applying some judgments. At the UV and shorter, I assumed that next-generation instruments would continue to be diffraction limited. At x-ray wavelengths, telescopes are limited by fabrication techniques, and I assumed modest improvements in quality. At higher energies, where area detectors are used, I assumed no improvement in angular resolution. At these higher energies, I did assume a drive toward 10-m arrays (100-m^2 collecting area), since measurements tend to be background limited and improve only by the square root of the collecting area.

Regarding other system parameters, I note the following:

1) Telescopes tend to be long, but of low weight. AXAF (Advanced X-ray Astronomy Facility) and HST are about 10 m in length, as limited by the Shuttle. The length tends to scale with the focal length.

Table 2 Detection characteristics of astronomical radiation

Discipline	Energy Range	Primary Interaction	Detector
X-ray astronomy	1-100 KeV	Photoelectric effect	Gas proportional counter
Gamma-ray astronomy	100 KeV-5 meV	Compton effect pair production	Crystal/ scintillation counters
Gamma-ray astronomy telescope	10 meV-50 meV	Compton	Compton effect (plastic and crystal scintillation counters)
Gamma-ray astronomy	50 meV-200 meV	Pair production	Spark chamber
Gamma-ray astronomy	>100 meV	Pair production	Gas Cerenkov telescope

2) Area detectors tend to be thin compared with the size in the collecting dimension. Gamma-ray instruments tend to be heavy (~100 kg/m^2). Pointing stability requirements are comparable to the angular resolution.

There are innumerable details not represented above. For example, there are what might categorized as environmental issues, paramount of which is the choice of orbit, which affects the radiation environment and the residual perturbations. Fortunately, a low-Earth orbit is acceptable for almost all astronomical missions, with the majority opting for low inclination to minimize the effect of precipitating particles associated with the South Atlantic anomoly and the aurora. As noted earlier, a polar, sun-synchronous orbit is desirable for certain missions.

Another aspect of the "environment" will be the effect of the space station itself, notably chemical and dust contamination and mechanical perturbations.

Unique Aspects of the Space Station

The space station, as a system, provides more than just an attachment point for astronomical missions, but rather a radically different capability compared with what US scientists have had available.

These differences are as follows:

1) The space station will be manned and serviced on a permanent basis. By this is meant a routine schedule of visits by ground-launched vehicles such as the Space Shuttle.

2) The United States is committed to support and exploit the space station as a major space facility. This means that a variety of support systems (tugs, service bays, power systems...) will be developed for use in conjunction with the space station.

These attributes of the space stations allow entirely new capabilities to the scientific community. These include:

1) The space station as an observing site. Facilities may be permanently attached to the space station while they are used to conduct scientific

observations. This can be done now on the Space Shuttle, but only for limited periods of time.

2) The space station as a service base. For facilities that operate remotely, servicing means the ability to rendezvous with the space station for repair, replenishment of expendibles, and other activities requiring material and human intervention. This can also be done with the Space Shuttle, as was noted earlier in the case of the SMM repair. However, the space station allows virtually unlimited time to conduct such operations. (The SMM repair came close to failing because of the short mission duration). Also there may be specialized vehicles (tugs) to assist in the rendezvous operation.

3) The space station as an assembly base. Space missions are limited in size by the largest available boost capability: currently, the Space Shuttle. It is no accident that the focal length of the HST and AXAF mirrors are in the range of from 8 to 10 m. These are the largest lengths that can easily be accommodated in the Shuttle bay. Assembly in space allows the use of multiple launches to build up very large observing facilities or the bringing up of such assemblies that are reassembled in orbit into a shape totally incompatible with stowage in the launch vehicle as a completed entity. Two crucial factors that emerge in the new capability afforded by the

Table 3 Capability of present and future missions

Current Capability				Future Missions?	
Discipline	Mission	Aperture	Angular Resolution	Aperture	Angular Resolution
Infrared	IRAS	<1 m	arc min	1-3 m	arc sec
				~10 m	
Optical/UV	HST	2.4 m	0.1 arc sec	5-10 m	0.01 arc sec
X-ray	AXAF	1 m	0.5 arc sec	1 m	0.1 arc sec
				1-3 m	1-10 arc sec
X-ray	XTE	1 m	0.1 deg	10 m	0.1 deg
Gamma-ray	GRO	1 m	~1 deg	10 m	~1 deg

space station are time in orbit--there is now almost unlimited time available to carry out various tasks--and a kind of conservation law: nothing carried into orbit need ever be discarded.

Let me return to each of these. The space station as an observing site allows for instruments known in Shuttle jargon as "attached payloads." These would be attached to the structure of the space station, outside and far-removed from any of the working modules comprising the laboratories and crew quarters. There should be unlimited area for such mounting. On the Shuttle, as an example, there is only 50-100 m^2 of area available for mounting. The baseline space station configuration is about 150 m long and shows 600 m^2 of solar panels. Thus, mounting area of ~1000 m^2 could be present. Consequently, space on the Shuttle should not be an issue in mounting experiments. Rather, bringing instruments into orbit is likely to be the limiting factor.

Regarding the requirements to be placed on the space station by such instruments, in my opinion, the best strategy is to require them to be self-contained aside from the bare essentials, such as power, data lines and certain essential supplies. The scientific community has proven itself remarkably adept at developing the capability to build sophisticated instruments for remote operation. Since weight and size constraints should be considerably relaxed compared with current practice, it should be possible for experimenters to develop, independently, the computers, control systems, the pointing capability, etc., they need to operate their instruments. NASA's experience in trying to develop common capabilities or to integrate groups of experiments has not been entirely positive.

By the time space station becomes operational, NASA will have a large inventory of instruments that will have flown on the Shuttle. These will be strong candidates for use on space station. One such set of instruments has already been identified, namely, the Solar Optical Telescope. This facility comprises a meter mirror, focal plane instruments and complementary instruments--perhaps 5 to 10 distinct instruments that would make up a high-quality solar observatory in space. Other instruments now under development and working in almost every wavelength range, will be available for use on the space station.

The Solar Maximum repair mission demonstrated to the whole world the effectiveness of servicing in space. The astronomical community need no longer watch their observatories become expensive space junk because of loss of expendables or trivial component failures. Furthermore, it should be possible to upgrade instruments, add new capabilities, change focal plane instruments, etc., for any instruments, either attached to the space station or accessible via a space "tug."

There are important caveats that need to be added here. Certain astronomical instruments, notably optical telescopes, have demonstrated long useful lives (up to 50 years). That is not universal; major radio facilities have been closed down. Also, it is a common problem that individual electronic components have a short lifetime in terms of the availability from a manufacturer, making it extremely difficult to repair electronics more than a few years old unless an adequate supply of spaces was provided initially.

Thus, while servicing is powerful on paper, it may be that the actual applicability will be limited. Nevertheless it is possible to plan for virtually unlimited life for certain facilities, and routinely it should be possible for years of operations. Limits on operational life will no longer be based on simply the fact that the instruments happen to be in space.

Fig 2 The Multiple Mirror Telescope in the final stages of assembly. The telescope is a joint project of the University of Arizona and Smithsonian Astrophysical Observatory. Photograph courtesy of the Smithsonian Astrophysical Observatory.

Finally, one comes to what should represent the key, new attribute of the space station--the basis for assembling large instrument configurations in space. No longer will the launch vehicle itself be the limiting volume and weight constraint for the instrument. The space station itself comprises hundreds of meters of structural elements. One need only glance at any large, ground-based telescope to recognize that the dominate feature is a large open structure. And scientists have proven themselves remarkably adept at constructing very large instruments of great complexity without exorbitant funding. The instruments available to astronomers, high-energy physicists, the plasma community, etc., bear witness to this. Figure 2 illustrates one such ground-based astronomical facility, the Multiple Mirror Telescope on top of Mt. Hopkins in Arizona. The telescope is shown during the final phase of construction. As one involved in the development of this facility, I can attest that in the late stages of the development, there were never more than a few individuals involved--the labor-intensive parts of the work were in the areas of detailed design and the manufacture of the various subassemblies. Futhermore, some of the most serious technical problems (and much of the cost) had to do with the fact that the facility was on the ground; namely, coping with gravity and with the weather.

The challenge to NASA will be to convert this potential to reality. At the present time, the limitations to orbiting instruments is the cost of the hardware and the availablility of launch space. For certain instruments, the latter will be greatly relieved because of the possibility of higher packing density (followed by assembly in space). At a minimum, the cost of the hardware will be reduced because of the longer lifetimes and the possibility of reuse.

Potential New Astronomical Missions

A number of astronomical missions have been identified that could either benefit significantly from the space station or could only be conducted because of the presence of the space station. I will describe them in terms of the kinds of capabilities provided by the Shuttle.

Observing Facilities

Already mentioned was the Solar Optical Facility (SOT), a high-quality facility planned for the Shuttle that should be operational by about 1990. Along with SOT will be available a number of high-quality instruments, such as those being developed for Spacelab II. Used on the Shuttle, these facilities will require essentially a full launch, extensive crew training, and limited observing time. However, there is really little need to return these instruments to the ground. Thus, the space station offers the possibility of an enormous decrease in operational complexity (and cost) because of the elimination of the need of continually preparing for a new flight, along with an enormous increase in the potential scientific productivity because of the extended flight duration.

There are two technical problems that need to be solved. One is that the instruments must be pointed at the Sun, whereas the space station as a whole will be gravity-gradient stabilized. A solution may be to tie the solar observing facility to the solar cells. Another technical problem is that a means, preferably in a shirt-sleeve environment, must be provided to service and modify the facility since it is necessary to rearrange the instrument configuration to meet specific observational requirements. That is the essence of a facility.

Assembly in Space

Whenever the space station is mentioned, almost in the same breath one brings up assembly in space--the potential for erecting very large structures. Obviously the ability to construct the space station itself means that the ability will exist to do so. It is only necessary to describe a few such facilities. One is QUASAT,[+] a large radio telescope that would operate in conjunction with ground-based telescopes as part of an interferometric network. The space radio telescope, perhaps ~100 m in diameter, would be assembled near the space station and then sent into a highly elliptical orbit. Another is XLA,[+] a large-area (~100 m^2) x-ray detector array, similar, but much larger than the X-ray Timing Explorer (XTE), which will only be ~1 m^2. XLA would operate near the space station, possibly attached to it.

Both these facilities share a common feature, namely, large, flat (or nearly so) arrays. Thus, the basic erection technique may not be dissimilar from what will be

+ QUASAT and XLA are made up names.

employed to erect most of the space station. Thus, these facilities, and others like it, represent an extension of the space station itself.

Construction in Space

Obviously there are other configurations besides "flat" that potentially are needed for astronomical facilities. Most telescope systems are large, complex, three-dimensional systems. Because of their size it is hard to avoid the necessity for construction in space. By construction I mean the use of a variety of assembly techniques. Furthermore, current experience in space indicates that such construction would need to be carried out in a pressurized environment. In this respect it should be noted that one needn't look far in the scientific community to find examples of scientists developing large, complicated facilities for conducting their research--large optical and radio observatories, high-energy particle accelerator experiments, plasma fusion machines. The lesson is clear. If NASA provides a place to work in space and a relative inexpensive means to bring material into space, it is likely that any facility found on the Earth can be duplicated in space.

Summary

As noted, astronomers have been responsible for a bewildering array of instruments for space. These instruments have always been limited by the available capability as defined by some launch vehicle and the "one-shot" nature; namely, having been launched, the instrument can never be touched again. With the Shuttle and space station, that is all changing, and for astronomers, given their relentless search for greater sensitivity and resolution, should lead to a great variety of new hardware concepts. At the outset I quoted the Space Science Board to the effect that the Shuttle and other expendable launch vehicles are adequate for the purposes of space sciences for the next 15 years or so. This is simply a reflection of the fact that all of us have been forced to think only of existing launch vehicles in configuring missions. With the space station becoming a reality, an entirely new generation of missions will emerge.

The Space Station Polar Platform: Integrating Research and Operational Missions

John H. McElroy* and Stanley R. Schneider†
National Oceanic and Atmospheric Administration, Washington, D.C.

Nomenclature

ADCLS = Automated Data Collection And Location System
AMR = advanced microwave radiometer
AMSU = advanced microwave sounding unit
APACM = atmospheric physical and chemical monitor
ATN = Advanced TIROS-N
ATOVS = Advanced TIROS operational vertical sounder
ATSR = along-track scanning radiometer
AVHRR = advanced very-high-resolution radiometer
CZCS = coastal zone color scanner
DE = Dynamics Explorer
DMSP = Defense Meteorological Satellite Program
ERBE = Earth Radiation Budget Experiment
ERBS = Earth Radiation Budget Satellite
ERS-1 = Earth Resources Satellite (European)
GEOSAR = land-related synthetic aperture radar
GOMR = global ozone monitoring radiometer
HIRIS = high-resolution imaging spectrometer
HIRS = high-resolution infrared radiation sounder
HMMR = high-resolution multifrequency microwave radiometer
JERS-1 = Japanese Earth Resources Satellite
LASA = lidar atmospheric sounder and altimeter
LFMR = low-frequency microwave radiometer
MAPS = measurement of air pollution from satellites
MEPED = medium-energy proton and electron detector
MLA = multilinear array
MODIS = moderate-resolution imaging spectrometer
MRIR = medium-resolution imaging radiometer

---

*Assistant Administrator for Environmental Satellite, Data, and Information Services.

†Technical Assistant to the Deputy Assistant Administrator for Satellites.

MSS = multispectral scanner
MSU = microwave sounding unit
NROSS = Navy Remote Ocean Sensing System
NSCAT = NROSS scatterometer
OCI = ocean color imager
OLS = operational linescan system
SAR = synthetic aperture radar
SBUV = solar backscatter ultraviolet spectral radiometer
SEASAR = sea-related synthetic aperture radar
SEM = space environment monitor
SIR = Shuttle imaging radar
SISEX = Shuttle Imaging Spectrometer Experiment
SMMR = scanning multichannel microwave radiometer
SPOT = Systeme Probatoire d'Observation de la Terre
SSM/I = special sensor microwave imager
SSU = stratospheric sounding unit
TDRSS = Tracking and Data Relay Satellite System
TED = total energy detector
TM = thematic mapper
TOVS = TIROS operational vertical sounder
UARS = Upper-Atmosphere Research Satellite
WINDSAT = Doppler lidar wind sensor

Introduction

Astronauts have often given up sleep to watch the panorama of the Earth passing beneath them. The fascination that people experience for the spectacular views of the Earth from space stems from their practical value, the intellectual challenges they raise, the major scientific questions they help address, and simply the great beauty associated with an unparalleled natural vista. Beauty, intellect, and human security are the wellsprings from which Earth observations derives its deserved place in the activities of nations, whether they are spacefaring or not. Mankind will continue to examine the Earth from the vantage point provided by space systems using both human and robot eyes, and NASA's Space Station program will create some of the most important opportunities to gain the clearest and most comprehensive views.

This paper examines the role that the Space Station program can play in Earth observations, and particularly the role that the astronaut-tended platform in Sun-synchronous, near-polar orbit can play. This element of the Space Station program is referred to below as simply the "Polar Platform." The other elements of the program, the inhabited segment and the co-orbiting platform, can be used to make midlatitude measurements, to test experimental instruments and to excite public interest in Earth observations, as will be noted later, but the focus of this paper will be on the Polar Platform.

In the beginning, middle, and end of every space applications program lies an interwoven science program. Space applications programs in Earth observations are built upon the scientific results obtained in research missions. Data produced by applications or operational missions are also essential to numerous scientific investigations--some directed at understanding measurements that have already been declared "operational." At the end of one instrument generation in an Earth observations program come the scientific questions regarding what the next generation of instruments and measurements should be, to say nothing of the continuing retrospective use of data gathered by instruments long since superseded by more advanced models. Thus, science activities--and not simply scientific foundations--are woven into every aspect of an Earth observations program from the earliest conceptualization phase to the last use of data drawn from an archive years after the mission has been completed.

For the above reasons, when a given discipline area is discussed in the following paragraphs, research and operational needs are not rigidly separated, but examined jointly. No operational program can remain effective without a vigorous accompanying research program. Similarly, no research program in Earth observations should be conducted without an awareness of operational programs, their objectives, and the data they produce. The two aspects are mutually supportive and synergistic. Therefore, it is the authors' view that the Polar Platform will be an integrated research and applications Earth observations system.

In recent years, much attention has been paid to studying the Earth as a total, coupled, interacting system. NASA's Global Habitability Program,[1] the International Geosphere-Biosphere Program (now renamed "Global Change"[2]), and other efforts are aimed at complementing the traditional studies of the Earth (e.g., biology, geology, meteorology, oceanology, etc.) with an examination of the interactions among these specialized areas. For example, oceanography and meteorology are disciplines with long and distinguished histories, but it is well recognized that they are not mutually exclusive, independent fields of study. Rather, they are connected and have many strong interactions that cannot be ignored. The view from space is ideal to provide the global synoptic analysis of the coupled machine that makes up the Earth. Measurements from space are vital, and the Polar Platform is the ideal carrier for an integrated instrument suite that permits simultaneous, multidisciplinary observations.

In the succeeding sections, the required observations will be characterized for each of the major scientific disciplines supported by Earth observations. The review of observations will begin with the solar-terrestrial environment surrounding the Earth and then move downward through the atmosphere to the hydrosphere, land, and biosphere. These discussions are not intended to serve as a summary of what is known or has been accomplished in each discipline using Earth observations techniques. They are only intended to state some of the principal views that will guide the selection of payloads for the Polar Platform. Further details about past and present measurements are given in a review paper prepared by one of the authors that will be published in 1985.[3] Note will also be made in each section of the synergism that measurements in one category have with those of other disciplines. These sections are followed by an examination of the necessary attributes of an Earth observations Polar Platform.

The material in this paper relies heavily on two documents.[4-5] Although these are technical reports that are not in the formal scientific literature, they are available from the authors.

Solar-Terrestrial Interactions

The Sun-Earth system is highly dynamic, with a varying solar flux of fields and particles that exert many different influences on the terrestrial environment.[6] The solar flux passes through the exosphere, past the thermopause, and on into the thermosphere (between 90 and 500 km in altitude). In this region lies the ionosphere, driven by the diurnal cycle of the solar flux. Below the thermosphere lies the middle atmosphere, bounded by the mesopause and tropopause and including the mesosphere and stratosphere. This section deals mainly with the exosphere, thermosphere, and upper sections of the mesosphere. The next section of this paper will discuss the stratosphere and the troposphere (the region below 10 km, where weather is produced).

The investigations of the space physics of the solar system have been among the most exciting intellectually of any of the space program. Yet, it would be incorrect to regard this field as one of pure research. Solar activity can have an adverse impact on long-distance and high-latitude radio communications, satellite operations, oil pipeline operations, electric power transmission systems, high-altitude aircraft flight, and human spaceflight. This has led to a program of solar event alerts, warnings, and forecasts that have been in existence for many years.[7]

The above-referenced program has in turn led to the requirement to fly space environment sensing systems on the civil environmental satellites (recently the TIROS-N and Advanced TIROS-N series) and the weather satellites flown under the Defense Meteorological Satellite Program (DMSP). The current Advanced TIROS-N (ATN) carries a space environment monitor (SEM) consisting of two instruments: a total energy detector (TED) and a medium-energy proton and electron detector (MEPED). The TED measures the total energy deposited by precipitating magnetospheric electrons and protons over a range from 0.3 to 20 keV. The MEPED uses four directional sensors and one omnidirectional sensor to measure protons, electrons, and ions having energies in the range from 30 to greater than 60 keV.[8-9]

The Block 5D-2 version of the DMSP carries five sensors in its space environment monitoring subsystem. They include a precipitating proton-electron cumulative-dose spectrometer called SSJ/4, a topside ionospheric plasma monitor called SSIE, a scanning gamma and x-ray sensor called SSB/A, an x-ray intensity detector called SSB/S, and a high-frequency receiver to monitor ionospheric phenomena called SSIP.[10]

The SSJ/4 measures the accumulated electron dose over a range of 1 to 10 MeV and protons for the range greater than 20 MeV. The SSIE has an electron sensor to measure ambient electron density and temperature, as well as the electrostatic potential of the vehicle. It has an ion sensor that makes the same measurements for the ion species as the electron sensor, but also measures the average ion mass.

The SSB/A measures x-ray intensity as a function of energy from nominally 2 keV to more than 100 keV. The SSB/S is a companion x-ray intensity detector measuring energies in levels of 25, 45, 75, and 115 keV. The SSIP monitors the ionospheric noise breakthrough frequency, a parameter used in ionospheric forecasting.

In addition to the above sensors, it is expected that the solar-terrestrial community will require both new research and operational sensors. An excellent illustration of the latter is the solar soft x-ray imaging sensor that is proposed for flight on either a civil or defense satellite.[11] This sensor would provide a greatly improved forecast capability, because it would provide a spatial characterization of the "source function" for insertion into the propagation models of NOAA's Space Environment Laboratory.

For the above reasons, it is assumed that a Polar Platform will carry an advanced SEM, which will include, at a minimum, the sensors currently carried by the ATN and DMSP and probably an x-ray imager as well. This operational set will be complemented by research instruments derived from the Dynamics Explorer (DE),[12-13] Solar Maximum Mission (SMM),[14] and other research programs.

In addition to the above instruments, there is the need to measure the total radiation budget of the Earth. The Earth Radiation Budget Satellite (ERBS) and the companion Earth Radiation Budget Experiment (ERBE) instruments that will be carried on a number of the ATN series are but the first step in this process. The objective of these measurements is to determine the monthly average radiation energy budget of the Earth on regional, zonal, and global scales. It is inevitable that it will be necessary to continue these measurements on the Polar Platform. The device or set of instruments is referred to below as simply the ERBI (Earth Radiation Budget Instrument).

To provide a timely, global view of the solar-terrestrial environment of the Earth, and to provide continuous observation of the Sun at x-ray wavelengths, it would be desirable to have two of the advanced SEMs in orbit at once. The needs of the research instruments would be addressed on a case-by-case basis. The ERBI requires a high solar angle and would, therefore, only be carried on a Polar Platform that crosses the equator near noon.

The Earth's Atmosphere and Meteorology

Space provides the best viewpoint from which to observe the atmosphere and its varying parameters. These variations involve changes in the constituents of the atmosphere over seasonal or longer time spans, as well as the rapid changes associated with the weather. The preceding section described the measurements that aid in understanding the flow of solar energy in all its forms into the atmosphere. The next step is to examine the effect of that energy flux.

The thermosphere was the principal subject of the preceding section, but the lower thermosphere and middle atmosphere are also the targets of NASA's Upper Atmosphere Research Satellite (UARS). The satellite will carry 11 instruments to measure on an integrated, global scale the concentration of ozone, as well as the parameters of the chemical species that affect ozone. UARS will also measure the energy balance and dynamics of the middle atmosphere. Thus, energy inputs, temperatures, and stratospheric and mesospheric winds will be examined.

The UARS measurements will be complemented by the continuing measurements that will be carried out on the operational polar-orbiting environmental satellites using the solar backscatter ultraviolet spectral radiometer, model 2 (SBUV/2). The device, flown for the first time on NOAA-9 in December 1984, is a spectrally scanning ultraviolet radiometer that operates over the range from 160 to 400 nm. It measures the solar spectral irradiance over that range, the total ozone concentration in the atmosphere with an accuracy of 1%, and provides data from which the vertical distribution of ozone in the atmosphere can be determined to an accuracy of 5%. A later generation of this instrument, termed elsewhere the global ozone monitoring radiometer (GOMR),[4] is an excellent candidate for international cooperation; more will be said of this later.

The above instruments will give rise to a series of research devices that will study the seasonal, annual, and multiyear variability of the constituents of the atmosphere. One of the questions that will be addressed is the relationship between man's activities and the ozone layer. NASA has outlined a proposed research instrument complement for the Polar Platform to address these issues[5] called the atmospheric physical and chemical monitor (APACM). Neglecting a Doppler lidar that will be discussed later in this section, the APACM includes an upper-atmosphere wind interferometer, tropospheric and upper-atmospheric composition monitors, and energy and particle monitors.

The upper-atmosphere wind interferometer uses a high-resolution, multipletalon Fabry-Perot interferometer to measure the Doppler shift of molecular and atomic absorption and emission lines in the atmosphere. The device provides a vector wind field from the tropopause to the stratopause with an accuracy of a few meters per second.

Tropospheric composition monitors will employ passive interferometers and spectrometers tuned to appropriate wavelengths, and other total column abundance measurements such as those carried out by the measurement of air pollution from satellites (MAPS) experiment carried by OSTA-1 on STS-2 in November 1981. Vertical profiles will be measured using lidar techniques. A laser atmospheric sounder and altimeter (LASA) has been proposed by NASA.[5]

Upper-atmosphere composition monitors will employ a variety of ultraviolet, visible, infrared, and submillimeter spectrometers and radiometers. They would be complemented by a microwave limb sounder.

The energy and particle monitors considered for inclusion with the APACM would properly be a part of the SEM package described in the preceding section and the Earth radiation budget instrument or ERBI that was discussed there as well. The reduced APACM (listed later in Table 1) has the Doppler lidar and energy/particle monitors removed.

In addition to the above measurements of the constituents of the atmosphere and their variations, it is also necessary to continue to monitor the dynamic patterns that are associated with the weather. The current ATN satellite carries an advanced very-high-resolution radiometer (AVHRR) that provides radiometrically accurate measurements in five visible and infrared spectral bands. A later version will add a time-shared sixth channel.[15] The AVHRR is used for local weather forecasting in some countries, and to measure snow cover, ocean and lake ice, and sea surface temperature. When prepared as a mosaic image, the data are also used to replace--partially--geostationary satellite imagery when those satellites fail unexpectedly. It has a current spatial resolution of 1 km,[8-9] but user demand in meteorology and other disciplines is likely to lead to a 500-m resolution in the future. Similarly, it is likely that the same demand will lead to a requirement for a ten-channel instrument, rather than the six-channel instrument that will meet the needs of the next series of ATN satellites.[4] Thus, the operational requirement will be a relatively straight-forward derivative of current technology. This sensor will be called the medium resolution imaging radiometer (MRIR) in the following text. The MRIR would have superior radiometric and spectral qualities--and comparable spatial resolution--to the operational linescan system (OLS), which is the primary imaging device on the DMSP.[10]

NASA has proposed a similar but more advanced instrument for the Polar Platform, called the moderate-resolution imaging spectrometer (MODIS).[5] It includes the ocean color imager (OCI) capabilities, an instrument that will be discussed in the next section. The MODIS is to provide visible and infrared radiometric imaging at a nominally 1-km resolution over land and 4-km resolution over the open ocean in perhaps 100 spectral bands. The MODIS would encompass the full capability of the MRIR and OCI instruments, and still further spectral channels as well. As a research instrument, it would provide direction for future improvements in the medium-resolution sensors. It is tempting to think of a merger of the MODIS, MRIR, and OCI functions, but it may be most cost-effective to initiate operations with the MRIR and OCI, add the MODIS when ready for a period of parallel operations, and then phase operations over to a MODIS-derivative at some appropriate future date. Because of the ocean color bands, the MODIS would be placed only on the afternoon platform.

The prime instrument on the ATN satellite is the TIROS operational vertical sounder (TOVS). The instrument provides vertical profiles of the atmospheric temperature and water vapor. Integrated with rawinsonde and surface observations, the vertical profiles or "soundings" establish the initial condition of the atmosphere from which the numerical weather forecasting models project forward to predict the future state of the atmosphere. The current TOVS includes a high-resolution radiation sounder (HIRS), a microwave sounding unit (MSU), and a statospheric sounding unit (SSU).[8-9] The SSU is supplied by the United Kingdom. The next-generation TOVS, called ATOVS below, will retain the HIRS for high-spatial resolution infrared soundings, but replace the MSU and SSU with a two-segment advanced microwave sounding unit (AMSU-A and AMSU-B). AMSU-A is an all-weather temperature sounder and AMSU-B is an all-weather water vapor sounder. The AMSU-B will be provided by the United Kingdom as a follow-on to their SSU work.

The DMSP satellites also carry sounding instruments to measure temperature and water vapor profiles. They will not be listed here as separate instruments because their capabilities would not augment those of a HIRS/AMSU (ATOVS) combination.

The above-listed sounding instruments would represent a part of the operational complement of the Polar Platform. As in the case of the imaging and other instruments, it is expected that the research community will propose new techniques as well.

In the discussion above regarding the APACM, the subject of a laser system to carry out direct tropospheric wind measurements[5] was deferred to the end of this section. Such a concept has been under investigation by NOAA's Wave Propagation Laboratory for a number of years[16] under the name WINDSAT. The motivation for a WINDSAT comes from the conviction of many researchers that improved wind data are essential to the continued improvement of numerical forecasting models, and that WINDSAT is a plausible means to obtain such data.[17] Thus, a lidar wind measuring system is a candidate to be a research instrument on the Polar Platform.

Weather forecasting requires frequent data to update the initial condition for the numerical models. Even more importantly, it must receive data reliably. This has led to the requirement for two polar-orbiting weather satellites, so that essentially every spot on the Earth is revisited with adequate frequency and so that data from at least one of the two could be assured at all times. Global synoptic times and data processing considerations have led to one satellite having an afternoon equator crossing time (nominally 1:00 to 2:00 PM northbound), and a second satellite traveling southbound over the equator early in the morning. These requirements would apply also to the Polar Platform and lead to the need to maintain two platforms. The purpose of the sensors discussed in this section is to maintain a continuing observation of the atmosphere from near the surface to the exosphere. These continuing observations must be done frequently enough to adequately characterize the changes in this rapidly changing medium. A two-platform observing system provides a maximum six-hour interval between measurements.

The next section of this paper will discuss the other great fluid medium that dominates man's existence, the oceans.

The Oceans and Coasts

When a map showing the measurement of surface and upper air weather conditions is examined, the most obvious characteristic is the sparsity of data over the oceans. One of the most fundamental reasons for the contribution that satellites have made to global weather forecasting is their provision of data from regions where no data were available. This was obviously most evident in the Southern Hemisphere, where the ocean area dominates that of the land. As weather forecasts extend further and further into the future, detailed information about atmospheric conditions over the oceans becomes more and more important. Likewise, as modeling becomes increasingly sophisticated, attention must be given to effects that were negligible in less detailed analyses. Atmosphere-ocean coupling is one such effect, and has important implications in both near-term and climate-scale forecasting.

The importance of understanding the dynamics of the ocean is self-evident, whether the objective of the understanding is to advance knowledge or to support maritime operations. Further, the technical means are at hand to carry out detailed observations that serve both objectives.

A number of instruments have been tested in space and found to be highly successful; they will find a permanent operational home in the space applications programs of the United States and other nations. They include synthetic aperture radars, radar scatterometers, radar altimeters, microwave radiometers, and ocean color imagers. When combined with the MRIR discussed in the previous section and a microwave imaging device called the special sensor microwave imager (SSM/I), carried on the DMSP satellites, a powerful suite of ocean instruments emerges. The role they can play will be discussed next.

It is useful to consider what information is desired by the maritime community as represented, perhaps, by the captain of a shipping vessel, a fisherman, or an oil-well drilling platform operator. It is obvious that the location of sea ice, ocean currents, regions of high biological productivity, cold upwellings, ocean temperature boundaries or fronts, and the range of wave heights and surface wind and wave direction and magnitude are some of the more elementary parameters that a maritime operator wants to know.[18] Further, because these are varying quantities, the sampling frequency must be commensurate with the significant period of variation. The instruments that will now be reviewed are aimed specifically at these data. In addition to their immediate importance to the person at sea, they also contribute to the improvement in weather forecasting mentioned above.

Two satellites paved the way in ocean observations, SEASAT and Nimbus-7. From these early experiments have come the plans for a series of US and foreign satellites that include GEOSAT, the Navy Remote Ocean Sensing System (NROSS), Topex/Poseidon, ERS-1, MOS-1, JERS-1, Radarsat, and others.[3] Among the most important observations was the synthetic aperture radar (SAR) carried by SEASAT and planned for Europe's ERS-1, Japan's JERS-1, and Canada's Radarsat. In ocean areas in general, and in high-latitude regions in particular, cloud cover is prevalent and negates the effective use of visible and infrared wavelength sensing systems. The SAR, by operating at microwave wavelengths, penetrates cloud cover and provides high-resolution imagery of the oceans, coasts, and, notably, sea ice.[19]

A SAR, based on SEASAT and the more recent shuttle imaging radar (SIR) experiments, is a high-priority passenger for the Polar Platform. Because much remains to be learned about optimum frequencies and look angles, the following discussion will make reference to a SEASAR (sea-related synthetic aperture radar) to denote a SAR that is specifically tailored to oceanographic and ice applications. This may or may not be an accurate assumption, but it will be tested by further research. The system would provide all-weather imagery at a ground resolution of 30 m. The SEASAR would be designed to meet the needs of both the operational community[4] and the research community.[5] In a two-Polar-Platform system, a SEASAR would only be required on one of the platforms. The platform allowing direct correlation with other ocean measurements would be chosen. This will be seen to be an afternoon platform later in this paper, even though scene lighting is not an issue with a microwave sensor.

A radar scatterometer was flown on SEASAT and is planned for the NROSS, ERS-1, and Radarsat satellites. The scatterometer provides the vector wind stress field from which surface winds and waves are determined. The NROSS scatterometer (NSCAT) will measure wind stress to an accuracy of 1 m/s in magnitude and 16 deg in direction and will be used as the model for a scatterometer for the Polar Platform. Because the scatterometer data are needed on a frequent basis to adequately characterize ocean dynamics, it is assumed that an NSCAT would be carried on both platforms in a two-Polar-Platform system. In this instance the same instrument serves both operational and research needs.

Radar altimeters have been flown on GEOS-3 and SEASAT, and will be flown on GEOSAT, NROSS, ERS-1, and other satellites. They provide the surface topography of the oceans and ice floes or fields, and measure as well the significant wave height of the ocean.[21-22] Radar altimeters would be installed on both Polar Platforms that would have the same precision as those on GEOSAT, namely 8 cm. Research uses of these data would involve such areas as ocean circulation studies and long-term studies of the changes in the polar ice fields. Operational uses would focus largely on the environmental data obtained from the wave shape of the radar return, which allows computation of the significant wave height (averaged over the area illuminated by the radar pulse).

Sea surface temperature is one of the most effective indicators of ocean currents, regions of cold upwelling waters (biologically rich), and other ocean features. The AVHRR that is flown presently on the ATN series provides sea surface temperature maps as one of its routine products, as will its operational successor, the MRIR discussed in the preceding section. Because these products are of high spatial and radiometric quality, they will continue to be in demand. As noted above, however, they are limited by cloud cover, because the MRIR operates in the visible and infrared wavelengths. Another unique instrument is the along-track scanning radiometer (ATSR) that the United Kingdom is providing for flight on Europe's ERS-1 satellite. The AVHRR and the MRIR provide a temperature accuracy of nominally 1K, and the expectation is that the ATSR will provide approximately a twofold improvement. The ATSR is also an infrared device and, therefore, suffers from the same limitation as the AVHRR.

Thus, while operational instruments such as the MRIR and research instruments such as the ATSR will continue to be flown, the user community desires all-weather capability. Current operational users must currently wait for days or even weeks to obtain data in some areas of the world.[23] It is this impetus that has led to plans for precision microwave radiometers that will be flown on the NROSS and ERS-1 satellites. The NROSS radiometer is called the low frequency microwave radiometer (LFMR).[24] The device will have a temperature accuracy of nominally 1K and a spatial resolution between 10 and 25 km. It is expected that there will be a continuing need for such data and that an outgrowth of the LFMR will be flown on the Polar Platform. It has been termed elsewhere as an advanced microwave radiometer (AMR).[4] NASA has proposed a research instrument called the high-resolution multifrequency microwave radiometer (HMMR)[5]; it seems likely that a single instrument can meet the needs of both the operational and research communities.

Both SEASAT and Nimbus-7 carried an experimental sensor called the scanning multichannel microwave radiometer (SMMR). The SMMR demonstrated the ability of a microwave instrument to measure sea ice (extent and age), sea surface temperature, precipitable water vapor, wind speed, and other atmospheric and oceanographic parameters. It has been followed by a microwave imaging device, the SSM/I, that will be carried on the future DMSP and NROSS satellites.[10] The SSM/I will continue and extend the SMMR measurements on an operational basis. Because the data products can be transmitted at a relatively low data rate and because they are of value to many in the maritime community, it is expected that this sensor will be employed well into the 1990's. It is important to note that the instrument provides wind speed, a scalar quantity, rather than the vector quantity of wind velocity. One of the data processing challenges will be the imposition of globally incomplete vector scatterometer wind fields on a much denser, and essentially complete, grid of wind speeds derived from multiple SSM/Is and the subsequent determination of unique solutions for the total global wind field. SSM/Is would be flown on both Polar Platforms to provide a sufficient sampling rate.

Nimbus-7 carried a coastal zone color scanner (CZCS) that has proven to be a great success in detecting near-surface phytoplankton biomass in the open ocean and in tracking the changes in mesoscale ocean features.[25] This experimental sensor, now long past its expected lifetime, is being used on a quasioperational basis. User community demand, both research and operational, will lead to a continuation of these measurements. NASA has proposed that the ocean color spectral bands be a part of the MODIS.
Because there is some technological uncertainty about the appropriateness or feasibility of incorporating such diverse requirements in a single instrument, it is suggested here that an OCI derived from the CZCS would be flown as an operational instrument until such time as the development and demonstration of the MODIS instrument is complete. Because the OCI requires a high solar angle, it would be placed only on the afternoon platform.

The next section of this paper will address the subject of remote sensing of the solid Earth and vegetation. Just as an overlap was observed between the sensors used for meteorology and oceanography, a similar overlap will be seen between those two areas and the land-related sensors.

The Solid Earth and Vegetation

One of the more surprising results of the meteorological satellite program has been the utility of the AVHRR in land remote sensing.[26] In spite of its coarse spatial resolution, its high spectral and temporal resolution have made it a very useful adjunct to higher spatial resolution sensors. Notably in agricultural applications, the AVHRR is an excellent, low-cost device for change detection.[27] Particularly favorable results have been obtained in monitoring African vegetation.[28] Thus, the MRIR and MODIS sensors discussed above with respect to their meteorological and oceanographic applications are also of value in the land sciences.

The multispectral scanner (MSS) and thematic mapper (TM) of the Landsat series have paved the way for precision multispectral analysis of surface features at spatial resolutions of 80 and 30 m, respectively.[3] The French

SPOT system will follow later in 1985. Multilinear array (MLA) detector technology is advancing rapidly. Even though it is currently found principally in the visible and near-infrared bands, it is expected to advance to new spectral regions before the deployment of the Polar Platform. At least one commercial proposal has been advanced that would provide 10-m spatial resolution, 20-nm spectral resolution, on-orbit spectral band selection (8 of 32), and along-track stereo and cross-track revisit pointing capability.[29] While the likelihood of this particular design being carried forward to operations is uncertain, it does give some measure of what industry feels confident in building. In the succeeding review, this instrument will be referred to as simply the MLA.

NASA has proposed a high-resolution device to complement the MODIS. The high-resolution imaging spectrometer (HIRIS) is an even more advanced concept than the MLA discussed above. It allows highly detailed spectroscopic analyses to be made of relatively localized areas, i.e., it is a directed sensor that is targeted on a particular area, rather than a sensor that is used for global "mapping" functions. The quotation marks indicate that the MSS and TM would fall in this category, as well as sensors of a more cartographic character. The HIRIS is derived from the proposed Shuttle Imaging Spectrometer Experiment (SISEX).[30] The SISEX provides 128 spectral bands over the range 0.4 to 2.5 μm, with a 10 to 20 nm spectral resolution. As with the MRIR and the MODIS, it seems likely that the MLA would be the near-term operational sensor until the HIRIS completes its development and produces well-understood results. The MLA sensor would be placed on a morning platform to retain continuity of scene lighting conditions with the existing and future archive of Landsat and SPOT data. The HIRIS could be placed on the same platform or, as proposed by NASA, on the afternoon platform. This would permit correlative research to be carried out between the HIRIS and the MODIS. For research and midlatitude measurements, the HIRIS could be carried on the inhabited module or co-orbiting platform.

Just as ocean and meteorological observations benefit from sensors having all-weather capability, land sensing experiences the same difficulties in many areas of the world. Further, sensors having the capability to penetrate vegetation or dry sand cover provide valuable insights into regional geology. This is particularly true of tropical rain forest and desert regions. Although SEASAT was directed at ocean applications, many of its scenes were of land. Those scenes were found to have great value. More recently, successors to the SEASAT SAR have flown on the Space Shuttle under the name Shuttle imaging radar (SIR). By providing variable incidence angles, polarizations, and frequencies, optimum sensor parameters are being developed for a number of applications. As of this writing it is not yet clear what the optimum combinations of parameters are. As noted in the section on oceanographic measurements, the best design for ocean applications may or may not be optimum for land applications. To allow for the possibility that the two designs are different, a SAR aimed at land applications is referred to below as a GEOSAR. As in the case of oceanography, a single instrument would serve both research and operational users. The GEOSAR would occupy the position on the morning platform that the SEASAR occupies on the afternoon platform.

This completes the review of the high-priority instruments for the Polar Platform. The next step is to discuss the arrangement of the instruments on the afternoon and morning platforms, and to discuss some of the general issues surrounding the Polar Platform.

The Polar Platform as a Vantage Point

The suite of instruments described in the last several sections of this paper represent the most powerful tools for the study of the Earth and its environment that have ever been conceived. A manifold, multidimensional view is attainable of the Earth and its radiation environment, atmosphere, ocean, land, and biota. This complex, interwoven, yet objective, view can only be obtained by means of a system such as the Polar Platform.

The observing system is too complex to assure reliable operation in the absence of servicing and the capacity to carry ample redundancy; the system is too costly an investment to be treated as an expendable commodity--its cost must be prorated over many years to reach cost-effectiveness.

Thus, a high-capacity, astronaut-serviced system is essential to the realization of the Earth observations capability described in this paper. It can therefore be stated without equivocation that the Polar Platform of NASA's Space Station program is mandatory for the full realization of the potential of Earth observations systems.

By this point the reader who is not intimately acquainted with all of the observing systems discussed above may find the numerous acronyms, and the sensors they represent, more than just a little confusing. Table 1 summarizes the discussion to this point and assigns each of the candidate passengers for the two Polar Platforms to their position. The Nomenclature provides the reader with definitions of the acronyms used with the instruments.

It is assumed that the platforms are placed in orbits commensurate with the current or planned polar-orbiting environmental satellites or future land remote sensing missions. This leads to Sun-synchronous, near-polar orbits at an altitude between 800 and 900 km, and with equator-crossing times of 8:00 to 10:00 am southbound and nominally 1:00 pm northbound.

From these vantage points, a comprehensive view of the Earth can be obtained with sufficient frequency to meet all meteorological and oceanographic forecasting requirements. Further, the sensor suites allow the correlative measurements that will be necessary to discover the subtle connecting mechanisms that tie the various spheres together.

Communications and a System of Global Services

The objective of the payload defined in Table 1 is to obtain a comprehensive view of the world and to revise that view on a frequent enough time scale to meet human needs.
This suggests that considerable attention be paid to data processing and distribution. The incorporation of operational meteorological and oceanographic sensors on the Polar Platforms places important boundary conditions on the data processing and communications subsystems.

Some processing must be done onboard, e.g., the generation of lower-resolution imagery or other products for direct broadcast. Several such direct links to and from the platforms are necessary. These include automatic picture transmission, high-resolution picture transmission, and direct sounder broadcasts for the international meteorological community (more than 1,000 receiving stations worldwide). They would also include similar downlinks for the oceanographic community, as well as transmissions to foreign ground stations of data from the land-related sensors. Further, to support US activities, the platforms would employ the Tracking and Data Relay Satellite System (TDRSS) for global data collection.

The principal focus of the above discussion has been on Earth observations, but there are two other important

Table 1 Payload for a two-polar-platform observing system

Category/Instrument	Platform Alpha (1:00 pm)	Platform Beta (9:00 am)
Solar-terrestrial		
Advanced SEM[a,c]	X	X
ERBI[a,c]	X	
Atmosphere/meteorology		
MRIR[a,c]	X	X
MODIS[b]	X	
ATOVS[a,c]		
HIRS[a,c]	X	X
AMSU-A[a,c]	X	X
AMSU-B[a,c,d]	X	X
GOMR[a,c,d]	X	
APACM[b]	X	
LASA[b]	X	
WINDSAT[b]		X
Oceans/coasts		
SEASAR[a,c,d]	X	
NSCATT[a,c,d]	X	X
Altimeter[a,c,d]	X	X
AMR/HMMR[a,c,d]	X	X
SSM/I[a,c]	X	X
ATSR[a,c,d]	X	X
OCI[a,c]	X	
Solid Earth/vegetation		
MLA[a]		X
HIRIS[c]	X	
GEOSAR[a,c,d]		X
Data Services		
Data collection and platform location[a,c,d]	X	X
Search and rescue[a,c,d]	X	X

[a]Principally operational instrument.
[b]Principally research or developmental instrument.
[c]Currently flying or scheduled to fly before the Space Station.
[d]Potential internationally-provided instrument.

sensor systems that will be carried on the platforms. They have already been included in the last category on Table 1. They are the data collection and platform location system and the search and rescue system.

The current polar-orbiting environmental satellites carry a data collection and platform location system provided by France and called the ARGOS system.[31] A receiver on the ATN satellites detects the signals from small transmitters that may be located on the land, sea, or in balloons. Land transmitters typically broadcast meteorological data, but have also been used in volcanology and seismology. Snow data and hydrological measurements for river basin management are other frequent uses for land transmitters. Sea transmitters are used for both oceanographic and meteorological applications and measure atmospheric pressure and temperature, ocean winds and waves, and marine pollution, salinity, and biological parameters. An upgraded ARGOS system is a very likely payload for the Polar Platforms. NASA has proposed a similar system called the Automated Data Collection and Location System (ADCLS), but only one of the two would be necessary.[5]

The second system is the international search and rescue system. The ATN satellites carry equipment provided by Canada and France that detects the faint emergency signals from crashed aircraft or ships in distress. Complementary equipment is carried on two Soviet satellites in a splendid example of international cooperation that has saved more than 300 lives in a little more than two years. Such equipment should be placed on any available civil satellite, including the Polar Platforms. The waiting time for a victim in an emergency situation is related directly the frequency with which an area of the Earth is observed, which is in turn related directly to the number of search and rescue equipped satellites in orbit.

For all of the above reasons, the Polar Platforms will be the hub of an international set of important services. It is essential that the importance of those services be understood and considered in the establishment of the capabilities and design constraints of the Polar Platform.

Platform Servicing

It is evident that the central capabilities that make the Polar Platform of value to the Earth observations community are the large and flexible capacity that is expected to be present and the capability to do on-orbit servicing. Capacity controls the amount of redundancy that can be employed and the servicing approach and schedule dictate the required design lifetime needed by a sensor and its various subsystems. It is these parameters that will determine the economics of future Earth observations.[4,32] The bounds within which favorable servicing costs must lie are well established, because the annual investment in operational environmental satellites and research missions is well known. If servicing can be shown to reduce those costs dramatically, or to allow greater capability within those costs, an astronaut-serviced Polar Platform will have the strongest justification possible. It is the authors' belief that a demonstration of this justification is within grasp of the Space Station program.

International Participation

Earth observations is a global science, and a science in which international cooperation and participation is not simply desirable but mandatory. The services resulting from this science are equally global in character. The relevant model for international participation in the Polar Platforms has been well-established through the activities surrounding the current ATN satellites. The satellites already have a strong international element, and efforts are under way to increase it even further.

The current satellites carry two instruments from France, one from Canada, and one from the United Kingdom. Discussions are under way with a number of countries, using a forum that developed under the Economic Summit process, to expand participation.[32] It is fully expected that substantial new projects will be proposed that may include as much as the provision by the international community of a Polar Platform. Even if this were not achieved, very large contributions of major science and applications sensors are inevitable. The listing of instruments in Table 1 has been annotated to indicate which instruments are either already being provided or could be provided by the international community as an outgrowth of current programs. Even some of those that are not annotated could be as well, but the more restrictive assumption was used to show the wide range of very substantive possibilities.

Conclusions

The Polar Platform of NASA's Space Station program can make possible the most sweeping examination of the Earth imaginable. Composite suites of sensors will allow correlative measurements to be made of the Earth's radiation environment, atmosphere, weather, ocean, land, and biosphere. The complex observation system described here can be created in no other way. A conventional, expendable satellite that does not employ servicing would be too costly and unreliable. Only an astronaut-serviced Polar Platform has the potential to bring this capability to reality--both technically and economically.

There is no doubt that the creation of the capability described in this paper is a major technical challenge. The difficult servicing and data management advances that it requires will not be easy. Yet, if this challenge is accepted, the entire world will benefit with improved understanding and vital services. That is an excellent objective for NASA's Space Station program.

References

[1] Malone, T. "Global Habitability: International Aspects and its Relationship to the International Geosphere-Biosphere Program," First International Academy of Astronautics Symposium on Global Habitability, Paper IAA-84-319, 35th Congress of the International Astronautical Federation, Lausanne, Switzerland, Oct. 8-13, 1984.

[2] Friedman, H. (chairman), "Toward an International Geosphere-Biosphere Program, a Study of Global Change," Report of a National Research Council Workshop, National Academy Press, Washington, D.C., 1983.

[3] McElroy, J.H., "Earthview--Remote Sensing of the Earth from Space," AIAA, New York, 1985 (to be published).

[4] McElroy, J.H. and Schneider, S.R., "Utilization of the Polar Platform of NASA's Space Station Program for Operational Earth Observations," NOAA Technical Report NESDIS 12, Washington, D.C., September 1984.

[5] Butler, D. (chairman), Science and Mission Requirements Working Group, "Earth Observing System," NASA Technical Memorandum 86129, Goddard Space Flight Center, Greenbelt, Md., 1984.

[6] Friedman, H., and Intrilligator, D.S. (co-chairmen), "Solar-Terrestrial Research for the 1980's," Committee on Solar-Terrestrial Research, National Research Council, National Academy Press, Washington, D.C., 1981.

[7] Federal Coordinator for Meteorological Services and Supporting Research, "National Plan for Space Environment Services and Supporting Research, 1983-1987," FCM-910 1983, National Oceanic and Atmospheric Administration, Washington, D.C., July 1983.

[8] Schwalb, A., "The TIROS-N/NOAA A-G Satellite Series," NOAA Technical Memorandum NESS 95, Washington, D.C., reprinted 1982.

[9] Schwalb, A., "Modified Version of the TIROS-N/NOAA A-G Satellite Series (NOAA E-J)--Advanced TIROS-N (ATN), NOAA Technical Memorandum NESS 116, Washington, D.C., February 1982.

[10] Fett, R.W., Bohan, W.A., Bates, J.J., and Tipton, S.L., "Navy Tactical Applications Guide; Operational Environmental Satellites; Polar-Orbiting Satellites, Geostationary Satellites; Spacecraft, Sensors, Imagery," Technical Report 83-02, Naval Environmental Prediction Research Facility, Monterey, Ca., June 1983.

[11] Suess, S.T., "Operational Uses for a Solar Soft X-Ray Imaging Telescope," NOAA Technical Memorandum ERL SEL-66, July 1983.

[12] Hoffman, R.A. (ed.), "Dynamics Explorer," Space Science Instrumentation, Vol. 5, No. 4, 1981, pp. 344-573.

[13] Spencer, N.W., and Nagy, A.F. (eds.), "Dynamics Explorer Results," Geophysical Research Letters, Vol. 9, No. 9, Sept. 1982, pp. 911-1299.

[14] Willson, R.C., Gulkis, S., Janssen, M., Hudson, H.S., and Chapman, G.A., "Observation of Solar Radiance Variability," Science, No. 211, 1981, pp. 700-702.

[15] Miller, D.B., and Sparkman, H.K., "Future U.S. Meteorological Satellite Systems," Paper 84-96, 35th Congress of the International Astonautical Federation, Oct. 7-13, 1984, Lausanne, Switzerland.

[16] Huffaker, R.M. (ed.), "Feasibility Study of Satellite-Borne Lidar Global Wind Monitoring System," NOAA Technical Memorandum ERL WPL-37, Boulder, Col., 1978.

[17] Select Committee on the National Weather Service, National Academy of Sciences, "Technological and Scientific Opportunities for Improved Weather and Hydrological Services in the Coming Decade," National Academy Press, Washington, D.C., 1980,

[18] Sherman, J.W., "Civil Oceanic Remote Sensing Needs," Proceedings of the Society of Photo-Optical Instrumentation Engineers: Recent Advances in Civil Space Remote Sensing, Vol. 481, Session 3, Oceanography, May 3-4, 1984, Arlington, Va., pp. 149-158.

[19] Thomas, R.H., "Observing the Polar Regions from Space," Proceedings of the Society of Photo-Optical Instrumentation Engineers: Recent Advances in Civil Space Remote Sensing, Vol. 481, Session 3, Oceanography, May 3-4, 1984, Arlington, Va., pp. 165-171.

[20] Li, F., Winn, C., Long, D., Geuy, C, "NROSS Scatterometer-An Instrument for Global Oceanic Wind Observations," Proceedings of the Society of Photo-Optical Instrumentation Engineers: Recent Advances in Civil Space Remote Sensing, Vol. 481, Session 3, Oceanography, May 3-4, 1984, Arlington, Va.,pp. 193-198.

[21]Patzert, W.C., "Spaceborne Studies of Ocean Circulation," Proceedings of the Society of Photo-Optical Instrumentation Engineers: Recent Advances in Civil Space Remote Sensing, Vol. 481, Session 3, Oceanography, May 3-4, 1984, Arlington, Va., pp. 159-164.

[22]MacArthur, J.L., and Brown, P.V.K., "Altimeter for the Ocean Topography Experiment (TOPEX)," Proceedings of the Society of Photo-Optical Instrumentation Engineers: Recent Advances in Civil Space Remote Sensing, Vol. 481, Session 3, Oceanography, May 3-4, 1984, Arlington, Va., pp. 172-180.

[23]Honhart, D.C., "Requirements for Space-Sensed Oceanographic Data," Proceedings of the Society of Photo-Optical Instrumentation Engineers: Recent Advances in Civil Space Remote Sensing, Vol. 481, Session 3, Oceanography, May 3-4, 1984, Arlington, Va., pp. 142-148.

[24]Hollinger, J.P., and Lo, R.C., "Low Frequency Radiometer for N-ROSS," Proceedings of the Society of Photo-Optical Instrumentation Engineers: Recent Advances in Civil Space Remote Sensing, Vol. 481, Session 3, Oceanography, May 3-4, 1984, Arlington, Va., pp. 199-207.

[25]Hovis, W.A., "Practical Applications of Nimbus-7 Coastal Zone Color Scanner Data," Proceedings of the Society of Photo-Optical Instrumentation Engineers: Recent Advances in Civil Space Remote Sensing, Vol. 481, Session 3, Oceanography, May 3-4, 1984, Arlington, Va., pp. 208-211.

[26]Schneider, S.R., McGinnis Jr., D.F., and Gatlin, J.A., "Use of NOAA/AVHRR Visible and Infrared Data for Land Remote Sensing," NOAA Technical Report NESS 84, Washington, D.C., Sept. 1981.

[27]Tucker, C.J., Gatlin, J.A., and Schneider, S.R., "Monitoring Vegetation in the Nile Delta with NOAA-6 and NOAA-7 Imagery," Photogrammetric Engineering and Remote Sensing, Vol. 50, No. 1, Jan. 1984, pp. 53-61.

[28]Tucker, C.J., Townshend, J.R.G., and Goff, T.E., "Continental Land Cover Classification Using Meteorological Satellite Data," Science, Vol. 227, No. 4685, 1985, pp. 369-375.

[29]EOSAT, "Earth Observation Satellite Company," Aviation Week and Space Technology, June 25, 1984, p. 111 (advertisement).

[30]Goetz, A.F.H., "The Shuttle Imaging Spectrometer Experiment," Proceedings of the Pecora VIII Symposium, Sioux Falls, S.D., Oct. 4-7, 1983, Abstract, p. 355.

[31]Service ARGOS, "Location and Data Collection Satellite System User's Guide," Centre National d'Etudes Spatiales, Toulouse, France, May 1984.

[32]McElroy, J.H., and Schneider, S.R., "Earth Observations and the Polar Platform," NOAA Technical Report NESDIS 18, Washington, D.C., January 1985.

Planetary Exploration in the Space Station Era

David Morrison *
University of Hawaii at Manoa, Honolulu, Hawaii

Introduction

Almost from the foundation of NASA, planetary exploration has been one of the most visible and productive elements of the U.S. space program. Initially, of course, the planet toward which most attention was directed was the Earth, and early studies from space provided the first perspective on our world as a planet. In the early 1960s our focus shifted toward the Moon, and much of the planetary effort during the next decade was carried out in support of the Apollo lunar landings. At the same time, the first robot spacecraft began direct studies of the inner planets: Mars, Venus, and Mercury. During the 1970s this planetary emphasis led to some of NASA's greatest triumphs of scientific exploration, culminating in the Viking landings on Mars and the Voyager missions to explore the outer planets.

The U.S. today is in a transition period from the "golden years" of planetary exploration (roughly the two decades from the first Venus mission in 1962 through the Voyager Saturn encounters in 1980 and 1981) to the Space Station era expected to begin in the early 1990s. This transition era also corresponds to a time of increasing interest, and capability, of other nations in planetary exploration. During the 1980s more planetary missions will be flown by the U.S.S.R. and ESA than by the U.S., and these are increasingly missions of substantial technical sophistication. Also, this is a time when there is great interest in international collaboration in planetary exploration, especially between the U.S. and the European nations. It is the planetary missions being planned now by NASA and

*Acting Vice Chancellor for Research and Graduate Education.

its counterparts in other countries that will provide the technical base and the cadres of trained scientists and engineers for a renewed exploration program in the 1990s and beyond.

As an Earth-orbiting platform, the Space Station (like the Shuttle) is primarily oriented toward our own planet and the uses that can be made of near-Earth space. It will play a major role in our continuing efforts to understand the Earth as a planet. It will also make possible laboratory studies in microgravity environments of significance for understanding some of the processes important to planetary science, and it will presumably provide a locus for continuing remote studies of the planets using Earth-orbiting astronomical telescopes. For many planetary missions, however, the Station will likely be irrelevant, at least through the 1990s, since the simplest interplanetary trajectories can be achieved directly from Shuttle launches using high-energy upper stages such as Centaur. Only for the largest and most challenging planetary missions of the 1990s will the Station play an essential, enabling role. Thus we can look forward to a mix of mission types, with most interplanetary spacecraft launched directly from the Shuttle while the largest and most complex, such as a possible Mars sample return, will demand the Station.

At present, the planetary science community has just begun to address the potential of the Space Station for the future of solar system exploration. Because of the severe impacts during the early 1980s of budget reductions and delays imposed by the Shuttle and upper-stage development timetables, NASA's planetary program has had to struggle simply to survive. Of necessity, we have had to think

Table 1 Core missions.

- Venus Radar Mapper
- Mars Geoscience/Climatology Orbiter
- Comet Rendezvous / Asteroid Flyby
- Titan Probe

• Mars Aeronomy Orbiter	• Saturn Probe	• Comet Atomized Sample Return
• Mars Probe Network	• Saturn Orbiter	• Multi Main-belt Asteroid Orbiter and Flyby
• Venus Atmospheric Probe	• Uranus Probe	• Earth Approaching Asteroid Rendezvous
• Lunar Geoscience Orbiter		

small, placing emphasis on less expensive and demanding missions. This philosophy is essential to the Core Program (Table 1) of planetary missions recommended to the Agency by the Solar System Exploration Committee (SSEC), which was charged with the development of a comprehensive NASA mission plan for the period 1985-2000. Only very recently has it seemed possible to give serious thought to more ambitious missions that might make use of the Space Station capabilities.

The second part of the SSEC report, dealing with larger and more technically challenging planetary missions that might be started before the year 2000, is to be released in 1985. At the same time, the National Academy of Sciences, through its Space Science Board, has begun a comprehensive two-year study of space science in the Space Station era (1990 through 2015), which includes planetary exploration in this expanded time-frame. The Academy report is not expected before 1987. Other, more limited studies are meanwhile directed toward specific potential uses of the Station, for instance in laboratory studies of planetary materials and processes. Since all of this work has just begun, the present chapter will necessarily be both tentative and somewhat speculative. At best, we can only outline some of the possible uses of the Space Station in the advancement of planetary science and exploration. A more mature perspective on these possibilities must await the results of other, more comprehensive studies involving the participation of many members of the planetary community.

The Objectives and Methods of Planetary Exploration

In its early stages, planetary exploration was limited by launch vehicle capabilities, which restricted potential targets to Earth and Moon. Even if higher launch energies could have been achieved, however, it is doubtful if spacecraft could have been built in the 1950s capable of survival for the months required to reach even the nearest planets. Even the beginnings of lunar exploration were depressingly dominated by spacecraft failures. Not until Mariner 2 made its historic flyby of Venus in 1962 did true planetary exploration begin, and it is well to remember that the first orbiter of another planet did not come until 1969, when Mariner 9 began its comprehensive survey of Mars.

Logically, the exploration of individual bodies in the solar system can be divided into four stages. Although there are exceptions, the history of our planetary program has generally followed this sequence.

1) Earth-based studies, carried out from ground-based and Earth-orbital telescopes. This phase covers the entire history of pre-spacecraft planetary research, but refers particularly to the much more detailed studies supported since the beginning of NASA's Planetary Astronomy program. Since Earth-based studies are much less costly than spacecraft missions, and they also can be extended over much longer time bases, it has always proved cost-effective to learn as much as possible about potential targets before missions are initiated, and to continue these studies in support of on-going planetary missions.

2) Reconnaissance, usually accomplished by fly-by spacecraft. These are the simplest spacecraft, placing smaller demands on navigation and on-board propulsion. In addition, we have learned how to use gravity-assisted encounters with one planet to speed the spacecraft on to a second target (for a total of four planets plus numerous satellites in the case of Voyager 2). At this stage, a characterization of global planetary properties can be expected. In terms of imaging, one might expect to map most of the surface to a resolution of a kilometer or two.

3) Extensive study, usually achieved with long-lived planetary orbiters and atmospheric entry probes. Penetrators implanted ballistically in the surface would also fall in this mission category, although they have not been used so far in planetary exploration. In the extensive study stage detailed investigation can be carried out of atmospheric structure and chemistry, while geological studies of surfaces can be based on global maps at sub-kilometer resolution, sufficient to reveal many of the processes at work. Orbiters can also provide a sufficient time base to begin to study variable phenomena, such as meterology and magnetospheric processes.

4) Detailed, targetted investigation. At this stage, sufficient is already known about a planet to ask specific questions and seek their resolution. The means employed might include long-lived landers, mobile surface laboratories, or return of selected samples to Earth. In the case of Mars, this stage was just being reached with the Viking landers and their biologically oriented scientific investigations, while the later Apollo missions to the Moon definitely pushed the study of our nearest planetary neighbor into this stage of exploration. Apollo and Viking, however, represented the high-point of this effort, and no missions of comparable complexity or scientific ambition have been carried out during the past decade.

In carrying forward the exploration of the solar system, NASA and the scientific community have always sought to balance the conflicting priorities of detailed investigation of a few bodies as against a broad reconnaissance of the entire planetary system. In general, emphasis since Apollo has been placed on achieving a broad, first-order look at the system rather than concentrating on any one planet. This balanced approach has been strongly endorsed by the National Academy of Sciences in a series of recommendations concerning planetary exploration. In its simplest form, the argument for a balanced program is simply that we cannot make a rational choice concerning where to concentrate our resources until we have some broader understanding of the potential targets available and the processes acting on them.

The National Academy has reinforced the need for balance by dividing the planetary system into three parts and avoiding any temptation to prioritize among them. These divisions are: The Inner Planets, The Outer Planets, and Small Bodies (Comets and Asteroids). The fact that each of these three areas has been treated as being of comparable scientific merit, however, has not resulted in equal numbers of missions. Because the inner planets are easy to reach, and also reflecting the exceptional interest always aroused by Mars, most of our effort has been spent on missions to the inner planets. In the 1970s, however, a major effort was made through the Pioneer and Voyager missions to carry out a comprehensive reconnaissance of the outer solar system, aided by a fortunate alignment of the outer planets. As of 1984, however, no U.S. missions have been authorized to the comets or asteroids, although there is considerable activity in this area abroad (especially for studies of Comet Halley in 1986). One major objective of the planetary program in the late 1980s and 1990s should be to correct this imbalance by moving ahead with small bodies exploration.

On a more fundamental level, the exploration of the solar system is motivated by a desire to find the answers to four fundamental classes of questions. The first three of these have long been recognized by NASA's planetary program, while the fourth was adopted by the Agency in 1983 at the recommendation of the SSEC. These basic objectives of planetary exploration are:

1) To determine the origin, evolution, and present state of the solar system; 2) To gain better understanding of the Earth by comparative studies with other planets; 3) To understand how the appearance of life relates to the chemical and physical history of the solar system; 4) To survey the resources available in near-Earth space in order to develop a scientific basis for future utilization of these resources.

The specific scientific and exploration objective of each NASA Planetary Mission can be related directly to these four fundamental questions. The first three have provided the scientific rationale for all of these activities to date. The fourth, with its more practical direction towards the possible utilization of space resources in the next century, could become of equal or greater importance if there is serious interest in committing the United States to an expansive human role in near-Earth space. This fourth question could supply the rationale for a series of resource-oriented missions to the Moon and to the nearest asteroids.

The SSEC Core Planetary Program

By 1981, less than 20 years after the beginning of the planetary program, NASA spacecraft had visited all of the planets known to ancient peoples, and in addition the close examination of several dozen planetary satellites, two ring systems, and several magnetospheres had greatly expanded the range of objects and phenomena under study. The result has been a revolution in our understanding of the solar system and of the Earth as a planet. However, this ambitious and highly successful program began a rapid decline in the mid-1970s, at about the time of the Viking Mars missions. The last launch of a planetary spacecraft took place in 1978, more than six years ago. The budgets for planetary exploration dropped to only 20% of their level a few years before. By the early 1980s, only one planetary mission - Galileo - was under development, and it was not scheduled to arrive at Jupiter until near the end of the decade. It was in this context that NASA asked the Solar System Exploration Committee (SSEC) to lay the groundwork for a renewed planetary program beginning in the late 1980s.

As described in the Introduction, the SSEC directed its original effort toward the definition of a minimum Core Program of planetary missions through the year 2000. In the view of the Committee the highest priority for the planetary program should be to implement these core missions with the vigor and continuity necessary to make major steps toward answering basic questions about the solar system. The Core Program should consist of two elements: the ongoing base activities, including basic research, missions operations, technology developments, and advanced planning; and the Core planetary missions. If the necessary continuity of the program and budget are achieved, the SSEC concluded that innovative approaches to spacecraft mission design could sustain the Core Program at a total budget level of between $300 and $400 million per year. At this funding level, planetary missions could be carried with a frequency that allows good use of spacecraft inheritance and commonality of systems and personnel. The SSEC considered that the Core Program would support a sufficient level of scientific investigation and accomplishments so that the United States could retain a leading position in solar system exploration.

The SSEC Core Program consists of fifteen recommended missions based on current assessment of technological readiness, launch opportunities, rapidity of data return, balance of disciplines, and various other programmatic factors. These missions are in addition to the ongoing Voyager mission and the Galileo Jupiter Orbiter and Probe, scheduled for launch in 1986. The Committee identified an initial sequence of four missions that should be given highest priority: The Venus Radar Mapper (Fig. 1), a Mars Geoscience Climatology Orbiter (Fig. 2), a Comet Rendezvous/ Asteroid Flyby Mission (Fig. 3), and a mission to the Titan system. These four missions will be briefly discussed in the following paragraphs.

The first mission — the Venus Radar Mapper (VRM) — is required to complete the first reconnaissance of the surfaces of the triad of the most Earth-like planets: Mars, Earth, and Venus. It is the highest priority core mission because of its scientific and exploration content, immediate

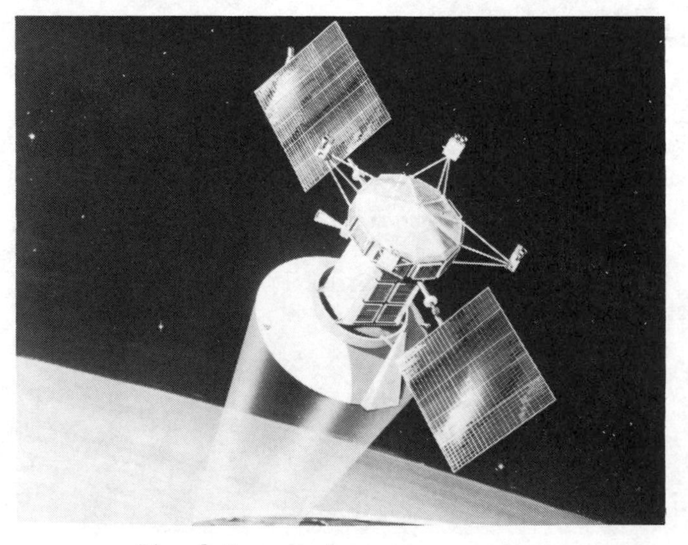

Fig. 1 Venus radar mapper.

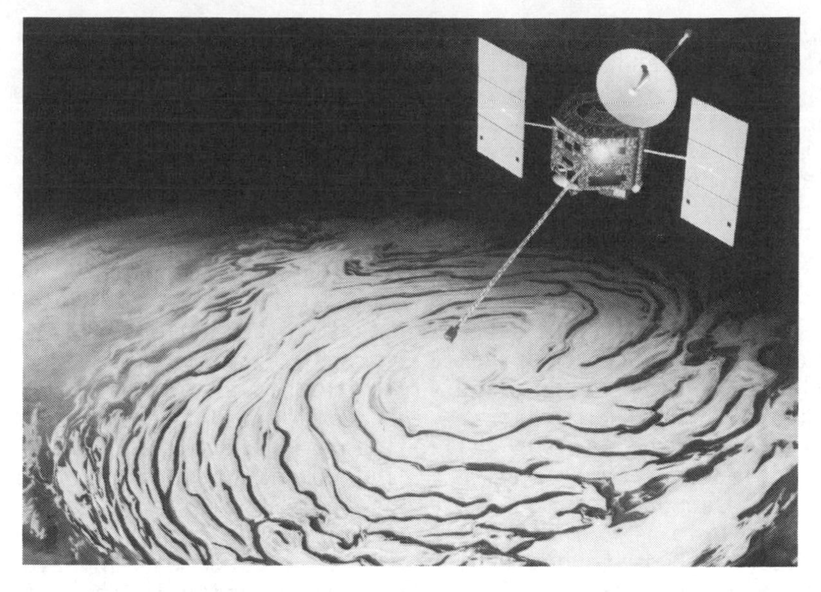

Fig. 2 Mars geoscience/climatology orbiter.

Fig. 3 Comet rendezvous/asteroid flyby.

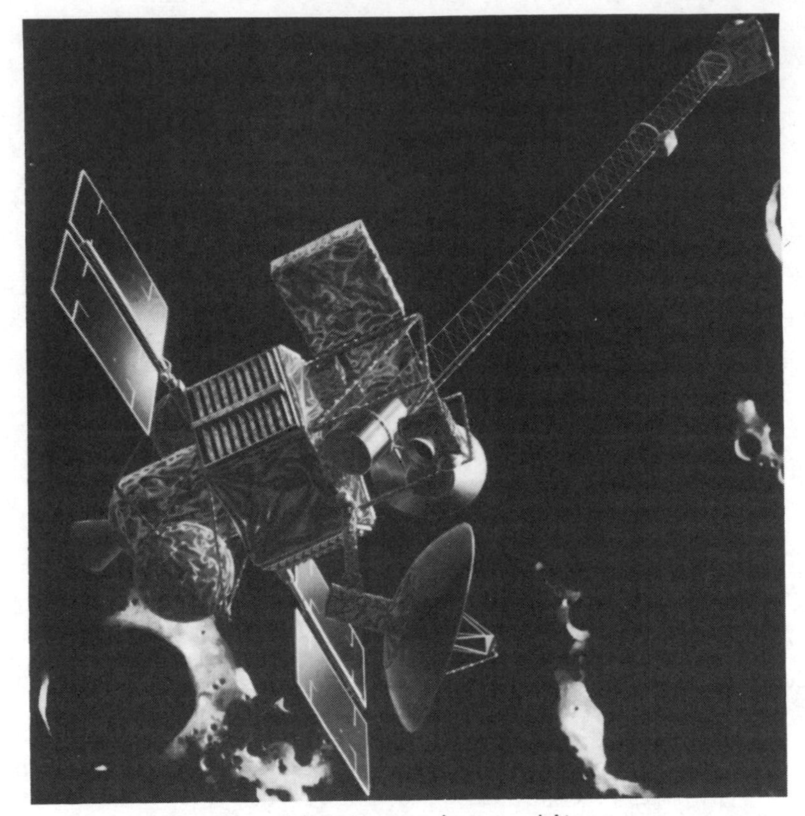

Fig. 4 Lunar geoscience orbiter.

technological readiness, moderate cost, and ability to return data within months after launch. The VRM was approved as a fiscal 1984 new start and is scheduled for launch to Venus in 1988.

The Mars Geoscience/Climatology Orbiter has a high scientific priority for resolving many first-order questions related to the evolution of the Venus, Earth, Mars triad. The cost of such a mission can be reduced substantially by taking advantage of the capabilities of Earth orbital spacecraft developed by the aerospace industry for commercial and scientific uses. Two fundamental objectives of Mars exploration are combined in this proposed Planetary Observer Mars mission: determination of the global surface composition, and determination of the role of water in the climate of Mars. This Mars Observer has been approved as a fiscal 1985 new start and is scheduled for launch in 1990.

The third of the high-priority SSEC Core Missions is the Comet Rendezvous/Asteroid Flyby. This mission (being

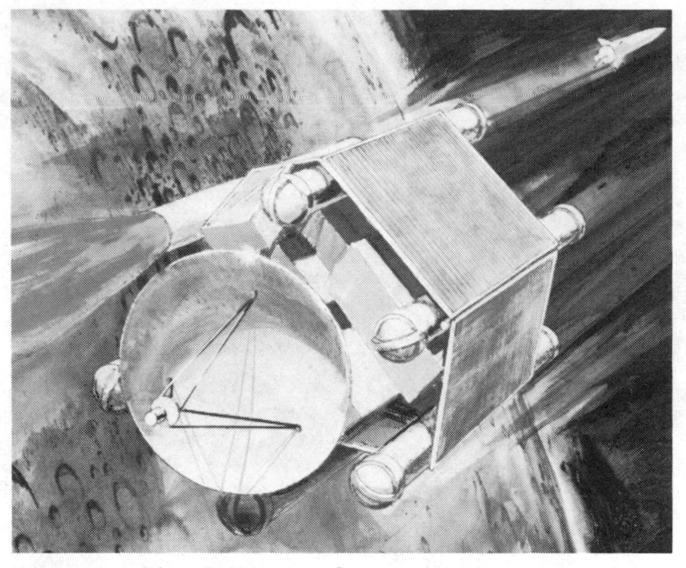

Fig. 5 Mars surface penetrator.

considered for a fiscal 1987 new start and launch in the early 1990s) will initiate the detailed exploration by U.S. spacecraft of the small bodies of the solar system. The comets and mainbelt asteroids include physically and chemically primitive objects whose study promises to provide profound insights into the formation and earliest history of the solar system. The availability of the Shuttle-launched Centaur upper stage, together with advances in spacecraft and instrumentation, bring within our capability exciting missions to these messengers from the distant past. An extended rendezvous with a short-period comet (probably Comet Kopff) will permit the detailed analysis of a cometary nucleus required for an understanding of its origin and evolution. En route to the comet the same spacecraft can provide a flyby encounter with a selected mainbelt asteroid.

The fourth initial mission will be to the Saturn system, with specific scientific focus on Saturn's large, cloud-shrouded satellite Titan. The atmosphere of Titan may yield insight into the chemistry of the prebiotic state of the Earth's atmosphere. Its cold, dense nitrogen atmosphere contains a substantial amount of methane that may well occur as methane rain, rivers, and seas, a variety of photo-chemically produced organic molecules, and a ubiquitous aerosol presumably composed of more complex compounds. A direct entry probe into the atmosphere of Titan is an essential element of this mission. Radar

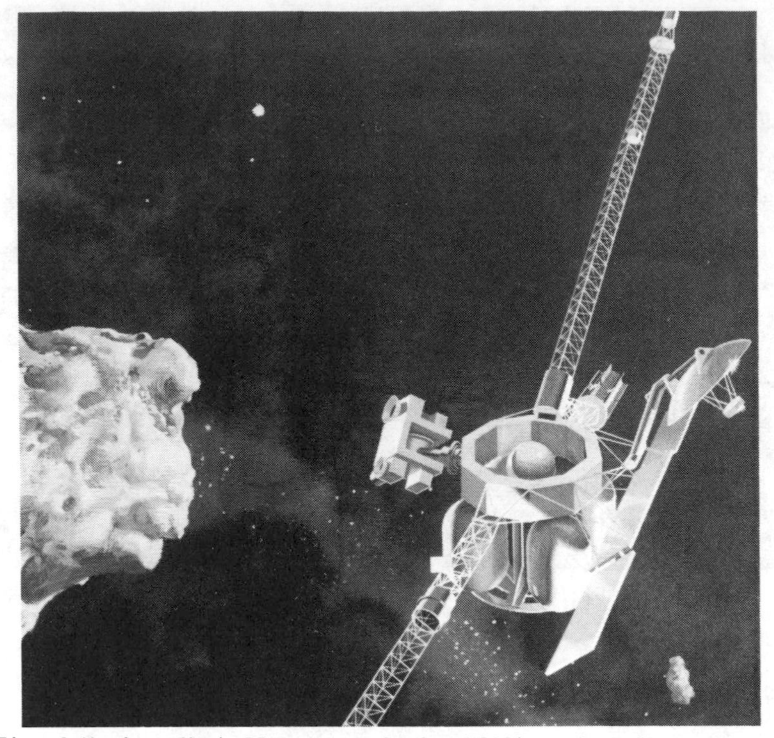

Fig. 6 Mariner Mark II spacecraft for NASA's main belt asteroid rendezvous mission.

studies of the surface of Titan also constitute an important scientific objective. Ideally, this Titan mission could be combined with a comprehensive Saturn System Orbiter resulting in an overall mission of comparable scientific yield to the forthcoming Galileo Jupiter Orbiter and probe. Studies jointly carried out by NASA and the European Space Agency hold promise for a joint U.S.-European mission to Titan and the Saturn system sometime in the middle to late 1990s.

The subsequent missions in the SSEC Core Program have not been prioritized. They can be conveniently divided into three groups: those to the terrestrial planets, those to the small bodies, and those to the outer planets. The proposed terrestrial body missions include a Mars aeronomy orbiter, a Venus atmospheric probe, a lunar geoscience orbiter (Fig. 4), and a Mars surface probe mission (Fig. 5). The subsequent core missions to comets and asteroids include a high-speed comet sample return, a multiple main-belt asteroid orbiter and flyby (Fig. 6), and at least one rendezvous mission with an Earth-approaching asteroid

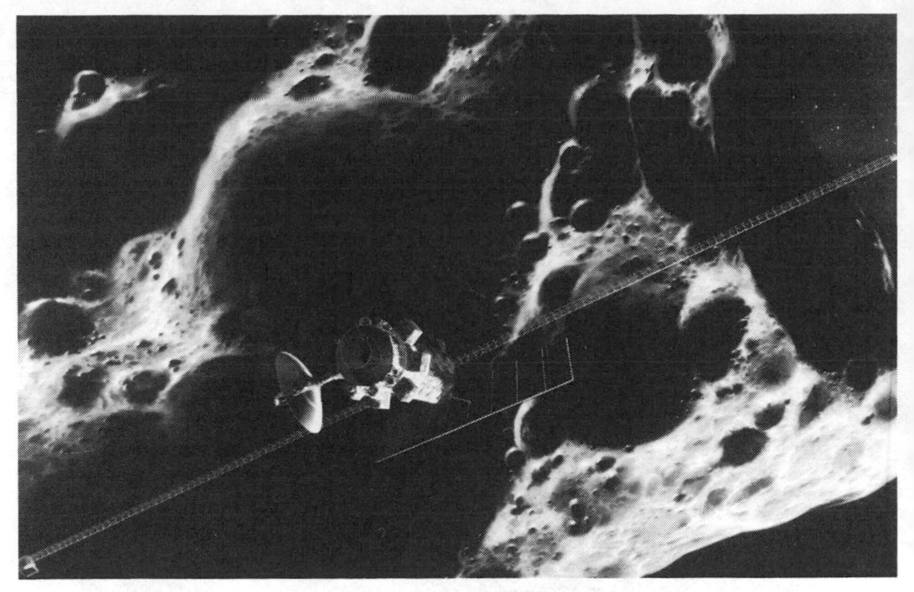

Fig. 7 Near-Earth asteroid rendezvous.

(Fig. 7). Additional missions to the outer planets include a Saturn orbiter, if this is not already included in the Titan mission, a Saturn probe, and a Uranus probe, each of the latter two to be deployed from scientifically instrumented flyby spacecraft. Most of these missions could be initiated by the year 2000 within a constant level planetary exploration budget of approximately $400 million per year.

Because of the emphasis in the SSEC Core Program on missions of low cost, no major enhancements in U.S. propulsion capability are required. These missions represent the kind of energetic scientific exploration that can be carried out with existing hardware and within the constraints of current NASA science budgets. In any discussion of a more ambitious planetary exploration program that might be initiated in the 1990s and extend beyond the year 2000, it is assumed that the low-cost missions of the SSEC Core Program will continue to provide a regular flow of new scientific data on the bodies of the solar system. Additional major initiatives requiring new technological advances will presumably be less frequent and will take place against the continuing background of the core scientific missions.

The SSEC Augmented Planetary Program

In its 1983 recommendation, the SSEC stressed that the Core Program should be augmented at the earliest opportunity by missions of the highest scientific priority that could not be included in the Core Program because of cost and technical challenge. In its subsequent studies, the SSEC has focused on two such advanced planetary missions: A Mars exploration program involving surface rovers and sample return to Earth, and a mission to a comet centering on return of a pristine sample of the nucleus for analysis in terrestrial laboratories. Of these two, the Mars program is particularly relevant to the present discussion because of its probable utilization of the Space Station and associated technological developments.

Missions in the Core Program will address many scientific objectives for the exploration of Mars, especially the global ones. However, the National Academy's objective of "intensive study of the chemical and isotropic composition and physical states of Martian minerals" requires return of an intelligently selected suite of Martian samples. Detailed study in terrestrial laboratories would allow the full range of the most sophisticated analytical techniques to be applied for the study of the chronology, elemental and isotropic chemistry, mineralogy and petrology of the Martian surface, and in searching for evidence of past life on that planet. Intensive study of local areas requires direct observations and measurements from a rover laboratory. The rover is also the best way to provide intelligent selection of samples for return to Earth (Fig. 8).

A sample return mission to Mars is very complicated, and there are many possible ways to carry it out. Based on the work of numerous NASA studies and science working groups, the following principles should be used to guide mission design: 1) The key geologic units to be sampled on Mars are the young volcanic material, intermediate age volcanic material, ancient cratered surface, and polar layered deposits; 2) Key materials to be sampled on Mars for return to Earth are: fresh igneous rock, weathered igneous rock, sedimentary rock, regolith, wind-blown dust, and the atmsophere; 3) The prime sampling objective of the mission is to select fresh and weathered samples of the most abundant material types in the near vicinity of the lander; 4) Special sampling tools and containers are required to acquire hard rock, regolith and atmosphere; 5) Some mobility and a surface exploration time of several months are required to properly sample a Viking-like land-

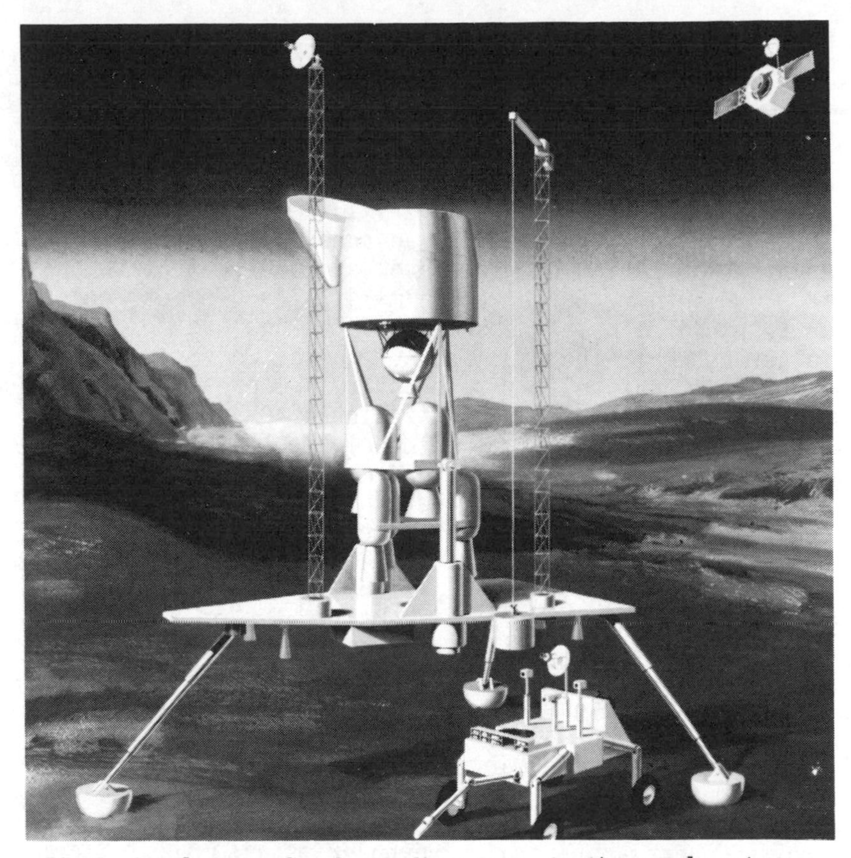

Fig 8 Sample transfer from a Mars rover to the sample-return canister.

ing site; 6) Sample return should maintain, as closely as reasonably possible, the conditions which the samples experienced on Mars. The samples should not be sterilized; 7) High resolution Viking imaging data obtained late in the mission are sufficient to identify suitable landing sites. No additional precursor mission is required; 8) The capability for a small landing error ellipse (10 to 20 kilometers) with fairly extensive sampling mobility (10 to 50 kilometers) would permit the sampling of several key terrains from a single landing site; 9) The capabilities required for a mobile sampler could also carry out the prime objectives of a rover mission.

Site selection for this Mars sample return mission can be based on existing Viking imaging data, and appropriate candidate sites have already been identified. Several terrains of different age should be within reach of the rover at any given site. Any Martian site visited will also contain a selection of wind-blown and impact derived debris which would increase the probability of sampling distant as well as local materials. Sampling near younger impact craters enhances the probability of obtaining material from the local subsurface in the form of rock fragments that have not been altered by long reaction with the atmosphere. Samples of atmospheric gases and soil volatiles can be collected at any site and will provide evidence on atmosphere-surface interactions, extent of planetary degasing, and degree of retention of primordial gases.

Ancient crustal material may be the most difficult unit to sample. Outcrops of this material cannot be identified on existing Viking images, and it is possible that samples may have to be obtained by moving to the nearest large fresh crater. Since such craters are sparse, sampling the ancient crustal materials may require in itself a 10-20 kilometer rover traverse over surfaces that could have many obstacles.

The total amount of material required to characterize and permit detailed study of individual rock or soil samples varies, but a few grams can be used effectively. Thus a sample collection strategy which maximizes the number of samples in the range of one to ten grams will also maximize the information content by making comparative studies possible.

Since this Mars mission would not be launched before the mid-1990s, we will be able to take advantage of the Space Station and developments related to it. By utilizing the Space Station capability and aerocapture at Mars, the proposed mission will enable both the landing of the rover and the return to Earth of a sample capsule. Previous detailed mission studies based on chemical retropropulsion and Viking-type entry systems at Mars required multiple launches, orbit rendezvous at Mars, and many more spacecraft modules. By utilizing the Space Station it will be possible to greatly simplify the mission design, thereby reducing both cost and risk.

The primary option being considered for this Mars mission involves the use of a refueled Centaur upper stage. The Centaur tanks could be filled to capacity by either fueling at the Space Station or by on-orbit assembly of a Mars spacecraft with a separately launched, fully fueled Centaur. The additional capability for aerocapture, aeromaneuvering, and terminal guidance at Mars reduces demands on initial spacecraft mass and enhances the probability of landing very close to the preselected targets.

Once the sample is returned to near-Earth space, either aerocapture or retropropulsion can be used to place the vehicle in Earth orbit. The capsule could be retrieved by an orbital transfer vehicle, by the Space Station itself, or by the Shuttle. Preliminary sample analysis at the Space Station might be used to minimize the risk of back-contamination and to maximize the scientific value of the samples. For example, a small portion might be sterilized and returned to Earth's surface for detailed study, while the major portion is retained in the Space Station. It is possible that political consideration will dictate a situation in which the only way an unsterilized sample of Martian material can be returned for Earth study is via a quarantine period in the Space Station.

A preliminary study has been made at the Jet Propulsion Laboratory of the characteristics of a two-vehicle Mars sample return mission for launch in 1996 and 1998 with the features discussed above. For study purposes this concept has been called Mission Columbus. The baseline design involves the use of the Space Station, full Centaur capability, aerocapture at Mars and Earth, aeromaneuvering and terminal guidance at Mars, orbital rendezvous, docking, and sample transfer at Mars, and semi-autonomous rover operations on the Martian surface. These concepts represent a number of technological challenges. They will extend our developing capability in robotics, autonomy, and artificial intelligence as well as in spacecraft technology. The existence of a Space Station is, of course, critical to this plan.

The second of the SSEC recommended missions for an augmented planetary program, the Comet Sample Return, has not been studied so extensively. The primary technology enhancement that appears to be required for such a mission is a high-capability low-thrust propulsion system, perhaps based on developing European ion-drive technology. It is not now clear whether on-orbit assembly or refueling at the Space Station are necessary to carry on such a mission. However, it seems extremely likely that the Comet Sample Return mission, like the Mars Sample Return, will involve retrieval and initial examination of the sample in a Space Station laboratory. Indeed, the techniques for sample handling and investigation for one mission should be directly applicable to the other. The Comet Sample Return mission could be initiated in the late 1990s, with return of a sample to the Space Station in the middle of the first decade of the next century.

Planetary Science On Board the Space Station

In addition to providing essential support for major planetary missions, the Space Station will also serve planetary science by providing a laboratory for microgravity and controlled gravity experiments. Plans for such studies are now in a very immature state, but the possibilities appear promising. The concepts briefly described here derive primarily from a workshop on "planetology experiments aboard the Space Station" held in Flagstaff, Arizona, in August 1984.

This workshop concluded that the Sation laboratory environment will provide a unique capability to stimulate many planetary processes, including: 1) impact cratering, 2) particle formation and interactions, 3) crystal settling and magma diffusion, and 4) studies of aeolian processes. In many of these experiments gravity is a critical term; however, there is no suitable means for duplicating a reduced gravity environment in experiments conducted in laboratories on Earth. On the Station, gravity can be reduced for a direct simulation of various free-space and planetary environments. Experimental results could therefore have a major impact on our understanding of a number of fundamental processes, especially those that are critical for modeling the formation and early evolution of the planetary system.

At the workshop, the experiment objectives and the level of sophistication of each topic varied widely among the science areas considered. In general, experimenters and discipline areas in which research programs already exist (such as in impact cratering) are more advanced than some of the other topics. Moreover, the topics considered are by no means exhaustive of all planetary-related experiments, nor are any of the ideas fully developed at this time.

The approach leading to the Space Station experiments involve intermediate feasibility studies. The first step is to establish (or continue where appropriate) laboratory experiments on Earth to develop general concepts. Feasibility studies also include experiments conducted on board the KC-135 aircraft (where reduced gravity conditions up to 40 seconds duration can be achieved) and on board Shuttle. The advantage of this phased approach (laboratory to KC-135 to Shuttle to Station) is that it provides the opportunity to produce useful science results and provides a test bed for engineering considerations.

The following is a summary of the objectives for the five general types of experiments considered at the workshop:

1. Formation and Interaction of Planetary Particles and Aggregates. The objective of these experiments is to study the behavior of particles under simulated free-space

and planetary conditions to determine the role of particles in planetary formation and in various planetary processes. Studies include: 1) refractory grain and ice condensation, 2) grain aggregation, 3) dynamics of low-velocity collisions, and 4) measurements of the physical and chemical properties and environmental affects on small-grain aggregate bodies which may be similar to cometary and protoplanetary bodies.

2. Impact Cratering. These experiments would be conducted to evaluate the effects of gravity on the general impact process by simulating specific low-gravity cratering environments. In particular, parameters controlling the dimensions and general morphology of craters would be investigated, along with study of regolith processes on various objects including asteroids and other small bodies. In addition, studies could be conducted of general planetary accretion and disruption by impact processes.

3. Aeolian Processes. These experiments would be conducted to determine the role of gravity in assessing interparticle forces during fluid threshold of wind-blown grains on planetary surfaces, as well as an assessment of low-gravity effects on particle trajectories and the resulting influence on flux and rates of erosion; in addition, experiments could be conducted to assess the role of gravity in the formation of dust devils.

4. Planetary Materials - Properties and Processes. The objective of these experiments is to carry out melting and recrystalization experiments of representative rocks, minerals, and ices under reduced-gravity conditions and atmospheres of various compositions. In addition, experiments could be conducted to assess the deformation of planetary materials under static and dynamic stress fields in the reduced gravity environment. These studies would have application to problems related to crystal settling and magma diffusion and other near-surface and deep-seated planetary processes.

5. Lithospheric Modeling. These experiments would involve the development of physical scale models to study global and local lithospheric processes. Previous experiments dealing with structural tectonic deformation of planetary lithospheres have been hampered by the inability to produce models of the appropriate strength, given the Earth gravity field. The improvement of various strength-scaling relationships by reduced gravity environments may lead to significant experiments to understand global and local lithospheric processes.

Further development of these and other ideas is to be expected. The Station will serve as an important laboratory for many investigations in basic physics and chemistry, and among these will certainly be research of great importance to the development of planetary science.

Conclusions

During the 1990s a reinvigorated NASA program of planetary science and exploration will make increasing use of the Space Station. The major Mars rover and sample return missions that are expected to be a central focus of the program require the Station for their success, both for on-orbit refueling and/or assembly and as a laboratory to which to return samples of Martian material. Comet sample return, expected for the next decade, also will utilize the Station. At the same time, smaller scale planetary missions will continue to be flown from the Shuttle or even from expendable launch vehicles. Also during the 1990s, a substantial beginning is expected toward the use of the Station as a laboratory for experiments aimed at answering basic questions about fundamental processes in planetary science which can not be addressed in the high-gravity environment of the Earth's surface.

As our perspective extends to more distant time frames, our predictions become less certain. Many people, both within and without the planetary science community, hope to see a major effort toward solar system exploration in the next century. Central themes of such a program might include establishment of a permanent manned lunar base, utilization of the resource potential of near-Earth asteroids, or initiation of manned exploration of Mars. All of these endeavors would make major use of the Space Station. In this vision of the future, planetary exploration would become the major theme of the U.S. space program, with the Station a necessary element of the capability required to support this mission.

The Next Steps in Satellite Communications

Walter L. Morgan*
Communications Center of Clarksburg, Clarksburg, Maryland

Space Stations and Platforms

For the purpose of this section, a **space station** will be defined as a large, manned, low-orbit spacecraft having, among other capabilities, the ability to assemble, store, and service satellites destined for other orbits.

A **space platform** is an unmanned satellite located in a low-, medium-, or high-altitude orbit. It may have various inclinations depending upon the mission. A second class of these satellites is located in the geostationary orbit (approximately 35,000 km above the Earth). While the principal function of a space platform may be telecommunications, it also will have capabilities that could be utilized for sensor data collection, onboard information processing, and research.

Space Station Communication Roles

Low-Orbit Assembly

The Space Station will provide unique opportunities for the construction of advanced space platforms. The absence of gravity permits consideration of new spacecraft designs. An example of this would be the use of extremely light solar arrays. At present, most solar arrays and antennas are constructed to withstand the force of 1 gravity (1g, see below) so that they may be tested on Earth.

The structure for a conventional satellite also has to withstand the currently high transfer orbit accelerations of several g. By changing the assembly location to the Space Station, where gravity is essentially zero, it may be

---

* President, Communications Center of Clarksburg.

possible to eliminate some of the constraints and therefore some of the mass and design limitations of satellites.

An alternative would be to construct the spacecraft on the ground, support it with foam or some interim structure, and carry it up to the Space Station. At this point, it would be reassembled using an ultralight structure that would be designed to only handle the transfer orbit loads. These loads, in turn, could be substantially reduced (to 0.1g) through the use of advanced propellants and a longer burn time, such as may be available from liquid stages.

Eventually it may be desirable to combined the construction of solar arrays (from raw silica or gallium arsenide) entirely in the Space Station. This would have the advantage of extremely low contamination levels, zero gravity (for crystal growing), and an extremely high vacuum for depositions of leads and grids.

Low-Orbit Tests of Satellites

As has become painfully clear in several instances, most satellites are not tested in orbit until they reach their final destination. This is especially true for geostationary satellites. By that time the satellite is literally beyond the reach of any human being, and the options for overcoming a problem may be extremely limited.

The Space Station offers an opportunity to make a complete test of a space platform prior to its launch from the Space Station. Almost all of the functions could be simulated in the zero-g high-vacuum condition of the low-orbit Space Station. The performance could be verified over a period of several weeks, (thereby eliminating many of the infant mortality failures).

The test could be carried out aboard the Space Station as a space platform drifts slightly ahead or behind the Space Station. The Space Station would be the testing device. An alternative method is to retain the space platform aboard the Space Station and put a measuring device (contained in a remotely controlled satellite) ahead or behind the Space Station. Antenna patterns, for example, could be verified by this method. Figure 1 shows the testing of a satellite aboard the Space Station.

The attitude control system could be referenced to the Earth and its performance verified. The thermal control subsystem, although in a different orbit with more frequent eclipses, could also be tested.

Solar arrays and antennas could be deployed and their operation verified. This would avoid the problems faced on two of the Insat satellites, where it became difficult or impossible to deploy certain portions of the spacecraft.

Low-Orbit Storage

Currently, it takes at least six months to replace a failed satellite. This assumes that a flight-worthy spare is available and the delay is entirely with the launch vehicle. Approximately 28 months were estimated to completely reproduce the Palapa B2.

It may make sense to store the replacement (either in a fully or partially assembled condition) aboard the Space Station. When an orbital emergency took place requiring the replacement satellite, it could be rapidly launched from the Space Station using the accumulated residual pro-

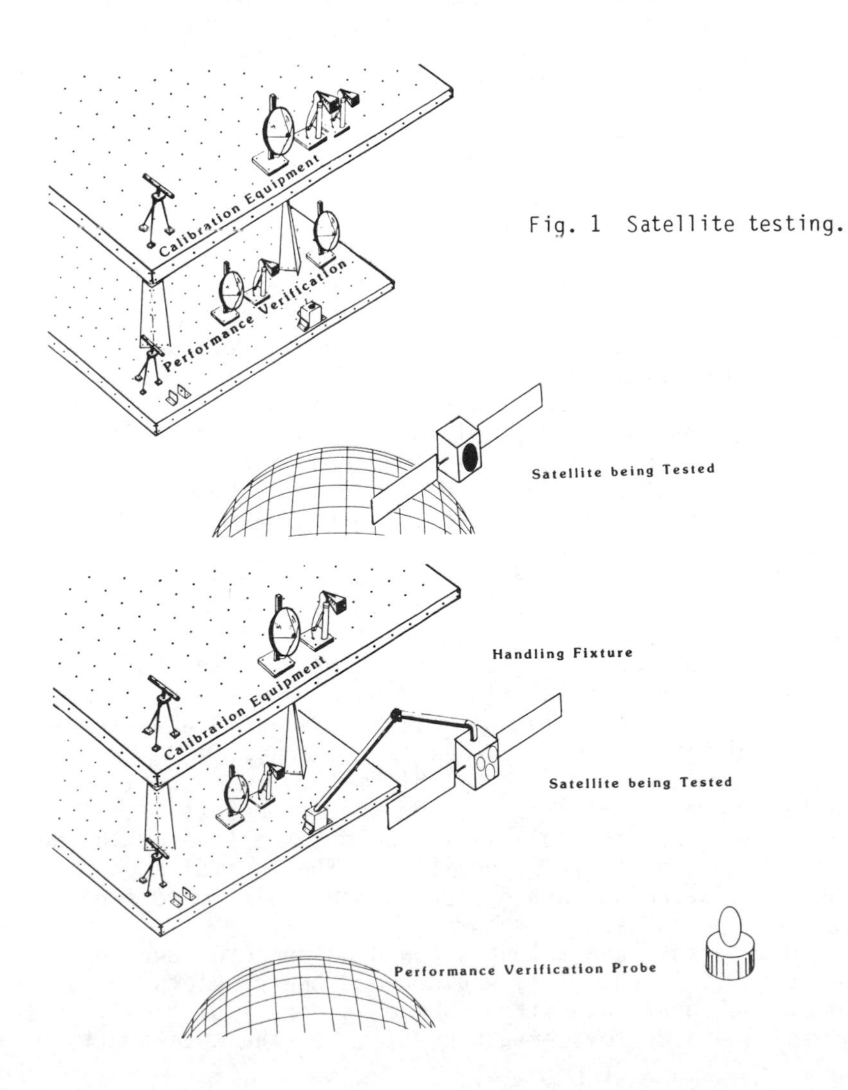

Fig. 1 Satellite testing.

pellants from the Shuttle. The replacement time, including test, could be measured in weeks rather than months.

This has a distinct economic attraction to the geostationary satellite owners, financiers, and to their insurers, as it reduces the duration of the service interruption. This, in turn, provides a higher level of assurance to the transponder owners and users. This capability has both cost savings and marketing advantages to the satellite owner.

The ability to store satellites in the low orbit on the Space Station also could permit reconfiguration of the satellite equipment (types of transponders, antenna patterns, interconnectivity, etc.) prior to the launch to another orbit. This would permit the custom design of a satellite for a particular orbit inclination and location, thereby providing additional freedom to the space platform owner. The Space Station could warehouse a collection of interchangeable modules, each with a different function. When it was decided that the next space platform was to be launched from the Space Station, a particular set of equipments would be selected, installed and tested to match the specific mission requirements.

Orbital Transfer Vehicle Requirements

Future advanced orbital transfer vehicles (OTVs) should have a thrust in the range of 0.05 to 0.1g or at a level in the range of the forces encountered in the most severe stationkeeping or stationchanging maneuver. This implies a liquid propellant. This concept may be compatible with the scavengering of unused fuel (H_2 and O_2) or decomposed from water (H_2O) that could be carried as ballast on the Shuttle flights that dock at the Space Station. Through the use a low-thrust OTV, the forces on the fully deployed satellite would be low enough so that it could be moved to the final orbit without any further steps.

The unpredictable reliability of solid propellants when combined with the high-g loads will decrease their attractiveness. Of still greater importance is the ability to combine the perigee, apogee, and stationkeeping management functions into an integrated liquid propulsion subsystem. The ability to start, stop, and restart liquid stages may permit additional flexibility. This may be attractive when the perigee and apogee functions are combined. Spacecraft designers are discovering substantial savings in special stages compared with the current PAM-D costs. Payload Assist Module D (PAM-D) is an upper stage designed to deliver up to 2320 lbs to a geosynchronous transfer orbit.

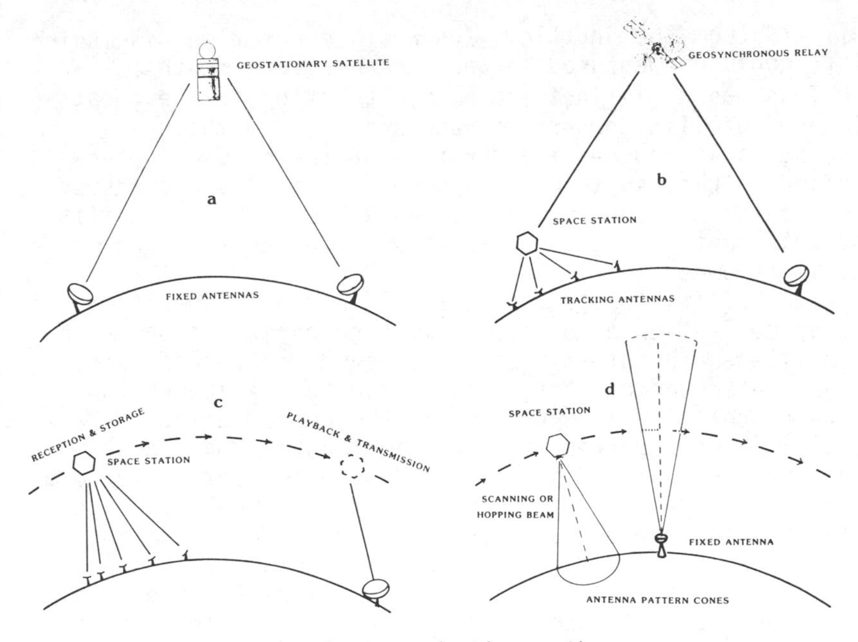

Fig. 2 Communications paths.

Table 1 Uplink characteristics

See Figure:	2a	2b	2c	2d
Uplink frequency GHz[a]	6	6	6	6
Earth station				
Antenna diameter, m	10	3	1	1
Tracking	No	Yes	Yes	No
Pointing loss, dB	0.2	0.2	0.2	4
EIRP, dBW	64	39	26	46
Transmitter, dBW	13	-2	-5	15
Transmitter, W	20	0.71	0.32	32
Uplink				
Distance to first satellite, km	35,000	2000	2000	300
Path loss, dB	199	174	174	158
Satellite antennas, dBi	30	10	10	45
Satellite noise, K	500	500	500	500
Peak bandwidth, MHz	2	2	0.1	200
C/N ratio, dB	30	30	30	30

[a]Although 6 GHz may not be the appropriate frequency for the low-orbit satellites, it has been used for consistency.

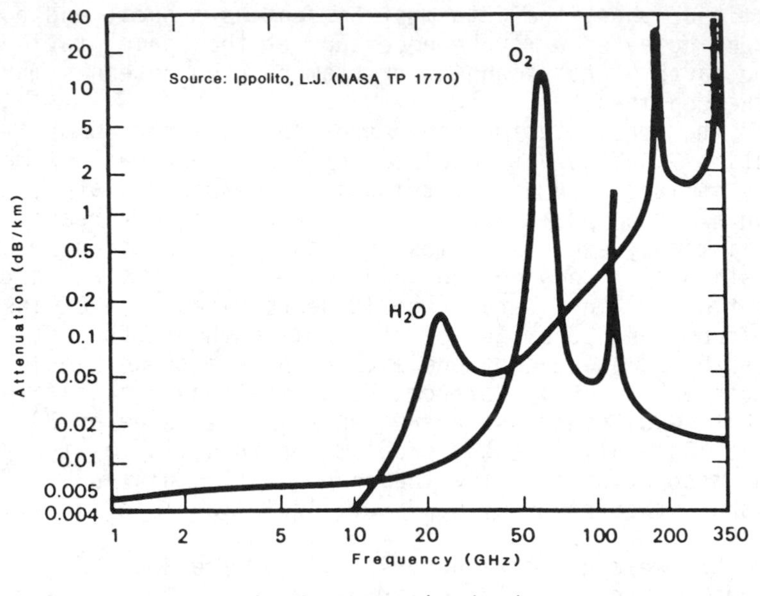

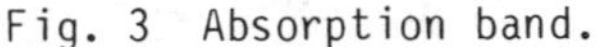
Fig. 3 Absorption band.

Table 2 Intersatellite frequencies and wavelengths

Allocation GHz	mm	Bandwidth MHz
22.55-23.55	13	1000
32.00-33.00	9	1000
54.25-58.20	5.5-5.1	3950
59-64	5.1-4.7	5000
105-130	2.6	25,000
170-182	1.7	12,000
185-190	1.6	5000
220-230	1.3	10,000
265-275	1.1	10,000

Space Station Communications Roles

The low-orbit Space Station could also provide the home for a series of communications links. These could receive sensory information from the ground and then relay it on to the geosynchronous orbit for transmission on the master earth station. The principal advantage of this particular type of link is that it is much shorter than going all the way to the geostationary orbit. Figure 2 and Table 1 show the differences in the two approaches. The time during which the Space Station is in view of the earth station is limited, thus, the data must be sent as a burst and a tracking antenna may be necessary at the Space Station or the earth. Beam scanning or beam hopping antennas may be other options.

The Space Station could provide links for communications that do not require continuous service. Examples may include document distribution, newspaper plates, audio/video update service, once-a-day account reconciliations, etc. At times when the earth station/Space Station link does not coincide with an earth station/geostationary link, service could be provided if the links to/from the Space Station and from/to the Earth continue to use the existing frequencies. A space station [in low Earth orbit (LEO)] to/from the geostationary satellite orbit (GEO) could use a frequency in the range of 30 to 100 GHz. Rainfall is not a problem for this link as the aerosol absorption is below the space station altitude. Table 2 shows the frequency bands available for intersatellite links.

Unlike a geostationary satellite, the Space Station moves with respect to an earth station. This provides a degree of security in these constantly moving Earth to/from Space Station links. Use of an absorption band (see Fig. 3) for the Space Station to/from geostationary space platform provides an extremely high security in this link from terrestrial eavesdropping. These links are also moving with time making interception even more unlikely.

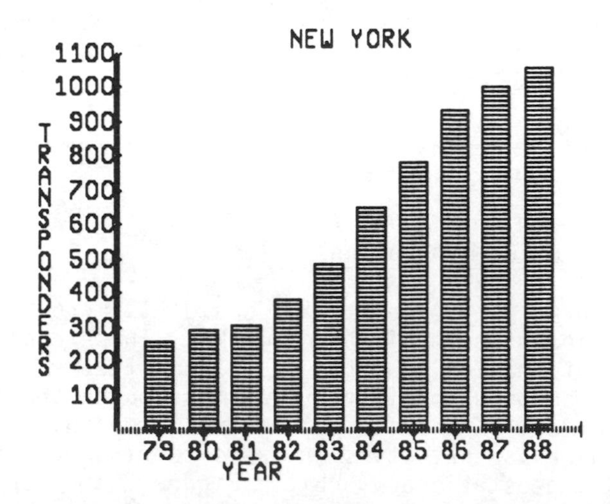

Fig. 4 Number of transponders visible from New York City (domestic US, Canada, and transatlantic).

Table 3 US fixed satellite service applicants

Applicants of the early 1970's	
Name	Result
Comsat General	
System A	Comstar D1-D4
System B	Not constructed
Fairchild	20% of Westar I-V
Hughes	Not constructed
MCI-Lockheed	Merged into SBS 1-3
RCA	Satcom I-IV
Western TeleCommunications, Inc.	Not constructed
Western Union	Westar I-V

Applicants of the late 1970's	
Name	Satellite name
Southern Pacific Comm.	Spacenet I-III
Western Union Space Comm.	Advanced Westar (not used)

Applicants of the early 1980's	
Name	Result
AT&T	Telstar 301-303
Advanced Business Communications, Inc.	Denied by FCC
Alascom, Inc.	Aurora I (Satcom V)
American Satellite Co.	ASC-1, ASC-2
GTE	Gstar I & II
Hughes	Galaxy I-III
Rainbow Satellite	Denied by FCC
USSSI	Denied by FCC

The Space Platform

Growth

Figure 4 shows the increase in the number of transponders visible from New York City. This capacity growth could propel the space industry towards the space platform as a logical means of meeting future requirements.

Satellite communications on a domestic basis started out in the mid-1970's as basically a telephony and television network distribution mechanism. Some of the original applications foresaw telegraphy and telex as being a major user, but it must be realized that little spectrum is required for these services. Another satellite network decided to base its fortunes on the coming digital revolution. Before many years were out, its principal revenues were coming from analog voice services.

In spite of errors in planning, these early services survived and eventually prospered. Just as soon as they were becoming profitable, a number of other entities (from outside the communications area) requested, and in some cases, were approved to provide competitive services.

Domestic satellites evolved into the means for connecting cable television systems. This was an economical approach that permitted the establishment of entirely new television networks based on the narrow-casting approach as opposed to the normal broadcasting. Narrow-casting examples included services devoted entirely to sports, religions, weather, women, and health. Had it not been for cable television, at least one, and perhaps two, of the domestic satellite systems could have floundered within the first few years.

The commercial television broadcast networks were rather slow to adopt this new mechanism for program distribution. It took nearly ten years before the first commercial TV networks really went to widespread program distribution by satellite, in spite of the Public Broadcasting System having shown that it could be done on a routine basis.

New Satellite Networks Versus Time

Table 3 shows the continuing growth of new satellite networks filing with the Federal Communications Commission. During 1983 alone, 15 new operators appeared (Table 4). It

Table 4 1983 Applications from new operators

Operator	Service type
Cablesat General[b]	Domestic FSS[a]
Columbia Communications[b]	Domestic FSS
Comsat General	Domestic FSS
Digital Telesat[b]	Domestic FSS
Equatorial Communications[b]	Domestic FSS
Federal Express	Domestic FSS
Ford	Domestic FSS
Martin Marietta	Domestic FSS
National Exchange[b]	Domestic FSS
Geostar	Domestic location
Mobile Satellite Corp.	Domestic MSS[c]
Skylink	Domestic MSS
International Satellite	Transatlantic
Orion Satellite	Transatlantic

[a]FSS = fixed satellite service.
[b]Denied by the FCC on July 25, 1985
[c]MSS = mobile satellite service.
Note: This list does not include applicants for the broadcasting satellite service.

is not expected that this rate will continue ad infinitum, or that each applicant will construct satellites, but it does illustrate that the industry is moving very rapidly. A number of new services may be emerging. Eventually satellite-delivered information services may provide instant access to the movie of your choice, access to the

Table 5 Dual-service satellites

Satellite	1	2	3	4	5	6	7	8	9	10	11
American Satellite Co.	X	X									
Arabsat	X		X								
Aussat		X	X								
Cablesat General	X										
Comstar D1-D4	X									X	
Sakura (CS)	X									X	
Ford Star	X	X									
Gorizont	X					X					
Intelsat V-F1 to F4				X	X						
Marisat						X	X				
Morelos	X	X									
Sirio-2									X	X	
Spacenet	X	X									

Column meanings

1 Domestic fixed satellite service (6/4 GHz)
2 Business satellite service (14/12 GHz)
3 Broadcast satellites (UHF, 2.5 & 12 GHz)
4 International fixed satellite service (6/4 GHz)
5 International fixes satellite service (14/11 GHz)
6 Military and government satellites
7 Mobile satellite service
8 Tracking satellites
9 Weather satellites
10 Experimental
11 Amateur satellites

Table 6 Triple-service satellites

Satellite	1	2	3	4	5	6	7	8	9	10	11
Gorizont later series				X	X	X					
Insat	X		X						X		
Intelsat V-F5 to F8				X	X		X				
TDRS/AW	X	X						X			
Telecom-1		X		X		X					

Column meanings

1 Domestic fixed satellite service (6/4 GHz)
2 Business satellite service (14/12 GHz)
3 Broadcast satellites (UHF, 2.5 & 12 GHz)
4 International fixed satellite service (6/4 GHz)
5 International fixed satellite service (14/11 GHz)
6 Military and government satellites
7 Mobile satellite service
8 Tracking satellites
9 Weather satellites
10 Experimental
11 Amateur satellites

Table 7 Quadruple-service satellites

Satellite	1	2	3	4	5	6	7	8	9	10	11
Intelsat V-F9 & up		X		X	X		X				
Intelsat VI		X		X	X		X				

Column meanings

1 Domestic fixed satellite service (6/4 GHz)
2 Business satellite service (14/12 GHz)
3 Broadcast satellites (UHF, 2.5 & 12 GHz)
4 International fixed satellite service (6/4 GHz)
5 International fixed satellite service (14/11 GHz)
6 Military and government satellites
7 Mobile satellite service
8 Tracking satellites
9 Weather satellites
10 Experimental
11 Amateur satellites

world encyclopedias, and custom searching of public data bases.

Multinetwork Spacecraft

When the author collaborated with Burton Edelson (now Associate Administrator of NASA) on the original concept of a large multinetwork satellite (then called the Orbital Antenna Farm), the paper was met with substantial interest, but most felt that the OAF was a long way off. Since 1977, there has been a quiet but discernible trend in the direction of having multiple networks on the same space frame. Tables 5 to 7 list satellites that are already in orbit (or are being planned), showing that this trend is already upon us. Not only are there dual-network satellites, but some even have as many as four discreet functions present.

While the term geostationary platform is generally associated with a large structure having multiple missions, this is not necessarily the case for the future. In many cases, these missions could also be satisfied through the use of many small satellites connected together by ground radio-frequency (rf) links or by intersatellite links. These could share either a single or a group of nearby orbit locations.

A large space platform could carry a number of communications payloads and other services. Because of its investment, the structural and electrical design should be capable of supporting the various missions of the next 20 to 30 years.

How Will the Space Platform Evolve?

A space platform will congeal at one of several orbit locations from the elements that make up the individual networks. The locations most likely to see this type of design are at 19, 31, and 101 West longitude. The originally predicted location was 100 West.

The location of 101 West is allocated to satellites using 6/4 GHz, 14/12 GHz, 18 GHz/direct broadcasting satellite frequencies. It is also a logical location for 30/20 GHz and possibly 8/7 GHz services. Table 8 shows the particular applicants for use of this orbit location. Tables 9 and 10 provide similar information for 19 and 31 West longitude.

The 19 and 31 West locations are unique from 101 West in that the direct broadcast frequencies, polarizations, beam patterns, power levels, bandwidths, transmission characteristics, etc. have all been predefined for the

Table 8 Applicants for 101 West

Applicant	Service	Frequency (Downlink, GHz)
Advanced Comm. Corp.[a]	BSS[b]	12.2-12.7
Direct Broadcasting Satellite Corp.	BSS	12.2-12.7
Ford[a]	FSS	3.7-4.2 and 11.7-12.2
Hughes[a]	BSS	12.2-12.7
National Christian Net.	BSS	12.2-12.7
National Exchange[a]	BSS	12.2-12.7
Satellite Syndicated Systems	BSS	12.2-12.7
Space Communication Services	BSS	12.2-12.7

[a]Construction permit granted.
[b]BSS=broadcasting satellite service.
[c]FSS=fixed satellite service.
As of August 1, 1985.

Table 9 Authorized users of 19 West

Austria
Belgium
France (TDF-1)
West Germany (TV-SAT-A3)
Guinea
Italy (Olympus)
Luxembourg
Namibia (South West Africa)
Nigeria
Switzerland
The Netherlands
Zaire

Table 10 Authorized users of 31 West

Azores
Canary Islands
Great Britain (Unisat or Britsat)
Iceland
Ireland (Eiresat)
Ivory Coast
Liberia
Portugal
Portugese Guinea
Sierra Leone
Spain
Upper Volta
Vatican City

operators at the World Administrative Radio Conference of 1977. In theory, this should eliminate competition between users and it would permit either a clustering of individual satellites or the construction of one or several large space platforms consolidating the individual national missions aboard one space frame.

Frequency Reuse

As predicted in the original Orbital Antenna Farm paper, the extensive use of spot beams permits reuse of the frequency spectrum many times across a broad area. Within the footprint of a particular beam, the frequency may only be reused twice (in two orthogonal polarizations). Through the use of multiple beams and adequate separation between co-frequency/co-polar satellites reuses of a bandwidth of 500 MHz may actually yield many times this capacity on a national basis. The Martin Marietta domestic satellite, for example, envisions a nine-times reuse.

A Novel Satellite Concept

As the beam diameter shrinks, the antenna gain (in absolute terms) increases proportionally. If the effective isotropically radiated power (EIRP) is held constant (e.g., 50 dBW), the power gain may be interchanged between the satellite power amplifier and the antenna gain. In the extreme case, a satellite with a large number of very narrow beams would have such a high antenna gain (e.g., 70 dBi) that the power amplifier, would have a very low radio-frequency power rating (for example 15 mW). This would substantially reduce the total dc power requirements. The mass and thermal demands of the active repeater would be so low as to completely reorder the normal construction goals of a satellite. This satellite would be physically dominated by the multibeam antenna. The antenna's structure would support the small housekeeping, transponder, and solar array modules. Only the antenna would be attitude controlled. Radio-frequency sensors could be used because of their inherent accuracy.

This concept could yield tremendous frequency reuse. In 1984, this ultimate concept is not possible because the high-precision, large-diameter antennas and feed arrays have not been developed sufficiently. Eventually, actuators along the antenna ribs will actively control the reflecting surface, which will be continuously surveyed by a laser measurement system. A similar method is being used to dynamically shape a 300-in. optical telescope at the Kitt Peak Observatory using hydraulics.

The multi-mini-beam satellite is an attractive concept, as the mass required for a large number of low-power amplifiers and the power generation/storage/conversion subsystem is small. The mass in the antenna and feed system has a slow incremental growth as beams and capacity are added. Basically these are the one-time mass of the reflector. Additional feeds and milliwatt power amplifiers have a small incremental mass. In a sense, this is the inverse of what is often thought of as a space platform in the geosynchronous orbit.

This space platform may have a series of these multiple beam antennas, along with onboard switching and regeneration to improve the signal as it passes through the space platform. This in turn permits the use of smaller and far less expensive earth stations.

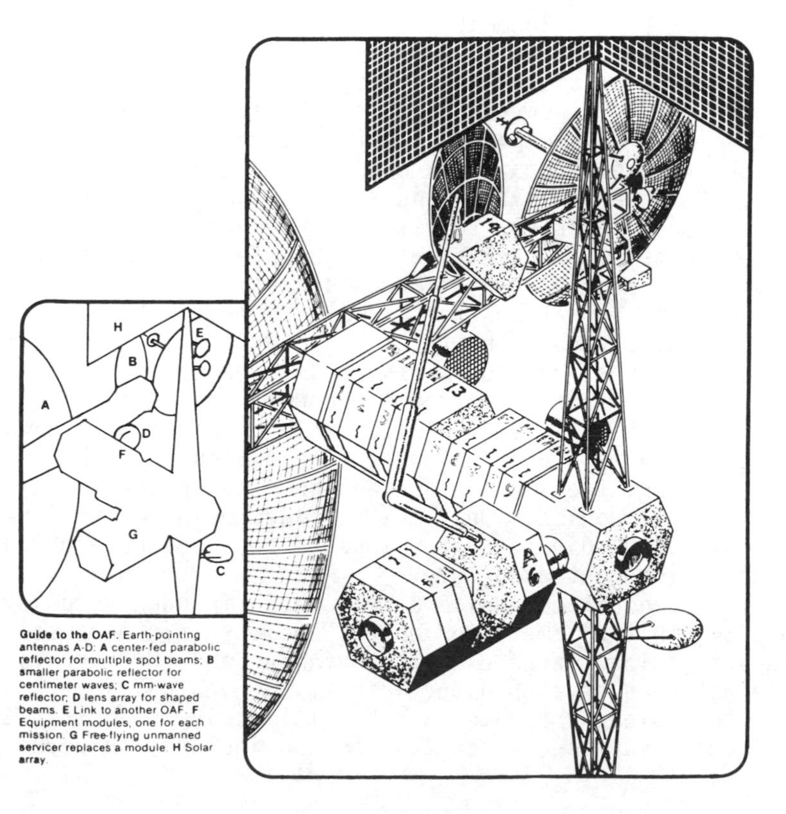

Guide to the OAF. Earth-pointing antennas A-D: **A** center-fed parabolic reflector for multiple spot beams; **B** smaller parabolic reflector for centimeter waves; **C** mm-wave reflector; **D** lens array for shaped beams. **E** Link to another OAF. **F** Equipment modules, one for each mission. **G** Free-flying unmanned servicer replaces a module. **H** Solar array.

Fig. 5 Space platform upgrade.

Of course, multiple networks may be aboard the same space platform, and therefore the total mass may climb back up but the per mission mass may still be lower than current designs. Eventually, this is the way the satellite may carve out its own niche of survival against fiber optics in point-to-point service.

Distribution Satellites

Spot beams provide very little advantage (if any) for wide-area distribution (point-to-multipoint services such as TV distribution). If this service is provided through such a satellite, many (perhaps all) of the individual beam patterns will be occupied simultaneously with the same video or audio information. Therefore, while there is a high theoretical reuse of frequency, the amount of intelligence transmitted has not improved at all. This is another reason why another class of satellites will evolve. These will be large, high powered, and massive. The ratio of point-to-point and point-to-multipoint satellites will change as a function of time. This is already happening, and fiber optics will alter the balance. Multipoint satellites should compete well with fiber optics in their market.

Refurbishment

For purposes of redundancy and absolute program continuity, two space geostationary platforms could be co-located in orbit. Under nominal conditions, each would carry approximately half of the traffic. While one of these space platforms is being refurbished and updated (through the automated replacement of some of its components from the Space Station) the traffic would be transferred to the other satellite. Because the space platform would be composed of many individual modules, the traffic diversion could be for a single network, whereas, the remaining networks could continue to use the satellite being upgraded. Figure 5 shows an early version of this change out. The structure was hexagonal in cross-section so that the maximum volume could be carried in the Shuttle cargo bay. Semihexagonal sections could be removed and replaced. In most cases the replacement is not for the purpose of repair, but rather for inserting a more recent module and upgrading the service quality or quantity. The central spine of the Orbital Antenna Farm contained dc and rf connections plus heat control paths.

Ownership

This is an area that concerned many people when the Orbital Antenna Farm was first proposed. Ownership of satellites had been on a single entity basis. In subsequent years, communications capacity has been leased (or sold) to others. Various forms of creative financing have evolved including limited partnerships, joint ventures, sale and leaseback, and consortia. Many of todays satellites (e.g., Galaxy I) have multiple transponder owners. It is anticipated that many of the ownership questions raised will become moot.

Conclusion

A space platform can provide:

1) Economic advantages of consolidated housekeeping functions on a common space frame.

2) Interconnection opportunities for various services and networks at the baseband level.

3) The use of high gain spot beams and milliwatt power repeaters with low incremental costs to add capacity.

4) The ability to refurbish the satellite, to change the capacity and to alter the service types as traffic evolves.

5) By increasing the communications capacity per orbit location, the orbital separation of satellites can be increased, thus allowing the use of simple and smaller earth stations.

Ownership of space platforms is a subject that lawyers will debate for years, but chief executive officers will settle within weeks.

Acknowledgments

Dr. Geoffrey Canetti and Dr. Ivan Bekey both provided useful review suggestions that have been incorporated. Portions of this chapter are based on Space Station studies conducted by the Communications Center of Clarksburg for Rockwell International and in turn for NASA.

The Potential of Materials Processing Using the Space Environment

James T. Rose* and Terrence D. Fitzpatrick†
McDonnell Douglas Astronautics Company, St. Louis, Missouri

Abstract

The potential for materials processing in space is enormous, but a number of problems--scientific, economic, structural, and political--need to be solved before this potential can be realized in commercial projects developed by private industry.

Basic materials processing research in space over the last 20 years confirms that a zero-gravity environment offers significant advantages that could lead to improved materials processing and new or improved products in at least four areas: biological materials (for pharmaceuticals), metal alloys and composites, semiconductor crystals, and glasses (for lenses and lasers). These improvements could take place on Earth, as a result of research in space, or in space, as a result of the development of commercial space projects.

The history of materials processing research in space is short, but it has started to expand in the last few years. This history is also international in character. In addition to American research that began with the Apollo program, the Europeans (especially the West Germans and French) are expanding their research through such programs as Spacelab, Texus, SPAS, and Eureca. The Japanese have budgeted a 15-year space program calling for over a billion dollars a year, with many millions marked for materials research. And the Russians, who began materials research in the early 1970's and are suspected of having

*Director, Electrophoresis Operations in Space.

†Technical Writer, Engineering Planning and Administration Department.

done more work than anyone, launched a prototype materials processing factory in 1983.

The only commercial materials processing project to date, however, is Electrophoresis Operations in Space (EOS), developed by the McDonnell Douglas Astronautics Company. The project aims at separating (in space) biological materials in the quantities and purities needed by pharmaceutical companies to develop new and improved medicines. Early space tests have confirmed that the space process can separate over 700 times more material and achieve purity levels more than four times higher than possible in similar operations on Earth. Commercial manufacture of the first product, following clinical testing and FDA approval, is planned for 1988. The company hopes to have full-scale production plants operating on manned or unmanned space platforms by the early 1990's. The prediction is that space bioprocessing could become a multi-billion-dollar industry within the next 20 years.

But private industry in general has not yet seemed willing to invest money in commercial space projects. Longer investment pay-back periods (ten years or more rather than the usual three to five years) and the need to form partnerships are among the barriers to investing in commercial space projects that need to be overcome. And a new national space policy that stimulates commercial investment is needed to help materials processing in space realize its potential.

Introduction

The field of materials processing in space has enormous potential, but this field is today only in the embryo stage. It will need 20 years or so to grow to maturity.

More basic research remains to be done. All kinds of hardware, including manned and unmanned space platforms, need to be built. More extensive and effective arrangements between government and industry need to be developed. Much hard work will have to be expended on discovering the new processes and products that the space environment affords and then in finding ways to market them. Finally, incentives will have to be found to convince private industries to risk their money on projects that require large initial investments and that have long pay-back times.

	PRECIPITATION (BATCH PROCESS – LARGE AMOUNTS)	ELECTROPHORESIS (CONTINUOUS PROCESS)	CHROMATOGRAPHY (BATCH PROCESS – SMALL AMOUNTS)	ELECTROPHORESIS (STATIC PROCESS)
QUANTITY	1	3	2	4
RESOLUTION	4	2	3	1
YIELD	3	1	2	4

Fig. 1 Ranking of ground separation techniques.

In spite of these problems, materials processing in space will grow. The motives for creating new life are simply too complex and too strong: the desire of consumers for new products and processes, the search for new areas of profit, the lure of competition among companies and nations, the pull of technology, and the excitement of working in the most challenging new frontier of this century.

But no one yet knows when materials processing in space will come of age as a commercial activity. Nor does anyone really know how many areas of work will open up as a result of our new access to space, although most observers speculate that there are four areas of seemingly great promise: separating biological materials for use in pharmaceuticals, producing new metal alloys and composite materials, growing purer semiconductor crystals, and developing new glasses for lenses and lasers. And at this stage, it is impossible to know just how large any of these industries could become.

What we do know is that the microgravity environment of space gives us the opportunity to study material processes in ways that are impossible on gravity-bound Earth. What we are learning may help us to improve the way we process materials on Earth, develop new ways to process them better in space, and devise ways in space to process materials that cannot be processed on Earth either for reasons of economics or due to limitations caused by gravity.

Because materials processing in space is a new and ill-defined field of work, a few definitions are in order. As used in this discussion, the word "materials" refers to metals, crystals, glasses, and biological substances. "Processing" refers either to work done in space for the purpose of improving processing methods on Earth or to work done in space that results in products that can be sold on Earth. This definition excludes such work as welding materials together in space to build a space platform. Finally, "space" refers to near-Earth orbit, the distance the Shuttle can travel from the surface of the Earth. This excludes such work as mining on the Moon or on asteroids because work at these distances is not likely to happen within the 20-year framework envisioned in this discussion.

The Role of Gravity in Materials Processing

What makes processing in space such an attractive prospect is the possibility of escaping from the forces of gravity. One day we may discover other space phenomena of equal importance, but for the foreseeable future it is the "zero-g" environment of orbiting spacecraft that holds our attention.

On Earth, gravity creates a number of phenomena--hydrostatic pressure, buoyancy and sedimentation, convection--that both shape the ways we process materials and limit the effectiveness of our methods. For example, sedimentation and convection seriously affect the ways we can separate biological materials into their constituent parts.

As Fig. 1 indicates, the five major separation techniques available on Earth lead to undesirable trade-offs among the four issues that matter in biological separations: quantity or throughput, resolution or purity, recoverable yield, and simplicity of operation. The precipitation method, which takes advantage of sedimentation forces, separates the most raw material but it achieves the worst resolution. Static electrophoresis is a method that tries to eliminate both sedimentation and convection forces by embedding the material to be separated in an unmoving gel (see Fig. 2). When an electrical field is applied across the gel, the differently charged particles migrate laterally and separate into their component parts. This method achieves the highest resolution, but because only a small amount of material can be separated at one time (typically about 0.01 ml) it has the worst record for quantity. It also has the worst record for recoverable yield--the separated material remains embedded in the gel. For these reasons, static electrophoresis is useful for laboratory research but not for producing pharmaceuticals, in spite of its ability to achieve high levels of resolution.

But there is a larger issue here than just comparing the advantages and disadvantages of various separation techniques. The real questions are what materials do we want or need to separate, how much of them do we need, how pure do they have to be, and how much will it cost to get them? There is an extraordinary degree of interplay among human, purely scientific, and rational economic considerations standing behind these questions, as the following example may help to show.

For many years, medical scientists have been doing research on ways to treat, alleviate, or cure diseases. Some diseases caused by lack of specific proteins could be treated by replacement therapy. The problem, however, is to purify the proteins to avoid immune responses. Another problem is that the separation techniques available on Earth either cannot separate proteins purely enough or in large enough quantities.

There is also a historical situation at work here, one that is at work in many fields besides medicine. Just as medical understanding has advanced to the point that specific proteins are needed but difficult to get, technological advances in many industries have increased our needs for better materials. But as we are discovering the characteristics these needed materials should have, we are also discovering that the methods we use to process these

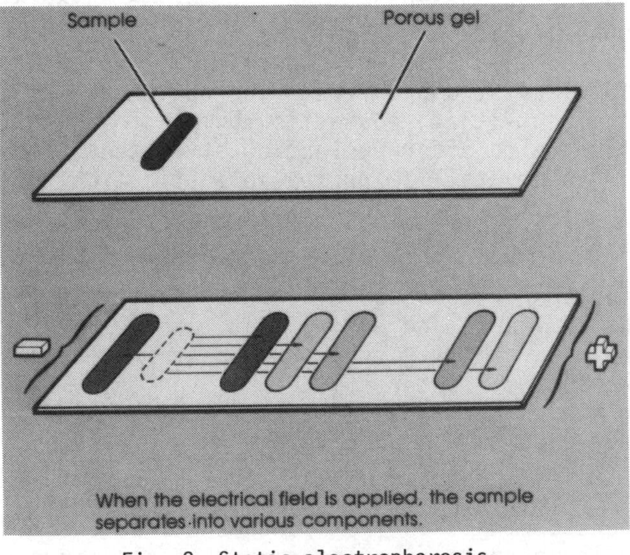

Fig. 2 Static electrophoresis.

materials cannot give us the needed characteristics. At the moment, we seem to have backed ourselves into a technological corner.

One obvious example of this bind, an example used by almost everyone in the field of materials processing, is the semiconductor industry. Our "ability to grow high-quality semiconductor crystals," an achievement driven first by military needs and then by the demands of commercial markets, has "revolutionized the electronics industry."[1] This advance has brought us computers of enormously increased speed, greatly reduced size and weight, and significantly lower cost. The problem, however, is that we have created a need for even better crystals. Among other things, we now need crystals that are free of impurities, crystals whose dopant distribution is extremely homogeneous, and crystals that are free of striations and other structural defects. And we need to develop methods for producing such crystals in large enough quantities to make them affordable for commercial markets.

This technology bind exists in other fields as well. For example, sedimentation, buoyancy, convection, and hydrostatic pressure all limit our ability to improve current metal casting techniques and to produce better alloys and composite materials. And in the field of glasses and ceramics, hydrostatic pressure is an important constraining force. Because it causes molten materials to deform under their own weight, we have learned how to cool (and supercool) materials in containers, but this method creates at least two problems. The container walls give off contaminants that reduce the purity of the solidifying glass. They also offer nucleation sites that allow some materials to crystallize during solidification, particularly metals that might otherwise produce glasses with unique properties.

Materials processing, of course, has taken place for thousands of years, and during that time we have made many advances in our ability to produce better materials. This is one reason we use terms like the Stone Age, Bronze Age, and Iron Age to mark advances in technology and civilization. But even though we have now entered the Silicon Age, it is important to remember that much of what we have learned has resulted more from minute increments of empirical understanding than from fundamental scientific breakthroughs in understanding the nature of the structure and properties of materials and of the processes that govern their formation.

John R. Carruthers, the former Director of Materials Processing in Space for NASA, has stated that advances in solid-state electronics since World War II, for example, "have caused a dramatic maturing of material science and deeper levels of understanding of the control of structure-property relationships in materials to meet increased performance demands. Important related advances in the characterization of the structure and composition of materials have also contributed to this increased sophistication."[2]

What he goes on to say, however, outlines quite well the nature of the problems that confront us in trying to produce even better materials:

"Unfortunately, our understanding of the materials formation processes themselves have not kept pace with these advances. The complexity of the processes used in the production of metals, chemicals, and crystals has made it necessary to understand them by modeling and simulating with simpler systems. Most major advances have come from the scaling up of production processes more by empirical engineering intuition than by fundamental contributions of science. The great cost of performing research on larger facilities precludes extensive work on them and thus further widens the gap between research and production. Many new high-technology industries spawned by new materials science are finding it difficult to sustain an adequate level of research activity and are rapidly reducing their edge in technology. All these pragmatic considerations serve as constraints to the involvement of materials and chemical scientists in basic studies of processing, whether on Earth or in space."[3]

These are excellent points, especially the connection that is made between scientific and economic constraints. Yet it is important to note that Carruthers made this argument several years before the Space Shuttle became operational. Since its advent, scientists have had the opportunity to do new kinds of basic research in materials processing.

The Space Shuttle is the pivot that could leverage materials processing to a new height. Just as other revolutions have been generated by a key event, just as the

Industrial Revolution was powered by the invention of the steam engine, so too may the Space Shuttle become the key event ushering in a new era for materials processing.

The Shuttle is a key event in the 5000-year history of materials processing because it gives us the opportunity to do both basic research and actual processing in a low-gravity environment. As the previous few pages have indicated, gravity-driven forces represent first-order constraints on many important formation processes. Working in an environment that effectively eliminates these forces, then, offers us two unique but very different kinds of opportunities. One is the possibility of improving Earth-based processing. This could take place in several ways. On Earth, forces like buoyancy convection are often so strong that they mask the operation of forces that are not caused by gravity, such as Marangoni flow (surface-tension gradient convection). With the disappearance of gravitational forces in space, it becomes possible to study the other forces at work in complex formation processes and to learn how to control them so we can improve our processing methods on Earth. Or by creating "prototype" materials and processes in space, we can also learn how to improve our Earth-based methods. One example of this possibility comes from low-gravity experiments on a KC-135 involving codeposition, the method of electroplating most affected by gravity. Even with gravity effects removed for so short a period as 20 seconds, these experiments revealed that a smooth substrate is extremely important in forming a good electoplated surface, a discovery that could lead to the development of better coatings.[4]

The other kind of opportunity comes from using the low-gravity environment of the Shuttle to produce new and better materials in space. The Shuttle could become a facility--a factory, if you will--for producing materials for commercial markets. With gravity removed, it should be possible, hypothetically at least, to improve any material whose formation is strongly influenced by gravity-driven forces. Processing in the Shuttle, then, could enable us to produce materials that cannot be produced on Earth, to produce other materials better, and--strangely enough--to produce some materials less expensively.

For example, the McDonnell Douglas Astronautics Company has estimated the relative Earth and space costs of separating enough biological materials to produce a single pharmaceutical that could satisfy the annual needs of patients in the United States. The cost of building separation facilities on Earth capable of providing even this much material would be more than 50 times greater than the cost of building a single facility with an equal capability that would fly on the Shuttle. And the yearly cost of operating the ground facilities would be more than four times higher than operating the single facility on the Shuttle. The cost of launching things into space is high, but the enormous increase in the amount of material that can be separated in space more than offsets the development and launch costs of the space project.

So it is access to a gravity-free environment that makes the Shuttle a key event in the history of materials processing. But such access is not the only fact about the Shuttle that makes it important to both materials research and production. The duration of Shuttle flights, up to ten days at present, gives us more time to work continuously in zero-gravity conditions that we can in drop-tower tests (about four seconds) and in parabolic flights of planes like the KC-135 (about 20 seconds). The size and configuration of the Shuttle, particularly its large payload bay, give us the room to do research and production work that was impossible to do in the cramped quarters of earlier manned spacecraft. NASA's decision to use what it calls mission specialists, people dedicated to performing research, gives us the ability to perform more extensive experiments than we could when astronauts had to divide their time with operating the spacecraft. The Shuttle's reusability gives it cost and scheduling advantages over ways of getting into space to do work. And when, in the near future, unmanned research and production platforms are operating in space, the Shuttle promises to be an excellent way to ferry people and materials back and forth (with decided advantages over the earlier Skylab project).

There are problems with the Shuttle, of course, but they can be solved. The major problem is that the Shuttle was planned and designed as a multipurpose vehicle, which means that NASA, the military, and private industries have to share its payload space. Likewise, NASA may not be able to schedule enough Shuttle launches to satisfy the growing needs of customers. One proposal to solve these problems has been to build a fifth Shuttle that could be leased or sold to private industry. Whatever the solution, though, the point is that the problems are political and economic, not technological. A good, safe way of getting into space to do work has been built and tested. What remains is to get on with the work.

A Short History of Materials Processing in Space

First, the history of materials processing in space is short, not much more than 20 years. Second, there is controversy about the value of the research that has been done in space. Some people regard many of the experiments that have been done as "show biz"; others claim that previous experiments are so conclusive about the advantages of working in a zero-g environment that industry should begin massive production efforts on all fronts. It is not our intention to enter into the details of this controversy. Our view, as members of the aerospace community, is that much of importance has already been learned and that much more remains to be learned. The last general thing to say about this short history is that it is not just an American history. The Russians have been doing materials research for a number of years and have probably been doing it on a scale that surpasses our own. The Europeans, particularly the French and the West Germans, have entered this field strongly in the last few years and have extensive plans for more work. The Japanese have also started to become active in this field and will become more so in the next few years. The growing international character of materials processing in space should give rise to both increased cooperation and competition, creating a push-pull environment that could help speed this many-sided field closer to maturity.

For all practical purposes, materials processing in space (MPS for short) began as the stepchild of the manned space program in the late 1950's. The need to design controllable propellant and other fluid systems for spacecraft gave rise to the need for research in fluid mechanics. Questions about solidification phenomena had to be answered before designing heat storage systems that depend on phase changes in materials. And the possibility of brazing and welding large structures in space depended on studying the flow of liquid metals in zero gravity.[5]

The need to solve these problems led to a number of experiments conducted in drop towers, planes flying parabolic trajectories, sounding rockets, and the last several

Apollo flights. Among the most important were the successful dispersion and solidification of materials (gallium and bismuth, for example) that are immiscible on Earth, composite castings in a furnace that showed that materials do not unmix as they do on Earth because of sedimentation, fluid experiments to determine the effect of weightlessness on heat flow and convection, and electrophoresis experiments designed to study what effects the elimination of buoyancy-driven convection would have on the separation process. These early experiments also provided important experience to help scientists design better experiments and apparatus used in later missions.

The long duration of the Skylab missions in the early to mid-1970's presented scientists with their first opportunity to do extensive research in space. Experiments were conducted in three main areas of MPS interest: processing of semiconductor materials, metallurgical processing, and the behavior of fluids. Several Skylab experiments were conducted to see if growing semiconductor crystals in zero gravity would minimize impurities and structural defects as well as make it possible to control the distribution of trace impurities or dopants deliberately added to change characteristics. In metallurgical processing, experiments included melting to study joining and casting, exothermic brazing, formation of immiscible alloy compositions, and formation of spheres by containerless melting and solidification. Fluids experiments included diffusion of radioactive tracers (which showed that gravity-driven convection creates about 50 times more transport than does simple diffusion caused by volume changes), liquid floating zone technique to create high-quality silicon, growing crystals from an aqueous solution, and separating charged particles in liquid buffers (isotachophoresis).

After Skylab came the joint US-Soviet mission, Apollo-Soyuz, in July 1975. This was a nine-day flight and a number of MPS experiments were carried out. Among them were a repeat of a Skylab vapor-growth crystal test (showing the same results: improved structure and growth habits), zero-g processing of magnets, alloy formation experiments to prevent the separation of materials with different densities, tests to determine the effects of surface-tension convection (Marangoni flow), and two kinds of electrophoresis tests to separate cells from other biological materials. The last two tests were shortly to prove of particular interest to the McDonnell Douglas Astronautics Company.

During the years between Apollo-Soyuz and the first launch of the Space Shuttle in 1981, there was of course no opportunity--in America, at least--for manned experiments in space. But the Space Processing Applications Rocket (SPAR) program did allow scientists to continue some kinds of work. Because the low-gravity part of a SPAR flight lasts only about five minutes, this program was not really useful for crystal growth and biological separation, but a number of fluids and solidification experiments were performed. While these experiments increased our knowledge to some degree, their real value was in giving scientists and engineers experience in designing and operating automated experiments.

At this point, it might be useful to take stock of what we learned from these early experiments. The reason for doing this now is that results from many Shuttle experiments and from the first Spacelab mission in late 1983 are not as fully known and clearly understood. According to Robert J. Naumann and Harvey W. Herring,[6] significant findings in the area of crystal growth include the following hopeful conclusions:

1) Macrosegregation in melt growth can be controlled, resulting in uniform dopant distribution over a short growth distance.

2) Microsegregation due to growth rate fluctuations can be eliminated, making it possible to grow some higher composition alloy semiconductors.

3) Containerless processing of crystal seeds was demonstrated, yielding crystals with extremely flat surfaces and substantially fewer defects than those grown on Earth.

Hopeful findings in the field of metallurgy include:

1) Buoyancy effects can be eliminated in producing composites with uniformly dispersed second phases.

2) The dominant role of gravity-driven convection in dendrite multiplication was confirmed.

3) Containerless melting and solidifying of metals was demonstrated (and recent tests show that extreme undercooling can also be achieved).

Findings in the field of fluid phenomena are somewhat conflicting, but some of the major conclusions are that:

1) Most experiments on manned missions were not adversely affected by vehicle accelerations or crew activity.

2) There was no evidence of surface-tension-driven convection in crystal growth experiments in which the solidified crystal pulled away from the container wall.

3) Free-column electrophoresis was demonstrated successfully, and methods for controlling nongravitational flows due to electro-osmosis were developed.

These conclusions that can be drawn from the first generation of space experiments, only a part of the conclusions that Naumann and Herring discuss, indicate that there are no fundamental, scientific reasons we cannot produce or process some kinds of new or better materials in space or that we cannot use what we have learned in space to improve our processing methods on Earth. Of course, it is impossible to prove a negative, but it is also a human need (and therefore a need of private industry) to eliminate as many negative possibilities as we can before acting. This is one major reason why scientists are continuing basic research as we enter the era of the Shuttle. Not only do we need to extend and deepen our understanding of the basic formation processes, we also need to eliminate more areas of scientific uncertainity before large numbers of private industries will begin committing research and development funds for commercial applications.

Which is how we would characterize much of the MPS research that has been going on in the Shuttle the last two years and that is being planned for the next few: basic research aiming slowly in the direction of commercial applications. However, there is no real value in detailing the completed experiments because all of the results are not yet known. What we do know in general is that they tend to confirm the promise of using zero gravity for research and production in the four areas mentioned at the outset of this chapter: separating biological materials, producing new alloys and composites, growing purer semiconductor crystals, and developing new kinds of glasses.

What does need to be mentioned here is that fact that NASA has developed "vehicles" other than the Shuttle itself to stimulate MPS and other kinds of research. There are four such vehicles, ways that groups ranging in size from private citizens to large corporations can work with NASA. One of these is the so-called Getaway Special: small, self-contained experiments can be carried on the Shuttle on a space-available basis for a fee determined by size and weight but not to exceed $10,000. As of mid-1983, more than 300 reservations for Getaway Specials had been made, and about 140 of them involved MPS research.[7] The other three vehicles, arranged in ascending order of the cost and size of research work, are:

1) Technical Exchange Agreements (TEA) are designed for companies interested in the application of zero-g technology but not yet ready for specific space experiments or ventures.

2) Industrial Guest Investigators (IGI) are scientists who collaborate, at company expense, with a NASA-sponsored Principal Investigator (PI) on a specific flight experiment.

3) Joint Endeavor Agreements (JEA) are arrangements whereby NASA provides, among other things, free flight time to a company conducting proof-of-principle testing of a project designed for commercial purposes.

Industry participation in these NASA programs has been very limited so far for a number of reasons: limited R&D funds, risk-aversion corporations adopting a wait-and-see attitude, doubts about NASA's ability to provide enough flights and space on the kind of schedule commercial projects will need, doubts about the existence or profitability of space-based projects. As of mid-1983, three companies had signed TEAs to do MPS research, one company that signed a JEA in 1982 dropped out in 1983, another company was near to signing a JEA, and one company that signed a JEA in 1980 was proceeding on schedule with the space testing of its Electrophoresis Operations in Space (EOS) project. The McDonnell Douglas Astronautics Company developed this project, and we will discuss it as a separate subject in the next section of the chapter.

The most ambitious and costly project to date is Spacelab, a self-contained laboratory built by the European Space Agency (ESA) that flew in the Shuttle payload bay for the first time in October 1983. It is a multinational project that has cost over a billion dollars. The initial idea was to operate it on some 50 Shuttle missions over a ten-year period. At present, only three more missions have been scheduled to take place by mid-1985, but a number of others are in the early planning stages.[8]

Because Spacelab is reusable, the scientific experiments and hardware designed for one mission can be modified for use on later missions. Spacelab is also designed to allow scientists of average health to work in their shirt-sleeves, a decided advantage over having to use superbly conditioned astronauts bundled up in spacesuits.

Spacelab is not designed merely for MPS research, but such research is an important part of misson plans. Spacelab Mission 1 carried three furnaces for experiments in alloy solidification, casting and brazing, producing new glasses, crystal growth, and lubrication studies. The second Spacelab flight, actually numbered Mission 3, will focus on MPS experiments involving crystal growth. The experiments for Mission 2 in March 1985 have not yet been determined, but the West German D-1 microgravity mission in June 1985 will be heavily oriented to MPS research, reusing the Materials Science Double rack that flew on Mission 1.

Spacelab is not the only project being worked on by the European Space Agency. Since 1982 they have been developing a free-flying platform that can be launched from and retrieved by the Shuttle. Called the European Retrievable Carrier (Eureca), its first mission is planned for 1987. Among the MPS experiments planned for this flight are metallurgical processing, crystal growth, and biological and biochemical studies. Because of its long-duration flight (up to six months) and its high-altitude orbit, Eureca should prove an excellent laboratory for MPS research.

At this point, MPS efforts by individual European nations are led by West Germany and France, the Germans stressing industrial participation and the French focusing on more basic scientific activity. The German government began its own MPS program in 1977, spending nearly $58 million through 1980,[9] and developing its own sounding rocket program called Texus (similar to America's SPAR). It should also be mentioned here that private industry in

Germany has developed a reusable satellite called the Shuttle Pallet Satellite (SPAS) that flew on the Shuttle in 1983, carrying two MPS experiments.

The activities of the French Space Agency (CNES) in materials processing have not been as extensive as those of the Germans, perhaps because of the amount of money they have been spending on the Ariane expendable launch vehicle program. They flew some experiments on the Russian Salyut 6, will fly others on Spacelab, and are reportedly considering purchasing Getaway Specials on the Shuttle. It has also been reported that their MPS budget in 1980 was about $1 to $2 million with plans to rise to about 5 million by 1983.[9]

The Japanese have been developing their own sounding rocket program for materials research and have launched two flights. The Japanese Space Development Program has a 15-year budget plan calling for expenditures of almost $1 billion per year. Not all of this money will be spent on MPS research, of course, but a good deal of it will, especially when, as some people predict,[10] Japan enters the MPS field after it has developed more fully.

Finally, there is the subject of Russian activity. In 1981, according to unofficial sources, they were spending three to four times more for MPS research than the US, which--if true--put their budget in the $60-80 million range per year.[10] Whatever the funding levels, however, many observers feel the Russians have been engaged in MPS research for years, perhaps as early as 1971 when they launched their first space "station" (Salyut 1). Since then, official TASS reports of results from experiments on Salyut 6 (launched in 1978) and Saylut 7 have confirmed that the Russians are working in many MPS areas--new glasses, containerless processing, metallurgy, crystal growth, new alloys and composites, and biological processing, among others. And in 1983 it was reported that Saylut 7 had linked up with a prototype materials processing module (Cosmos 1443), which may be the first Russian factory for manufacturing in space. Certainly there is no doubt that the Russians consider MPS a field of both scientific importance and economic potential. In fact, a recent TASS report (Sept. 29, 1983) claimed that the value of space manufacturing might reach "50,000 million roubles by the 1990s."[11]

We have deliberately ended this "history" of MPS research on the notes of cost and economic potential. We have done this rather than end by assessing the current state of MPS research because we believe that cost and economic potential are the issues that will ultimately determine just how large an industry materials processing in space will become. So on these notes we will turn our attention to the one MPS in the United States that was conceived as a commercial venture and that is proceeding rapidly toward commercial production.

The EOS Project: Electrophoresis Operations in Space

We have chosen to write about our own project in the context of a general discussion of materials processing in space for what we feel are a number of compelling reasons. First, our pride in and optimism about the EOS project play no small role in shaping our feeling that MPS activity has enormous commercial potential. Second, our experience in developing this project over the past eight years has given us insights into ways that high-technology research can be turned into viable commercial projects. We hope to draw on these insights for the reader's benefit in this discussion of the EOS project and in the final section of the chapter, where we discuss what we think needs to be done to help MPS research achieve its commercial potential. Finally, the fact that EOS is the only commercial MPS project alive today makes it--inescapably--a model for any discussion of materials processing in space. We hope it is a positive model, but that is not for us to decide.

What, then, is the EOS project? Simply stated, it is the use in space of a process called continuous flow electrophoresis to separate biological materials in quantities and purities that are unattainable on Earth. With large quantities of purified materials at their disposal, pharmaceutical companies can then proceed to manufacture, test, and market new medicines that could achieve breakthroughs in a number of diseases that afflict millions of people.

The EOS project is also a carefully planned business venture. Years of careful research took place simultaneously on both business and technology issues before any large amounts of money were authorized. Candidate commercial products were thoroughly studied, and a partner with expertise in pharmaceutical products was sought and found. Agreements that allowed us to share some of the risks of developing a new process in space were forged in talks with NASA, and in 1980 we signed the first Joint Endeavor Agreement ever written. Finally, we set up a program for developing the necessary space hardware that would proceed in careful (and affordable) steps from a small unit located in the middeck of the Shuttle (first flown in 1982), to a much larger prototype production unit situated in the Shuttle payload bay (scheduled to fly in 1985), to an even larger commercial production unit that we plan to be flying in some kind of "free-flyer" or space station by the late 1980's or early 1990's. And by 1995 or so, we hope to have a number of separation factories in continuous Earth orbit that will be visited every few months by Shuttle crews who will deliver raw materials and pick up separated products.

That is a quick overview of the EOS project. What it leaves out is the actual starting point, the reason we need to spend the money and energy to go into space to separate biological materials. Looking again at Fig. 1, you can see that there is a nearly inverse order among the five separation techniques in terms of quantity and resolution (or purity). Precipitation produces the greatest quantity but the lowest purity, while static electrophoresis achieves the least quantity but the highest purity. If you need both (as well as usable product or yield), you might choose continuous flow electrophoresis as the best available compromise. But it all depends on how much material you need and how pure it has to be.

What's interfering here with the emergence of a "perfect" technique is gravity. In the precipitation method we try to take advantage of gravity by adding some substance that will cause the biological material we want to aggregate to precipitate out of the original mixture. But the aggregation process is not very selective and is certainly not a way to achieve purity.

In static electrophoresis, on the other hand, we try to "overcome" the effects of gravity by using a technique that eliminates convection currents. As shown in Fig. 2, the biological material to be separated (the sample) is placed on a stationary, or static, medium like a porous gel plate. An electrical field is then applied across the plate long enough for the differently charged particles to migrate to separate areas. Although excellent resolution can be achieved with this method, only a very small amount of sample can be separated at one time (typically about 0.01 ml). And because the sample is embedded in the gel, it cannot easily be removed for use. For these reasons, static electrophoresis is useful for laboratory analysis but not for production purposes.

On Earth, the method best suited for separating commercial quantities of material--hypothetically, at least--is continuous flow electrophoresis. In this method (Fig. 3), a stream of sample material is continuously injected into a flowing buffer solution that carries the sample from the bottom to the top of a long, rectangular chamber. As the sample flows through the chamber, an electrical field is applied across the flow. This causes the differently charged particles in the sample to move laterally at different rates, splitting the sample into separate particle streams that exit through an array of collection outlets at the top of the chamber.

As we said, this is the method that should be best suited for separating large quantities of highly purified material. The use of an electrical field should allow the process to yield the highest purity levels, and the continuous injection of materials should allow it to separate very large quantities. The problem, however, is that gravitational forces cause problems that seriously limit the effectiveness of the process.

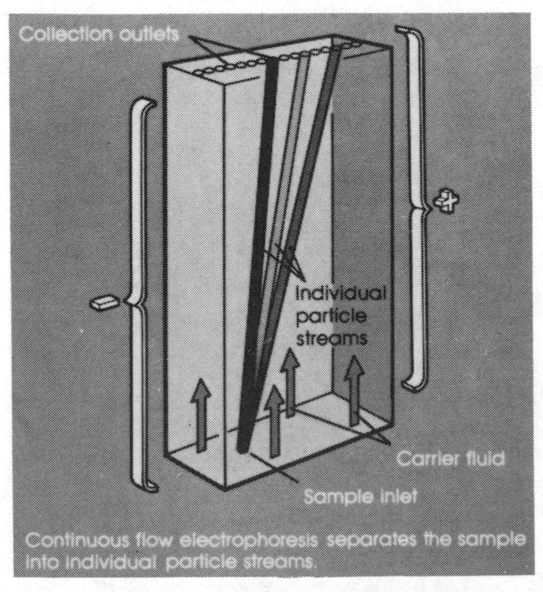

Fig. 3 Continuous flow electrophoresis.

One such problem, shown in Fig. 4, is sample collapse. For continuous flow electrophoresis (CFE) to work properly, the sample must be light enough to flow with the buffer solution in a steady, continuous stream. Because the sample is typically much heavier than the buffer, gravity causes the sample stream to collapse around the inlet port. To avoid this problem the sample must be highly diluted (to about two-tenths of 1% concentration), but this dilution has the unwanted effect of limiting the amount of sample that can be separated.

Another problem caused by density differences between the sample and the buffer is called bandspreading, but this problem affects both quantity and purity. When the density differences are slight (as opposed to large, which causes sample collapse), the heavier sample lags behind the surrounding buffer (Fig. 5). This means that the particles near the center of the stream spend more time in the chamber. Subjected to the electrical field for a longer period, these inner particles move greater lateral distances, spreading away from the trailing edge of the stream and pushing out the leading edge (Fig. 6). This widening phenomenon keeps the streams from separating completely, seriously limiting the purity of the separated material (Fig. 7). The solution, as with sample collapse, is to dilute the sample to equalize its density with that of the buffer, but this reduces even more the amount of material that can be separated.

Fig. 4 Sample collapse.

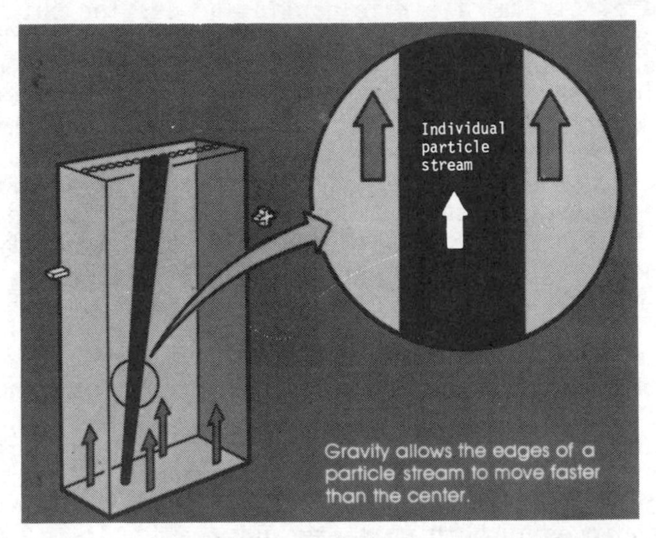

Fig. 5 Product/buffer density differences cause velocity gradients.

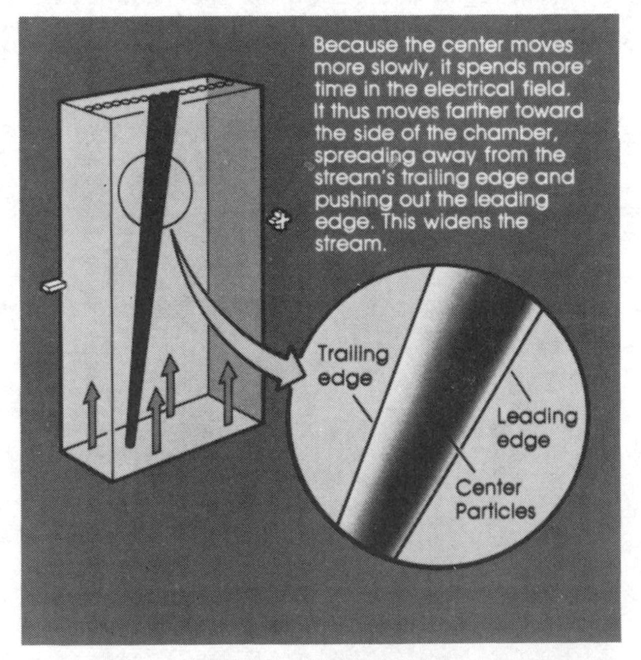

Fig. 6 Velocity gradients widen streams.

The other major problem is caused by convection. When the electrical field is applied across the chamber, the buffer solution heats unevenly. Gravity causes the warmer buffer to rise and the cooler to sink, creating convection currents that make the sample streams waver back and forth (as shown in Fig. 8). This causes the streams to exit through each other's outlets, a result that degrades purity. To minimize the disruptive effects of these convection currents, it is necessary to use a very thin separation chamber (1.5 mm). Unfortunately, a thin chamber restricts the size of the sample inlet, which in turn limits the amount of material that can be separated.

On Earth, these problems can be minimized but not overcome. In space, they virtually disappear. With their disappearance, continuous flow electrophoresis is able to achieve its potential for full-scale production. As Fig. 9 shows, CFE in space ranks first among the separation techniques in both quantity and yield. It is second only to static electrophoresis in resolution.

These comparisons are based on the results of actual space tests of the EOS separation device on four Space Shuttle flights. During these tests we demonstrated that

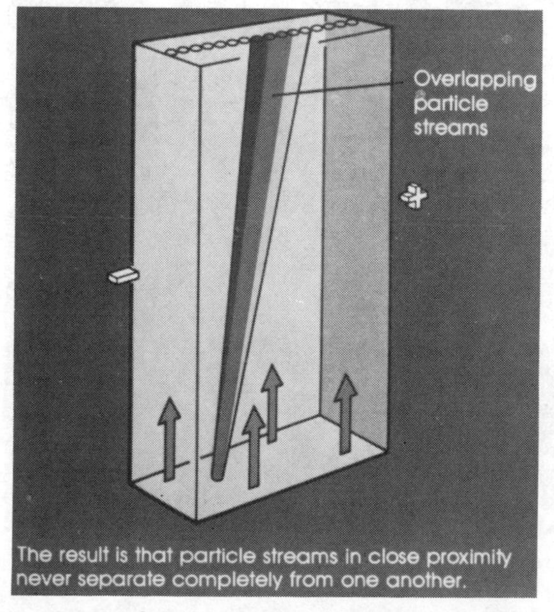

Fig. 7 Overlapping streams decrease separation.

the EOS space device can separate over 700 times more material and achieve purity levels more than four times higher than is possible in the same amount of time in the EOS ground device, one of the best CFE systems on Earth today. In the absence of gravity, the sample concentration can be increased over 130 times and the flow rate of the sample can be increased over five times. These two improvements yield an increase of 718 times.

Space also affects the level of resolution or purity that can be achieved. On Earth, the biological mixture containing the candidate material was separated and collected into 66 outlet tubes, but the material of interest collected into just four tubes. About 66% of the material in these four tubes was the material of interest. In space, the same mixture separated into 188 tubes, but the material of interest collected into nine tubes. About 92% of the material in these nine tubes was the material of interest. Dividing the unwanted material in the tubes on Earth (34%) by the unwanted material in the tubes in space (8%) yields an improvement ratio of 4.25.

When we began work on this project in 1975 through 1976, we of course did not know that conducting CFE in space would yield such significant improvements. We knew that two basic experiments in electrophoresis conducted on Apollo-Soyuz in July 1975 has shown that it seemed to work

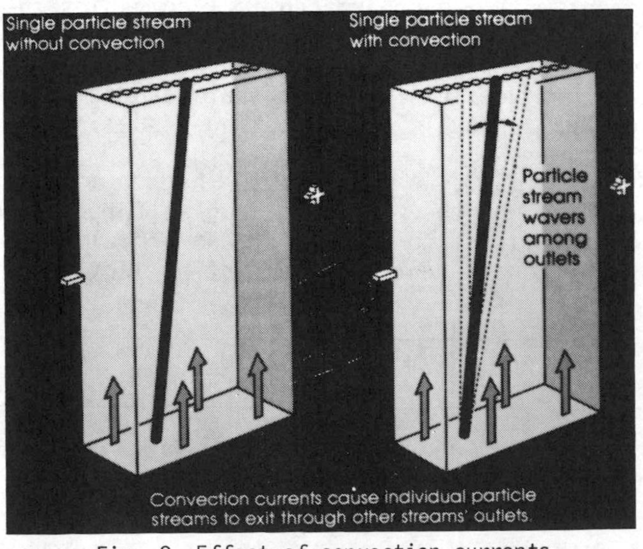

Fig. 8 Effect of convection currents.

better in zero gravity. We had already done some preliminary studies on the effects of gravity on electrophoresis and other materials processes in anticipation of a NASA program to build MPS hardware, so we put together a project team to begin studying the economic and marketing issues.

When we began this project, we had a number of assets that helped us carry through what was then a risky project in an unknown field. First, we had a good deal of experience in aerospace engineering (a point to which we will return in the final section). We also happened to have a small staff of biological scientists, people working for McDonnell Douglas on the Mercury, Gemini, and Skylab programs to develop microbial monitoring systems, that we could commandeer to work with our engineers on a project that weds biology, fluid mechanics, computer technology, and mechanical systems. Finally, we had a management that not only prided itself on its high-technology reputation and ambition but also demanded hardheaded business analysis.

During 1976 and 1977, we were allowed to proceed on a limited basis. We built several rudimentary electrophoresis devices to characterize the process and to confirm that gravity-related problems were first-order phenomena. We began to develop ways to minimize the problems, the biologists working on sample collapse and band-spreading, the engineers on convection. And we began discussions with many pharmaceutical companies to find out their needs, their attitudes about biological processing in space, their sense of which products might be profitable to produce this way, and their assessment of testing, FDA approval, and marketing issues.

By early 1978, we felt that we had to meet five major objectives before the McDonnell Douglas Corporation would fully commit itself to an untried commercial project costing millions of dollars and lasting many years before

	SPACE ELECTROPHORESIS (CONTINUOUS FLOW)	PRECIPITATION (BATCH PROCESS – LARGE AMOUNTS)	GROUND ELECTROPHORESIS (CONTINUOUS FLOW)	CHROMATOGRAPHY (BATCH PROCESS – SMALL AMOUNTS)	GROUND ELECTROPHORESIS (STATIC)
QUANTITY	2	1	4	3	5
RESOLUTION	2	5	3	4	1
YIELD	1	4	2	3	5

Fig. 9 Ranking of separation techniques (ground and space).

there would be any returns on investment. First, we needed an associate from the pharmaceutical business. Second, we needed to identify at least one product for commercial development. Third, to lower our financial risks to an acceptable level, we needed NASA support to verify and demonstrate the space technology. Fourth, to show that there would be a clear economic advantage to operating in space, we needed to optimize the ground process to establish its capabilities for use as a baseline. Finally, we needed to develop a realistic, comprehensive business analysis showing that the EOS project would be financially attractive.

By late 1978, our first objective was met when the Ortho Pharmaceutical Corporation, a subsidiary of Johnson & Johnson, joined us as an active associate in the project. We divided our tasks as follows. In the experimental stage of the project, we were to develop the separation process, characterize the products, and conduct space verification and commercial feasibility tests. Ortho was to identify candidate products, do market research, conduct a product development program--including animal and human clinical testing--and gain FDA approval for new products. In the commercial phase, we are to develop and operate the commercial space plants, while Ortho is to manufacture, package, market, and distribute the products.

Our second objective was also met by this agreement with Ortho. In 1980, after an extensive 16-month market survey, Ortho identified the first candidate product. This product is proprietary, so all we can say is that it is a natural biological material that will provide a breakthrough treatment for a common, chronic ailment that afflicts millions of people.

In January 1980, we met our third objective. We became the first private company to take advantage of NASA's new incentives policy by signing a Joint Endeavor Agreement. Under the terms of this agreement, NASA is providing us with the flight time on eight Shuttle missions to verify our process and equipment and is also allowing us to protect our data. We agreed to make sure that the results of our technology reach the commercial market with all practical speed. We also agreed to provide NASA with the opportunity to use our process for some of their own basic scientific experiments. In this agreement, no funds are exchanged.

It took us a little over three years to meet our fourth objective, but by June 1980 we finished optimizing the ground process. In doing so, we significantly advanced the state of the art for preparative biological separations. Using empirical studies and mathematical modeling, we designed a series of increasingly efficient separation devices until we reached the optimized device, one in which the potential for separation is six times greater than in previous electrophoresis devices. This advance in technology is now being put to use through our support of continuing research at a number of universities and biological laboratories.

We met our final objective in June 1981 when EOS was upgraded from a research program to a commercial project. Buttressed by having met the first four objectives, we were able to show that the project could be rationally ordered, providing the corporation with major milestones by which to measure progress, assess risks, and make decisions about continuing investments. The reader will appreciate the fact that we cannot divulge the details of our business plans, but two major points can be safely made. First, we devised a plan that would allow growing development costs (as we moved from a small, experimental device on the Shuttle to a dedicated space factory in continuous orbit) to be offset by modest levels of profit

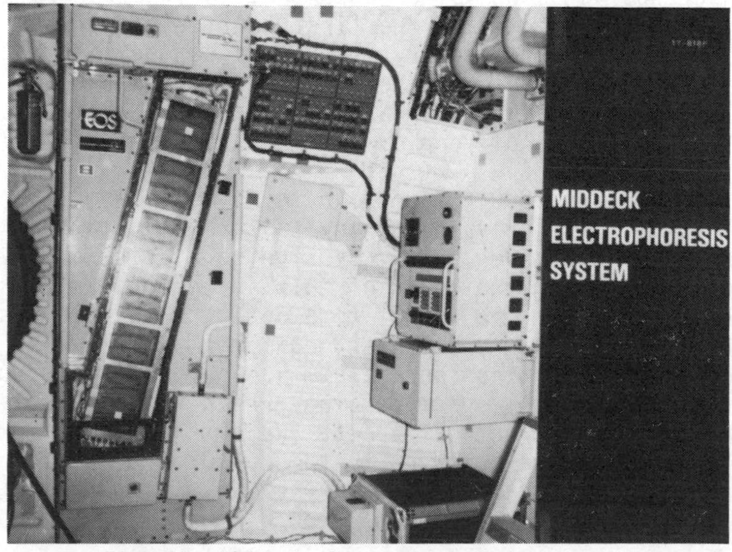

Fig. 10 Middeck electrophoresis system.

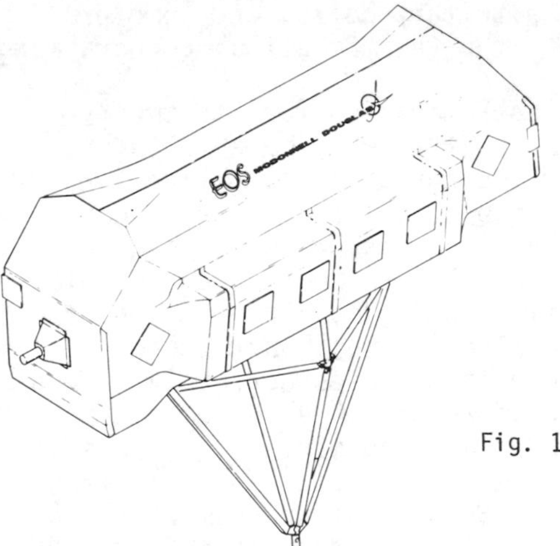

Fig. 11 Production Prototype Unit.

from products developed along the way. Second, we were able to show that the project would be economically viable even when operated in a sortie mode on the Shuttle. Just by operating during the seven-day missions of the Shuttle, we could separate enough of the first product to satisfy the needs of a few percent of the potential US market. This would be a large enough market to get the project to the break-even point. With an orderly transition to long-duration flight and operation, we could produce about ten times more material. And with the addition of more plants during the 1990's, we could meet the worldwide demand for our first product.

In June of 1981, however, this goal was still a long way off. But in June of 1982 it moved one step closer to reality when our experimental electrophoresis device was flown on the Shuttle for the first time. This device, shown in Fig. 10, is called the Middeck Unit because it is located on the middeck below the flight deck in front of the payload bay. The large, slanted piece of equipment on the left is the separation chamber.

The Middeck Unit, modified as time goes on, had flown on the Shuttle four times by the end of 1983 and will fly twice more during 1984 and 1985. As we mentioned earlier, it has demonstrated its ability to separate materials in the quantities and purities that are needed for commercial production. Among the materials separated so far are hemoglobins, two kinds of albumins, dyed latex particles (for a NASA experiment in fluid mechanics), pancreatic beta cells, kidney cells, pituitary cells, and some proprietary material for the Ortho Pharmaceutical Corporation.

The Middeck Unit has been modified for its two flights in 1984 and 1985 so that it will be able to operate continuously for about 100 hours on each flight (rather than the 12-14 hours on each of the first four flights). With these production runs, we will be able to separate enough material for Ortho to begin their clinical testing program. In addition, NASA has agreed to let one of our engineers, Charles Walker, ride aboard these flights to ensure that the unit operates correctly. When Walker flew on the Shuttle in August 1984, he became the first person representing private industry to fly in space.

The second stage of the EOS flight program will begin in November of 1985 when a Prototype Production Unit (PPU) will fly in the payload bay of the Shuttle. This unit will operate continuously and automatically throughout the length of the Shuttle orbit, and it will have a production capability 24 times greater than the Middeck Unit. With this kind of production output, we will be able to provide Ortho with the quantities of material they need to finish their clinical testing program and to gain FDA approval for a new product, hopefully by 1987.

This unit is shown in Fig. 11. Because we plan to fly the PPU in the Shuttle's payload bay on a commercial basis for several years, we have designed it for maximum long-term efficiency. Its weight and size also allow it to fit nearly anywhere in the payload bay, a fact that provides flexibility in scheduling flights and that gives us the opportunity to take advantage of any last-minute openings on the Shuttle. This mode of operation will allow us to separate limited amounts of material for a single product. It will give us, in fact, enough material to be able to carry on a viable commercial program into the early 1990's.

Space Stations and the EOS Program

To expand the EOS program, either to separate more material for a single product or to separate a number of materials for different products, we will need some kind of a permanent space station. Only a space station offers the kinds of technical and economic advantages that make expansion of the program possible.

Perhaps the most important issue driving the need for a space station is the need for increased operating time. For example, Shuttle manifesting will probably limit us to less than 30 days on orbit per year, and Shuttle's power capabilities will not allow us to operate at full capacity. These constraints limit us to processing only enough material of a single product for a few percent of the US market.

A space station, manned or unmanned, would allow us to operate all year round. Being able to do this would have two great advantages. Production would obviously be greatly increased. And cost effectiveness would be greatly improved, not only because production runs would be more efficient but also because the most expensive step in getting equipment into space--the launch--would occur only once.

A space station has other important advantages besides increased operating time and improved cost effectiveness. Compared with the payload bay concept, whose size depends on the fixed size of the Shuttle, the space station factory concept has almost unlimited growth potential. As more production capability is needed, more factory modules could be attached to an existing space station or more space stations could be placed in orbit.

The design of the Shuttle also bears on another issue, operating power. Operating power on the Shuttle is limited and must be divided among Shuttle systems, crew member needs, and military and commercial operations. A space station can be designed to provide greatly increased amounts of power and it can be designed to dedicate that power to a single user. For a system like EOS, increased power is nearly as important as increased operating time to achieve economies of production.

All these advantages of a space station--increased operating time, production cost-effectiveness, expandability, and increased power--will allow us to greatly shorten the time it takes to bring a new product to the commercial market. Operating on the Shuttle, it could take as much as several years to bring a new product on the market. Operating on the space station, we could greatly reduce this time, giving us the chance to develop more products.

We are currently discussing the design and production of space stations with a variety of companies interested in business ventures in space. Although both manned and unmanned space stations offer distinct advantages over Shuttle operations, at this point we prefer the manned concept for several reasons.

First, humans are still needed to monitor and repair time-critical operations like biological separation. This fact was proven once again on Flight 41-D of the Shuttle in August 1984. It was the presence and know-how of our engineer, Charles Walker, that made it possible for our operation to continue when the equipment malfunctioned. Until the day comes when we can design foolproof automated space equipment, man-in-the-loop is a necessity.

Man-in-the-loop is not only a necessity, but also a positive advantage. A second reason we prefer a manned space station to an unmanned one is that with humans aboard, two kinds of work can go on simultaneuously. While the equipment is separating production quantities of biological materials, people on board the factory can also be conducting research and preparation work on new materials. This will allow us to develop new products more quickly than we could in an unmanned operation. For example, we estimate that we could reduce the time needed for both product characterization and production of clinical materials from one or two years to a few months. And when this time reduction is added to the fact that having humans on board makes it possible for many products to be evaluated or produced in short runs in the same time frame, we estimate that 15 new products can be developed in ten years on a manned space station as compared with three new products on an unmanned one.

It is clear to us that space factories, manned or unmanned, are necessary for space bioprocessing to reach its full commercial and human potential. What that potential is can be indicated fairly quickly. When we began the EOS project, we investigated over 200 specific materials within the broad categories of hormones, enzymes, other proteins, and living cells that seemed amenable to space bioprocessing. We narrowed this list to 12 materials that fulfilled our criteria for near-term development: commercial potential, difficulty in developing adequate products on Earth, no anticipated improvements in Earth-based processing methods within five years, sufficient data to show biological efficacy, suitability of material for electrophoretic separation, and data available for comparative feasibility studies. Among the 12 materials that met these criteria were urokinase for treating blood clots, antitrypsin for emphysema, interferon for viral infections (including some kinds of cancer), pituitary cells for dwarfism, and lymphocytes for antibody production.

A conservative estimate of the number of people who could benefit from products derived from the 12 materials exceeds 17 million in the United States alone. Worldwide, the number exceeds 250 million. Besides the humanitarian value of bringing new medicines and treatments to this many people every year, patient populations of this magnitude translate into a commercial market running into billions of dollars.

Realizing the Potential of Materials Processing in Space

As we argued at the outset, the commercial potential for materials processing in space is enormous. Biological processing in itself could become a multi-billion-dollar industry within the next two decades. The question is, however, how can we turn this potential into reality?

Although basic MPS research remains to be done in a number of areas, we believe that what has been done so far confirms that materials processing in space offers decided advantages in four areas: pharmaceuticals, alloys and composites, semiconductor crystals, and glasses. Now that this is known, private industry needs to begin research and development programs for specific commercial applications. Stated another way, materials processing in space will not begin to realize its potential until private industry begins investing money in actual commercial projects.

Given that industry in general has so far adopted a "wait-and-see" attitude about commercial space projects, it seems to us that the real barriers to developing commercial MPS programs are economic and political rather than technical, issues of attitude rather than science. But attitudes can have the force of facts, so the real question is how to change the "wait-and-see" attitude that is blocking the development of materials processing in space for commercial purposes.

It may be that only a commercial success will change this attitude. If this is true, we hope the EOS project will soon become the example of commercial success that stimulates others to enter the new field of materials processing in space. But until that happens, we can do our share to help change attitudes by offering some general ideas as well as some recommendations that we think should be included in a new national policy for space commercialization.

One general idea is based on a simple proposition: To learn new ways, it is sometimes necessary to unlearn the old ways. Learning how to think about weightlessness, for example, takes setting aside not only our experience of life in a gravity field but also our hard-won understanding of many phenomena. Likewise, thinking about the costs and risks of developing a commercial space project means setting aside the normal business time frame. On Earth, we think in terms of three to five years as an acceptable time to wait for returns on investment. In space, we need to learn to think in terms of eight to ten years, perhaps longer. It will not be easy to make this adjustment in thinking (and planning), but it must be done.

The second idea is really a corollary of the first. Private industry is used to thinking and operating competitively. Developing commercial projects in space, however, requires sharing roles and responsibilities. The costs and risks in this new field are so high and the nature of the work so different from what we know that we need to learn new ways to work together. Our own experience proved to us that we could not have developed the EOS project without pooling knowledge and resources with pharmaceutical companies and university researchers.

How individuals and individual companies interested in commercial space projects might think through the specifics of these general propositions is not our business to say. Furthermore, the agreements we reached with Ortho and NASA that made the EOS project possible spell out roles and responsibilities that might be unworkable for other groups with other projects in mind. But we do know that joint ventures and long development periods are likely to be necessary conditions for successful space projects for a number of years to come.

There is one issue, however, on which we feel we can be specific. This is the necessity of a stable, long-term national policy on space commercialization. Earlier in this chapter we argued the view that the Shuttle is a key event in the history of materials processing. Now we would like to amend that claim: The Shuttle is also an event necessary to the future of materials processing. Without a national policy that "guarantees" the availability of a space transportation system, few businesses are likely to commit themselves to commercial space projects. Given this likelihood, the McDonnell Douglas Astronautics Company believes that any future national space policy should include the following:

1) Formal recognition by the Administration and Congress that the Shuttle is a national resource. This recognition would include long-term support of space commercialization because of its social and economic benefit to the nation.

2) Assurance that sufficient Shuttle capacity will be available to support commercial activity as it develops.

3) Setting of long-term pricing policies for use of the Shuttle. Current policy sets prices for only three years, and most serious commercial programs will require 8 to 12 years to yield favorable financial results.

4) Follow-through on plans for a fifth Orbiter and a space station to provide needed capacity and facilities.

Two other recommendations, addressing issues that are more structural and political in nature, are:

5) Direction to NASA to expand its current programs (e.g., Getaway Specials, JEAs, etc.) for encouraging space projects.

6) Direction to other Government agencies to review and adjust their policies to support space commercialization (e.g., tax incentives, depreciation schedules, etc.).

The point behind these recommendations, of course, is for the Government to develop ways to stimulate private investment in a new field of work that will benefit everyone. Precedents for this kind of Government action exist. Neither the railroads nor the interstate highway system would have come into existence without Government support.

But Government support alone will not transform the potential of materials processing in space into reality. That will happen only when private industry commits itself to commercial projects. As pioneers in this new field, we are finding its open spaces fertile. We look forward to the day when it is settled.

References

[1]Naumann, R. J. and Herring, H. W., Materials Processing in Space: Early Experiments, NASA, Washington, D.C., 1980, p. 15.

[2]Naumann, R. J. and Herring, H. W., Materials Processing in Space: Early Experiments, NASA, Washington, D.C., 1980, p. vi.

[3]Naumann, R. J. and Herring, H. W., Materials Processing in Space: Early Experiments, NASA, Washington, D.C., 1980, p. vii.

[4]Savage, D., "Perspectives: Experiments Test Weightless Process," High Technology, Vol. 3, No. 9, September 1983, pp. 72-73.

[5]Naumann, R. J. and Herring, H. W., Materials Processing in Space: Early Experiments, NASA, Washington, D.C., 1980, pp. 33-103.

[6]Naumann, R. J. and Herring, H. W., Materials Processing in Space: Early Experiments, NASA, Washington, D.C., 1980, pp. 105-106.

[7]Merrion, P., "Space: The New Business Frontier," Fact Magazine, April 1983, p. 18.

[8]Bulloch, C., "Spacelab Off At Last?", Interavia, October 1983, pp. 1105-1108.

[9]Waltz, D. M., "Is There Business in Space? Outlook For Commercial Space Materials Processing," AIAA Paper 81-0891, Frontiers of Achievement Anniversary Celebration, Long Beach, Calif., May 12-14, 1981.

[10]Meermans, M. J., "Materials Research in Space," Materials Engineering, October 1981, pp. 32-37.

[11]Aerospace Daily, October 17, 1983, p. 246.

Chapter IV. Space Station and Space Platform Concepts: A Historical Review

Space Station and Space Platform Concepts: A Historical Review

John M. Logsdon*
The George Washington University, Washington, D.C.
and
George Butler†
McDonnell Douglas Astronautics Company, Huntington Beach, California

Abstract

The design of the United States space station system, which will include both manned and unmanned elements, can draw upon over 75 years of thinking and planning with respect to a permanent human outpost in space. This paper reviews the evolution of space station and space platform concepts, from the ideas of space pioneers, such as Tsiolkovsky, Oberth, and von Braun, through the increasingly more detailed engineering studies of the post-Sputnik period. The various space station and space platform studies sponsored by NASA from the early 1960's until the early 1980's are summarized.

Introduction: The Pioneers

When the US Space Station begins operation in the first half of the 1990's, it will be drawing upon over 75 years of thinking and planning with respect to a permanent human outpost in space. As long ago as 1903, the Russian, Konstantin Tsiolkovsky, discussed the concept, including the energy for power and the potential of a closed ecological system for life support (Fig. 1). In 1923, Tsiolkovsky wrote of staying "at a distance of 2000-3000 versts [a Russian unit of distance equal to 0.6629 miles] from the Earth, as its Moon. Little by little appear colonies with supplements, materials, machines, and structures brought from Earth."[1] Although this chapter

---

*Director, Graduate Program in Science, Technology, and Public Policy.

†Director, Advanced Space Programs.

does not discuss the development of the space station concept in the Soviet Union, it is clear that there is a direct line between the thinking of pioneers such as Tsiolkovsky and Soviet space plans. In the United States, Robert Goddard in 1918 visualized a nuclear-propelled ark carrying civilization from a dying solar system toward another star, and in 1920 he suggested the use of extraterrestrial resources to manufacture propellants and structures for space operations.

Another of the fathers of the space age, Hermann Oberth, was the first to describe an orbiting manned satellite as a "space station." In his 1923 book Die Rakete Zu den Planetenraumen (The Rocket into Interplanetary Space) and subsequent works, Oberth suggested that such stations could be useful as communication links and refueling stations for outward-bound spaceships and for observing the Earth. He wrote that if one were to place rockets "around the Earth in a circle they will behave like a small moon. Such rockets no longer need to be designed for landing. Contact between them and the Earth can be maintained by means of smaller rockets so that the large ones . . . can be rebuilt in orbit better to suit their purpose."[2] Oberth discussed the construction in orbit of a 100-km solar mirror to concentrate sunlight; such a mirror could be used to illuminate the Earth's surface at selected locales. If the beam were of sufficient power, noted Oberth, it could sterilize areas infected by insects,

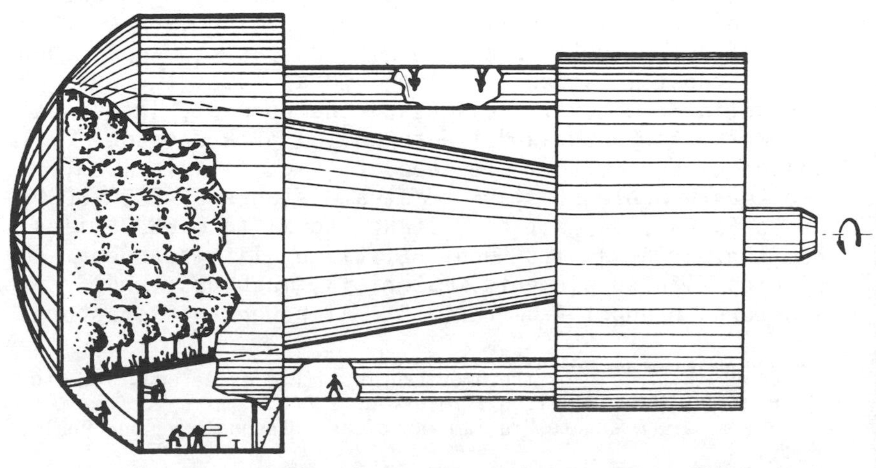

Fig. 1 Tsiolkovsky space station concept, 1903.

drain marshlands of surplus moisture, or perhaps melt away portions of polar ice.[1]

In 1928, two other visionaries published proposals for space stations. Baron Guido von Pirquet suggested in the journal Die Rakete that three stations be established, one in a 470-mile orbit to observe the Earth below, one in a 3100-mile orbit to serve as a launch platform for interplanetary spaceships, and a third in an elliptical orbit intersecting the orbits of the first two stations. (The problem of differing missions requiring differing orbits has been around from the start.) Herman Noordung (a pen name for an Austrian Imperial Army officer named Potocnik) prepared a detailed study of a three-element station consisting of a "Wohnrad" ("living wheel") crew module, a power-generating module, and an observatory (Fig. 2). The Wohnrad was shaped like a doughnut and was designed to rotate around a central hub to create artificial gravity on the perimeter. Power was to be obtained from solar heating through a system of collecting mirrors, boiler tubes, and condenser pipes. The station was to be located in geostationary orbit, and the Earth-oriented observatory was to have instruments powerful enough for mapping the Earth, warning ships of approaching icebergs, predicting the weather, and observing military conflicts.[2]

From 1930 until after World War II, few other original space station ideas emerged in the literature. The attention of space-oriented engineers was instead

Fig. 2 Noordung's "Wohnrad" concept.

Fig. 3 von Braun 1952 station concept.

focused on developing rocket engines and on ballistic and aerodynamic missiles primarily for military applications. For one thing, it was clear that space stations could only be dreams until the rockets to launch them existed. With the World War II demonstration of the feasibility of key elements of rocket technology, the stage was set for a postwar surge of interest in orbiting satellites, both manned and unmanned.

Early Postwar Thinking

For example, in a January 1949 paper in the Journal of the British Interplanetary Society, H. E. Ross noted that "the advent of powerful military rockets is certainly a useful step forward in the progress of astronautics." Ross's article was titled "Orbital Bases" and sketched a concept for a large rotating station to be used for meteorology and astronomy, for research related to zero-gravity and high-vacuum conditions, for cosmic and solar radiation studies, and for communications. Ross's ideas were stimulated in part by Noordung's work and in part by Arthur C. Clarke's observations regarding the particular advantages of the geostationary orbit. The Ross station was a large installation requiring a 24-person crew consisting not only of engineers and scientists but also, in an attempt to extend the style of the British Empire into orbit, "two cooks" and "four orderlies."[3]

Perhaps the most articulate and certainly the most publicized advocate of the values of a space station during the early 1950's was Wernher von Braun, who had come to the United States after the war and was working for the US Army at the time. Some form of large space platform was a central part of von Braun's thinking, and his ideas got wide dissemination in a Collier's magazine special section of March 22, 1952, titled "Man Will Conquer Space Soon." Von Braun's contribution to the magazine was an article called "Crossing the Last Frontier," in which he claimed that "scientists and engineers now know how to build a station in space that would circle the Earth, 1,075 miles up. . . . If we do it, we can not only preserve the peace but we can take a long step toward uniting mankind." Von Braun's plan called for a triple-decked, 250-ft-wide, wheel-shaped station in polar orbit that would be a "superb observation post" and from which "a trip to the moon itself will be just a step" (Fig. 3). The main station would be accompanied by a free-flying unmanned but man-tended astronomical observatory.

Von Braun noted that the station would not be alone in space; "There will nearly always be one or two rocket ships unloading supplies near to the station." "Space taxis" or "shuttle-craft," as von Braun described them, would ferry both men and materials from the rocket ships to the station itself.

Von Braun noted a number of uses for a space station: "a springboard for exploration of the solar system"; "a watchdog of the peace"; a meteorological observation post; a navigation aid for ships and airplanes; and, "a terribly effective atomic bomb carrier."[4]

The launch of Sputnik I on Oct. 4, 1957, transformed engineering speculations into potential Government programs. Even before the start of the space age, however, most of the possible uses of a space station had been identified, and many of the uncertainties related to the concept, such as whether artificial gravity was required for long-duration manned flight, had been discussed. The debate since 1957, in the United States at least, has centered on how to transform the visions of Tsiolkovsky, Oberth, von Pirquet, Noordung, Ross, and von Braun, and of their colleagues, into reality.

The Earliest Years: Planning a Post-Mercury Space Program[5]

Even before it opened its doors for business on Oct. 1, 1958, the National Aeronautics and Space Administration (NASA) had been assigned the responsibility for manned space flight and had developed a one-man, short-term experiment, Project Mercury, as its initial manned activity. The NASA leadership set as a high-priority task developing a long-range plan for the agency's first decade. A space station was a leading candidate for a post-Mercury goal. The House Space Committee in early 1959 concluded that stations were the logical follow-on to Mercury, and von Braun (then still working for the Army) presented a similar view in his briefings to NASA. At this time, the German rocket team had developed an elaborate scheme, called Project Horizon, for Army utilization of space, including military outposts on the lunar surface.

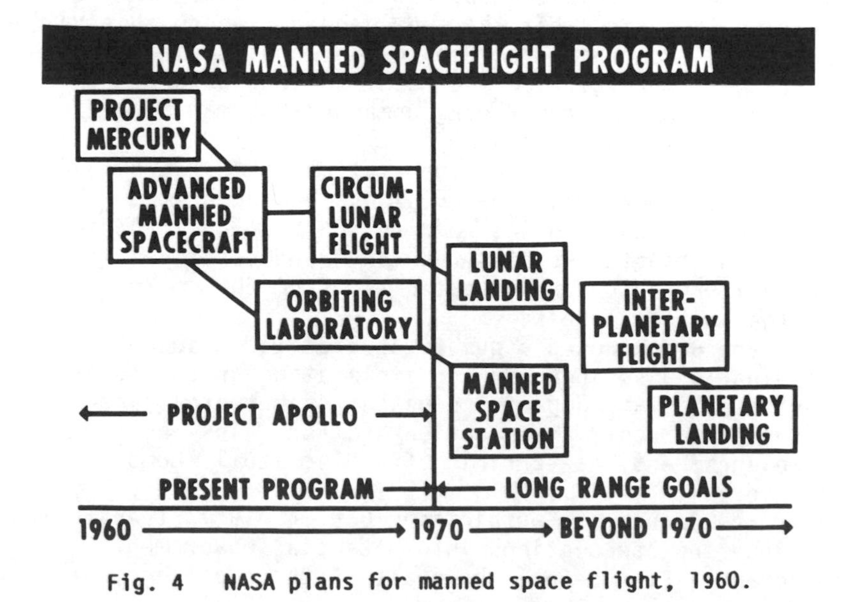

Fig. 4 NASA plans for manned space flight, 1960.

In the first half of 1959, NASA created a Research Steering Committee on Manned Space Flight, chaired by Harry Goett. At the first meeting of this committee, members placed a space station ahead of a lunar expedition in a list of logical post-Mercury steps. In subsequent meetings, the debate centered on the research values, especially with respect to biomedical studies, of a station versus the excitement of a lunar landing goal. While some members of the committee argued that "the ultimate objective of space exploration is manned travel to and from other planets," others argued for an interim step, since "in true space flight man and the vehicle are going to be subjected to the space environment for extended periods of time and there will undoubtedly be space rendezvous requirements. All of these aspects need extensive study. . . . the best means would be with a true orbiting space laboratory that is manned and that can have a crew and equipment change."[6] Ultimately, the Goett committee recommended that a lunar landing be established as NASA's long-range goal, on the grounds that it was a true "end-objective" requiring no justification in terms of some larger goals to which it contributed.

These recommendations were not immediately accepted. In 1960, Robert Gilruth told a space station symposium that "it appears that the multimanned Earth satellites are achievable. . . , while such programs as manned lunar landing and return should not be directly pursued at this time." At an August 1960 industry briefing on NASA's future plans, George Low presented a scheme in which a manned lunar landing and creation of a space station were given equal treatment as long-range goals of the NASA program (Fig. 4). Low told the conference that "in this decade, therefore, our present planning calls for the development and demonstration of an advanced manned spacecraft with sufficient flexibility to be capable of both circumlunar flight and useful Earth orbital missions. In the long range, this spacecraft should lead toward manned landings on the Moon and planets, and toward a permanent manned space station."[7] Low also announced the name of the advanced spacecraft program, aimed both at the Moon and at space stations; it was to be called Project Apollo.

The debate was effectively terminated by John F. Kennedy's May 1961 announcement that "I believe we should go to the Moon." As NASA planners accepted Kennedy's challenge and chose lunar-orbit rendezvous as the approach to accomplishing a lunar landing mission before 1970, any chance of developing a space station in the 1960's disappeared. The rest of the decade was spent studying the concept, with the anticipation that it would be the logical post-Apollo program.

Space Station Plans During the 1960's

Throughout the 1960's, there were almost always funds available for engineering and design studies of future concepts. These studies were conducted both within NASA and by the various aerospace contractors (particularly those without a major role in Apollo), and they resulted in the examination of a wide variety of concepts, ranging from inflatable balloonlike structures, through the use of refurbished rocket stages, to very large stations requiring the use of Saturn V boosters to put them in orbit. Three NASA field centers, the Manned Spacecraft Center (MSC) in Texas, the Marshall Space Flight Center (MSFC) in Alabama, and the Langley Research Center (LRC) in Virginia, managed these in-house and contractor studies, and they were coordinated by the Advanced Missions Office of the Office of Manned Space Flight at NASA headquarters in Washington, D.C.

Indeed, space station concepts had been on the study agenda of NASA and major aerospace firms in the years immediately after Sputnik. One of the earliest space

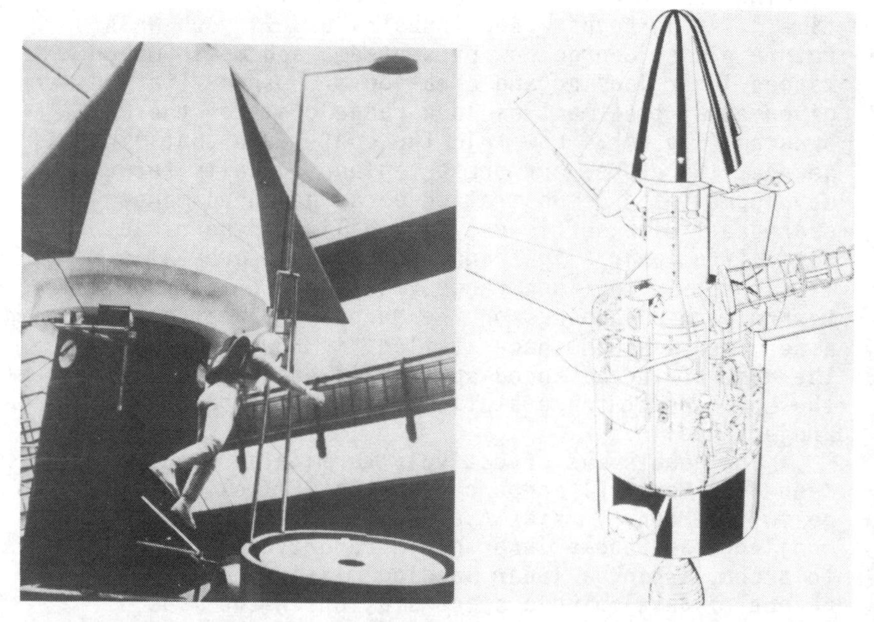

Fig. 5 London Daily Mail space observatory concept.

station concepts to be mocked up in full scale was developed by the Douglas Aircraft Company in 1959 for display at the London Home Show (Fig. 5). The theme of the show was "A Home in Space." A number of months before the show, a representative of the London Daily Mail had visited major aircraft companies throughout the world, soliciting ideas. Based on these visits, the Douglas Aircraft Company's concept for a manned space observatory was selected as being most appropriate for the theme concept, and Douglas was asked to design and build a mockup for display at the Home Show. The "Space Station" was designed to function as an astronomical observatory, the primary instrument being a Cassegrainian telescope with a 60-in. primary mirror. The "Space Station" was the second stage of a two-stage launch vehicle. During launch, this second stage contained oxygen and hydrogen tanks. When the vehicle reached orbit and the propellant was expended, the crew would enter the tanks (which would have been flushed and pressurized), and the scientific equipment and habitability features of the station would be assembled in orbit by the four-man crew. During launch, the crew would ride in a reentry vehicle on the nose of the rocket. When their mission was completed, they would use this reentry vehicle to return to Earth. The "Space Station" was to be launched in equatorial orbit from Christmas Island in the Pacific Ocean.

This study was notable for producing many of the concepts for zero-gravity operation that were used in later programs. The common bulkhead design between the hydrogen and oxygen tanks was similar to that later used in the S-IV stage of the Saturn rocket, and the live-in-tank concept was similar to the "wet" workshop concept that later evolved into Skylab. Many of the concepts for crew restraint systems, crew equipment items, and crew bunks also were similar to those later used in Skylab.[8]

At a 1960 space station symposium, a variety of station concepts was discussed. Lockheed proposed a modular station to be launched by a Saturn I rocket and assembled in orbit (Fig. 6). Each module of the station was habitable. Provisions were made for artificial-gravity living areas and zero-gravity modules for docking and experiments. The station was to be powered by a nuclear reactor. Other concepts examined during this period used smaller launch vehicles and were to be automatically deployed, rather than assembled, in orbit. Some were based on inflatable structures (Fig. 7); others were based on rigid but self-deploying designs.

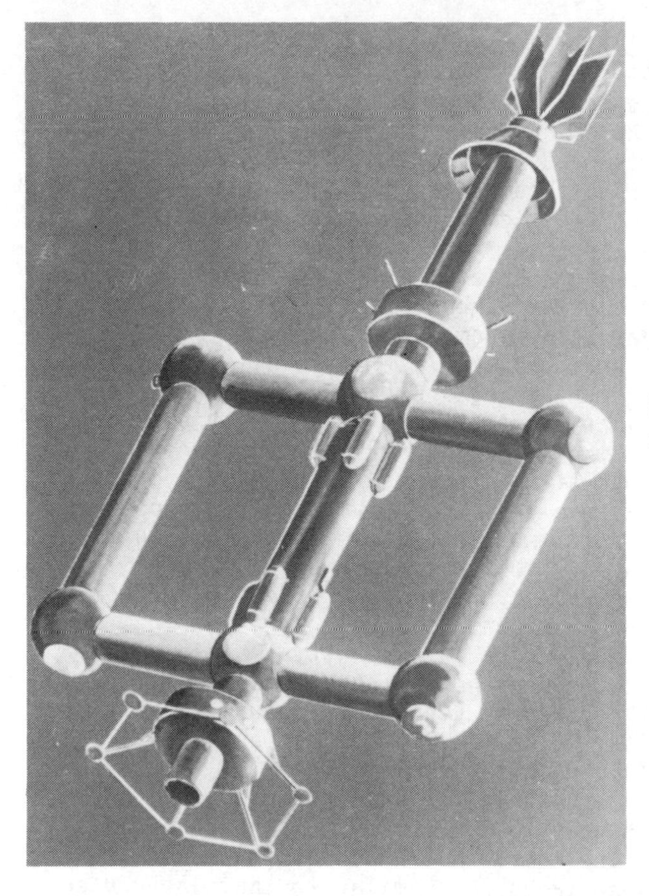

Fig. 6 Early space station concept, 1960

This latter approach was extensively studied by North American Aviation during 1961 and 1962[9] (Fig. 8).

As one participant in early space station studies has noted: "Virtually all of the space station concepts studied during this time were dominated by design considerations which would provide long-term artificial gravity. Stations were carefully configured to provide large radii. . . . The size of the launch vehicles that were to be available was uncertain. . . . As experience was gained in manned space flight, designers turned to considerations other than providing for Earth-like environment for space-flight crews. Emphasis shifted to the usefulness of the space station."[9]

Manned Orbital Research Laboratory

The NASA facility most active in space station studies at this time was the Langley Research Center.

Fig. 7 Inflatable space station.

Fig. 8 Self-deploying space station.

Their efforts dated from at least mid-1959, and by 1962 enough work had been done to form the basis for a space station symposium.[10] Langley researchers noted that "a large manned orbiting space station may have many uses or objectives." Among these they listed:

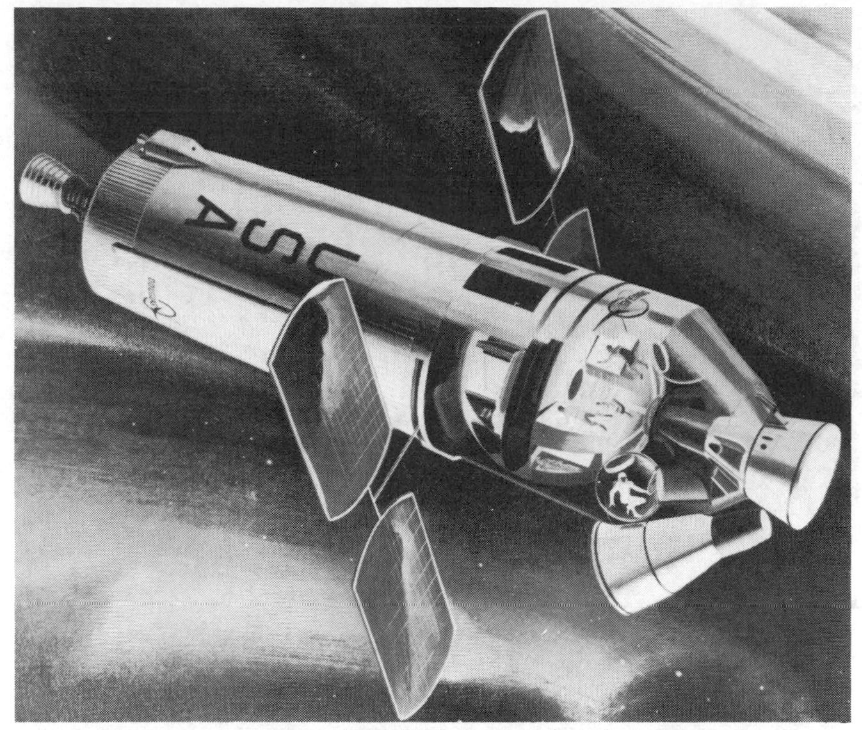

Fig. 9 Manned Orbital Research Laboratory.

1) Learning to live in space
    a) artificial-gravity experiments
    b) zero-gravity experiments
    c) systems research and development
2) Applications research
    a) communications experiments
    b) Earth observation
3) Launch platform experiments
4) Scientific research

With respect to launch platform experiments, Langley suggested that "the space station with its crew of trained astronauts and technicians should be a suitable facility for learning some of the fundamental operations necessary for launching space missions from orbit. The new technologies required for rendezvous, assembly, orbital countdown, replacement of defective parts, and orbital launch can be determined."[10]

Among the various space station studies carried out by Langley contractors during the first half of the 1960's, perhaps the most detailed was that of a Manned Orbital Research Laboratory (MORL) conducted by Douglas

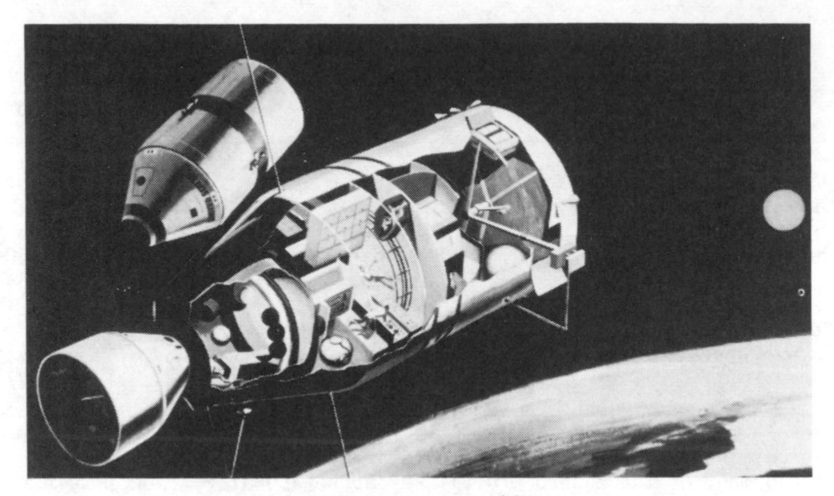

Fig. 10 MORL centrifuge.

Aircraft from 1963 to 1966[11] (Fig. 9). In this study, a baseline technical concept was established first, then the "utilization potential" of such a station was examined, i.e., design preceded requirements. When the original design was compared to various requirements, it was inadequate, and a larger station in a different orbit evolved as the final result of the study effort. The study found that the highest utilization potential came from "key engineering and scientific research studies augmented by specific experiments directed toward potential Earth-centered applications."

As the study effort proceeded, the MORL got steadily more complex and bigger, as there were no criteria established to limit the addition of new experimental requirements. Table 1 illustrates the growth and changes that developed as the concept evolved.

The MORL requirements study examined 1) Earth-centered applications, 2) national defense, 3) support of future space flights, and 4) the space sciences.

From this analysis, the study predicted the need for "hundreds of thousands of man-hours" in orbit to carry out all useful applications; this implied a long-range requirement for "near-permanent operations and support of probably several space stations." The study also noted, foreshadowing a future issue, that "the limiting factor on the number of such stations and on the crew size of each station appears to be the cost of logistic support."

Table 1 MORL concept evolution

	Sept. 1963	Nov. 1964	Feb. 196
Man-hours of experiments	9670	19,500	53,000
Crew size	4	6	9
Program duration	1 year	2-5 years	5 year
Orbit	28.72 deg 200 n.mi.	28.72 deg 200 n.mi.	50 deg 164 n.m
Launch date	1968	1970	1972
Cost ($ millions) w/o operations	$1050 (FY/'63)	$1566 (FY/'64)	$1939 (FY/'66)

Source: Douglas Missile and Space System Division, "Report on the Development of the Manned Orbital Research Laboratory (MORL) System Utilization Potential," SM-48822, January 1966, p. 14.

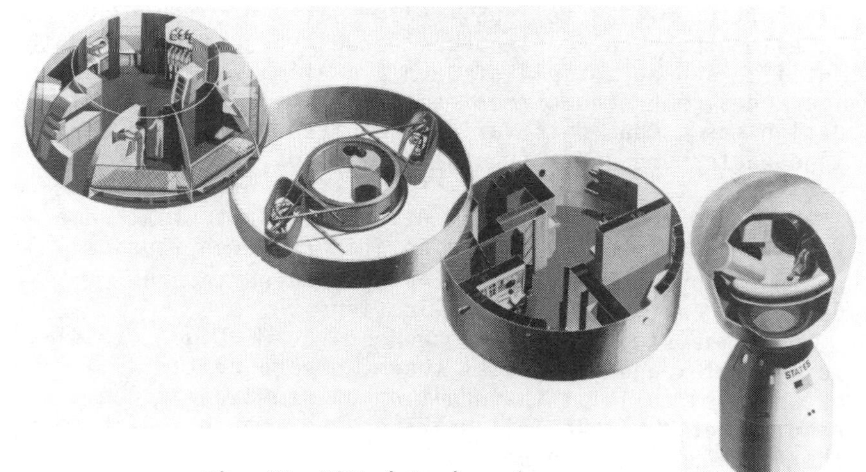

Fig. 11 MORL interior view.

The MORL was basically a zero-gravity station, but it had an onboard centrifuge for simulating reentry and testing physical condition, and for physical therapy in case zero-gravity conditions were debilitating to the crew (Fig. 10). This centrifuge was located between an upper deck, which contained crew quarters and a wardroom, and a lower deck containing control consoles and laboratory facilities (Fig. 11). Another compartment served as a hangar area in which supplies could be transferred from logistics vehicles to the station in a pressurized,

Fig. 12 Resupply of MORL.

"shirt-sleeve" environment. It was proposed that Gemini spacecraft be used as ferries for crew logistics, to be replaced later by Apollo spacecraft (Fig. 12).

The MORL study also examined artificial-gravity experiments in which the MORL was attached to a Saturn upper stage by connecting cables, and the configuration rotated. This configuration supplied 1/3/g at the laboratory.

The MORL was 260 in. in diameter and was sized for launch by a Saturn I-B launch vehicle; its crew would consist of six to nine people. Early planning assumed that the station would be powered by solar panels, but an isotope/Brayton cycle system was also examined. The MORL studies emphasized the need for a better understanding of just what was to be done in a space laboratory and why. To this end, NASA undertook several large studies of an Earth-orbital experiment program for a space station. The results of these studies were assembled into a set of documents called Blue Books; these provided reference payload programs for later space station studies.[12]

Other Space Station Studies

Although the MORL was the most extensively studied space station concept during most of the 1960's, both MSC and MSFC also sponsored a number of in-house and external

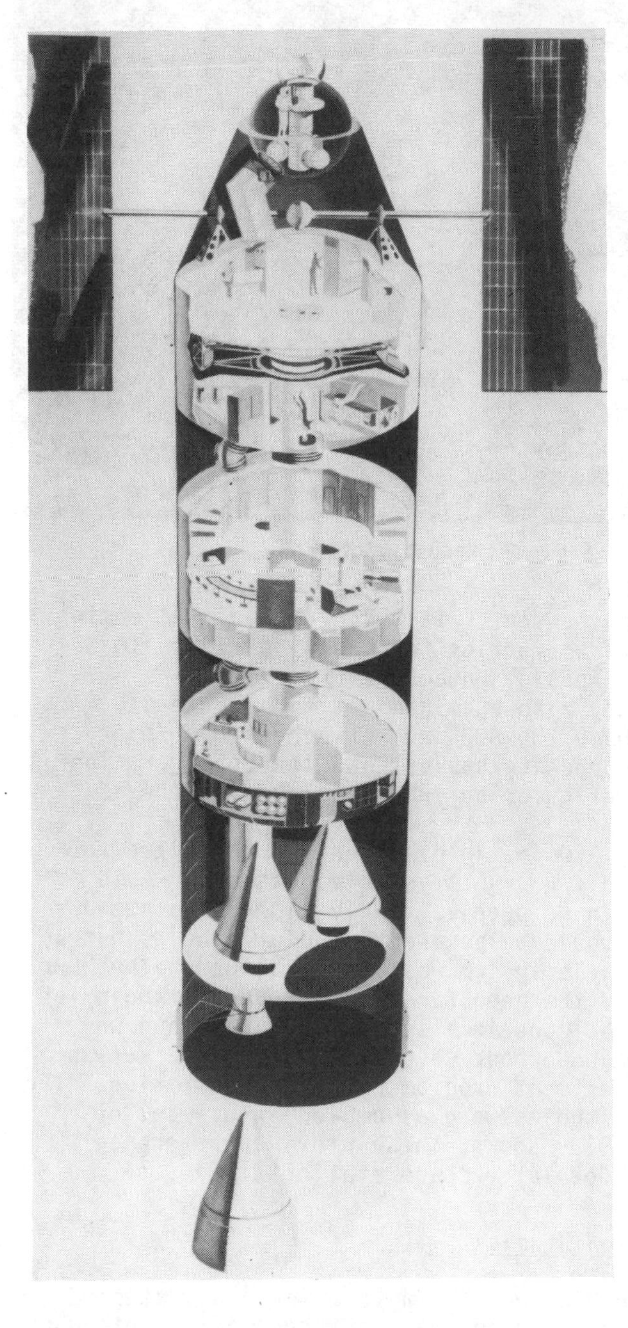

Fig. 13 Project Olympus concept.

studies. By mid-1962, MSC had defined Project Olympus, which called for as large a combined artificial- and zero-gravity rotating space station as could be launched by a two-stage Saturn V vehicle. The crew size was projected at 24, with a 12-person Apollo-type supporting logistics vehicle. This kind of large station concept dominated Houston's planning during the 1960's; a variety of configurations was examined, and both Earth-orbital and interplanetary missions were studied[13] (Fig. 13).

Most of the MSFC study effort concentrated on creating a space station inside a Saturn IV-B upper stage, which could either be launched "wet" (filled with fuel required for stage operation) on a Saturn I-B first stage and Saturn II second stage, or "dry" (without fuel) on a Saturn V first stage and Saturn II second stage. Crew sizes were smaller than those preferred by MSC, and a need for a precursor interim space station to try out various design concepts was identified. The MSFC studies also examined experiment modules that could be either attached to the station or operated in a remote mode[14] (Fig. 14).

On Dec. 10, 1963, during a news briefing at the Pentagon, Secretary of Defense McNamara announced the

Fig. 14 Orbiting workshop concept, 1967.

cancellation of a program to develop a manned spaceplane called Dynasoar and said: "We propose to substitute for it . . . a new program which we are calling a manned orbital laboratory, sometimes known as Gemini X plus a laboratory--Gemini X referring to a modification of the Gemini capsule that would be used in connection with what might be thought of as a trailer to follow behind it, which itself would be the manned orbiting laboratory."

In the course of this briefing, Secretary McNamara made the following comments: "The objective of the Gemini X manned orbiting laboratory program will be to explore operations in space using equipment and personnel which may have some military purpose. . . . we are proposing the manned orbiting laboratory, not for a precise, clearly defined, well-recognized military mission, but because we feel that we must develop certain of the technology that would be the foundation for manned military operations in space."

In January 1964, the Air Force awarded contracts for Orbital Space Station Studies (OSSS) to establish specifications and characteristics, including configuration and size, orbital life, crew composition, mission duration, support and facility requirements, orbital test equipment requirements, and recovery techniques necessary to demonstrate and assess quantitatively the utility of man for military purposes in space. In August 1966, Douglas Aircraft was awarded a

Fig. 15 Launch of Air Force Manned Orbital Laboratory.

contract to develop the Manned Orbital Laboratory (MOL) for the Department of Defense. This program was terminated in June of 1969 due to budget cutbacks and the lack of a defined Air Force requirement for a manned space facility. The MOL (Fig. 15) was to have been launched on a Titan III. It was to have been a manned test-bed that would sustain two men in a shirt-sleeve environment for 30 days while they conducted scientific experiments. Other objectives included determining biological responses of man while in orbit and providing data from design, ground test, and flight phases applicable to the design and development of future manned systems.

Management Attention to a Space Station

While NASA and contractor studies of space station concepts dated from the late 1950's and continued with increasing momentum through the 1960's, it was not until the Apollo program was well under way that the top levels of NASA began to pay much attention to the idea. By early 1963, NASA Associate Administrator and General Manager Robert Seamans called for study of an Earth-orbiting laboratory (EOL) from "an overall NASA point of view." Such study was needed, said Seamans, since an EOL had been studied and discussed "by several Government agencies and contractors" and because NASA and the DoD "are now supporting a number of additional advanced studies." Seamans ordered an agency-wide, four- to six-week high-priority study that would examine EOL proposals in terms of, among other factors, Department of Defense interest, international factors, and other Government agency interest.[15]

Little came of this study effort, for neither 1964 nor 1965 were propitious years for proposing major new starts in the national space program. In 1964, a careful in-house examination of NASA's future options had recommended that NASA defer "large new missions for further study and analysis."[16] There was concern within NASA about maintaining an adequate work load for both NASA centers and NASA contractors as the development phase of the Apollo neared completion, and an evolutionary approach from Apollo to more advanced missions was one way to meet this need, given the low probability of a major new start on a post-Apollo program. Thus, in 1965, NASA initiated the Apollo Applications Program (AAP).

With the Apollo Applications Program under way and Apollo itself pushing toward an initial lunar landing in 1968 or 1969, by mid-1966 NASA top-level management recognized that it was "timely that we update our studies of a permanent manned station." This agency-wide study was to be divided into two parts; one effort would focus on system design, the other on the "need for, the requirements of, and the constraints on a space station" to support a number of objectives:

1) Astronomy with large optical and radio telescopes.

2) Geographical studies with emphasis on Earth resources.

3) Meteorological sensor development.

4) Biological research.

5) Aeromedical research and development.

6) General laboratory research and development in advanced technology.

7) Flight operations development directed toward more efficient resupply, service, and orbital operations.

8) Long-duration flight leading to a future mission capability of manned flight to the planets.

The study charter emphasized that "it is still a question whether a permanent space station is the best approach to achieving the envisioned mission objectives," and called for a review of both advantages and disadvantages of a station as the means for accomplishing them. Finally, the study charter indicated that the request for such a study "is in no way an indication of an agency decision to proceed with or propose a manned space station; it is very important that this point be clearly understood by all concerned with the study and by those with whom it is discussed."[17]

The requirements segment of the 1966 space station study was directed by Charles Donlan; the design segment, by E. Z. Gray. It was the Donlan group that assessed the conflicts between various experimental objectives and design requirements. The Donlan study found that "the most difficult problem to resolve is the matter of artificial gravity. If artificial gravity becomes a firm requirement for crew comfort," it noted, "its implementation can have a very large impact on space station design," since "zero gravity is mandatory for major portions of the experiment programs." Other conflicts included those between astronomical observations and Earth observations, which are "obviously in conflict in both direction and stabilization modes."

The conclusion of this assessment was that "the manned space station concept emerges as an unparallelled opportunity for learning how to exploit the features of space," and that "the prime justification for a manned station [is] associated with the potential the station offers for undertaking broad-based research and development programs in science and technology addressed to all the space objectives of the United States." The group considered both a single station in which there would be an attempt to integrate various conflicting experiments by sequential scheduling, and a concept of splitting the station into several components, each optimized for a particular set of activities. The study concluded that "a minimum station that would appear to satisfy much of the program needs would operate in 200-mile orbit at an inclination of 50 deg, be capable of operating for a five-year period, continuously or intermittently, and large enough to house a staff of from 8 to 12 people."[18]

Based on the 1966 space station study, other NASA and contractor studies, and the lead time required to get a space station program defined and implemented, NASA leadership in the fall of 1966 decided to go ahead with what the Donlan committee had defined as a minimum space station program, and requested $100 million in the FY 1967 budget for detailed definition (phase B) studies. However, the Bureau of the Budget allocated no funds for space station work. Throughout 1967 and 1968, NASA used the limited resources available to it for advanced mission studies to continue "phase A" preliminary conceptual space station studies, with the intent that during 1969 NASA would initiate two "phase B" detailed design studies. The anticipation was that the results of those studies would provide the basis for an agency and national decision to develop a space station.

Space Station Studies Go to Phase B

NASA was preparing in early 1969 to involve the aerospace industry in defining the program through two phase B studies. These studies were initiated in September 1969 and extended over most of the next two years.

Defining the Preferred Concept and Its Rationale

NASA was finding it quite difficult to tell prospective contractors what kind of station it wanted to develop and for what purposes. This was so even though NASA had been studying space station concepts throughout the 1960's. The basic requirements that had emerged from that study effort were 1) qualification of man and systems for long-duration Earth orbit flight; 2) demonstration of man's ability and functional usefulness in performing engineering and scientific experiments; and 3) periodic rotation of the crews and resupply of the space station.

The average crew size for this station was planned to be six to nine persons, with a two-year orbital lifetime design goal.[19] An Apollo command and service module launched by a Saturn I-B booster was to be the logistics vehicle for the station; the station itself was to be launched on a Saturn V booster.

Incorporating these requirements into an agreed-upon statement of work for the phase B studies proved difficult. In January 1969, NASA was still asking center directors rather fundamental questions:

1) Are the goals, objectives, and uses of the space station program stated adequately?

2) Does the statement of work imply a sufficiently forward-looking approach?

3) Is there proper emphasis on low-cost operations, particularly with respect to logistics transportation?

4) Is the space station approach as to size, flexibility, operating mode (zero gravity and artificial gravity, for example), and safety stated properly?[20]

Out of the responses to these questions and subsequent discussions within NASA came a rapid change from the earlier concept of a space station program. NASA administrator Thomas Paine heard, and agreed with, the advice of his center directors that NASA should be bolder in its plans. In February, Aviation Week reported that "all previous concepts have been retired from active competition in favor of a large station," with the focus on "a 100-man Earth-orbiting station with a multiplicity of capabilities" and the "launch of the first module of the large space station, with perhaps as many as 12 men, by 1975." Top NASA officials were reported to have rejected earlier space station plans as "too conservative."[21]

In February, NASA headquarters decided that "the definition study will concentrate on a Space Station Module applicable to the concept of a much larger space station but with independent capability for application and experiment activities with a 12-man crew starting in 1975."[22] Later in the month, NASA further refined its conception of the space station program to have four major elements:

1) A large space base with artificial-gravity capability and composed of modules launched by a two-stage version of the Saturn V.

2) A smaller space station, which would be the first element of the space base.

3) An intermediate logistics system that would be new, based on a Gemini capsule, available at the first launch of the space station, and would be launched on either a Titan III-M or modified Saturn V vehicle.

4) An advanced logistics system, which would be a low-cost reusable Shuttle.

Phase B Competition

NASA issued a statement of work for the Phase B Space Station Program Definition on April 19, 1969. Prospective contractors were ready; they had been following the rapidly expanding character of the program closely and were "already forming teams in anticipation" of the phase B competition.[23]

The work statement described the space station as "a centralized and general-purpose laboratory in Earth orbit for the conduct and support of scientific and technological experiments, for beneficial applications, and for the further development of space exploration capability" and noted that the work requested would include "the Space Base but will focus on the mid-1970's Space Station as the initial but evolutionary step toward the Space Base." The objectives of the space station program were stated as:

1) Conduct beneficial space applications programs, scientific investigations, and technological and engineering experiments.

2) Demonstrate the practicality of establishing, operating, and maintaining long-duration manned orbital stations.

3) Utilize Earth-orbital manned flights for test and development of equipment and operational techniques applicable to lunar and planetary exploration.

4) Extend technology and develop space systems and subsystems required to increase useful life by at least several orders of magnitude.

5) Develop new operational techniques and equipment that can demonstrate substantial reductions in unit operating costs.

6) Extend the present knowledge of the long-term biomedical and behavioral characteristics of man in space.

The initial space station was to have a crew of 12, and normally was to operate in a zero-gravity mode, but there would be during the early weeks of its mission an assessment of the effects of artificial gravity; a counterweight would be tethered to the station, and the configuration spun to provide the gravitational effect.

The station was to be 33 ft in diameter and was normally to operate in a 270-n.mi., 55-deg orbit, but also be capable of operating in polar and slightly retrograde orbits.

There were to be two parallel phase B contracts of $2.9 million each, one managed by MSC, the other by MSFC; NASA headquarters would organize a Space Station Task Force to coordinate the study efforts. The effort was to be divided among four study areas: 1) conceptual definition of space base (15%); 2) definition of space station (60%); 3) definition of an initial logistic system (15%); and 4) development of interface requirements for an advanced logistic system (10%‡).

The statement of work contained a "candidate experiment program" for the space station, but this was "a summary of an illustrative program . . . to assure the system has the inherent capabilities to support those specific experiments and other experiments not yet identified."[24]

Shortly after the original proposals in response to the April 19 statement of work were received by NASA, a new requirement was added to the phase B effort. Not only was the space station to be designed so it could be the core around which a space base could be developed; the station module would also be the core of a spacecraft designed for a manned trip to Mars. This requirement was a reflection of the high hopes for all NASA's future manned programs that were pervasive in the immediate aftermath of the first lunar landing (Fig. 16).

Phase B Contracts Let; Studies Begin

Three firms, North American Rockwell, McDonnell Douglas, and Grumman Aircraft, submitted proposals to NASA in response to the phase B statement of work, and on July 22, NASA awarded phase B contracts of $2.9 million each to North American and McDonnell Douglas. The studies were to run for 11 months beginning in September; MSC would manage the North American effort, and MSFC, the McDonnell Douglas study. Each contractor during the course of the study would invest significant amounts of its own funds to supplement the NASA contract, in the hope that it would ultimately be chosen to develop the space station. Each prime contractor was backed up by one or two major

---

‡The advanced logistic system was the subject of a parallel study effort that ultimately led to the definition of the Earth-to-low-orbit Space Shuttle.

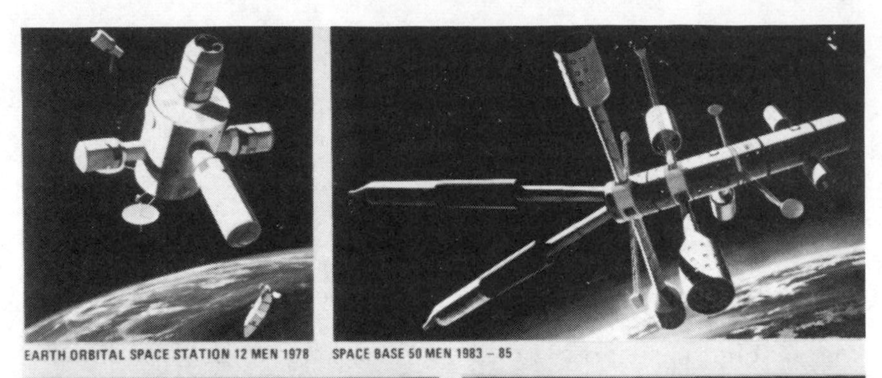

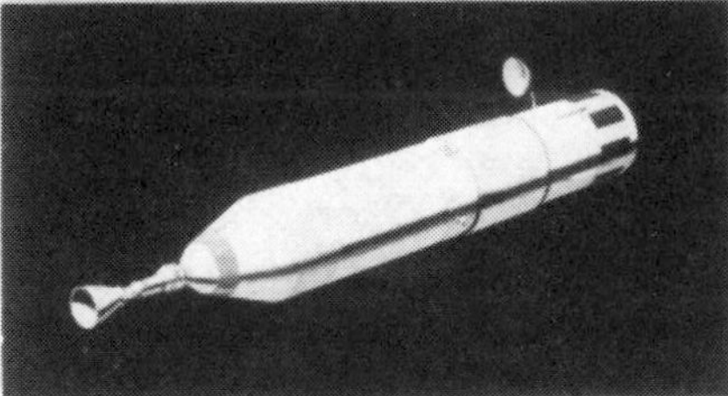

Fig. 16 Proposed space station missions, 1969.

subcontractors and a number of other subcontractors. In all, a substantial industrial effort was mobilized in support of the space station phase B studies.

Both centers organized in-house space station task groups to manage the contractor and parallel in-house studies; overall coordination and control of the studies was provided by an Office of Manned Space Flight (OMSF) Space Station Steering Group and Space Station Task Force, and an independent Space Station Review Group oversaw the technical content of the study effort. Quarterly reviews of the study effort were conducted by top NASA management and others outside NASA interested in the space station; representatives of companies from other countries were invited to participate in several of these reviews, with the hope that non-US firms could participate in the station development effort.

The phase B studies were extended for six months on June 30, 1970; by this time, the planning date for the first station launch had slipped to 1977. The cost of the program was now estimated at $8-$15 (FY '71$) billion, including both development costs and 10 years of on-orbit operations; this estimate did not include the cost of the Space Shuttle program. In addition to the technical design activities, NASA was undertaking a phase B effort to define experiment modules to be added to the core station, and planning a year-long study to involve potential users, both domestic and international, in the program as it was developing. A user's symposium to kick off this effort was scheduled for September 1970, and both study contractors were building full-scale mockups of the station. The space station was to be a cylinder 33 ft in diameter and 52 ft long, divided into four decks plus an equipment section and a 10-ft-diam central tunnel. Two decks were for living quarters and operations (each crewman had private quarters) and two for laboratories. The station was normally to be resupplied every 45 days, with crew rotation every 90 days (Figs. 17 and 18).

Each contractor had come up with a station concept that had met the requirements of the phase B statement of work, but there were a number of differences between their two concepts in areas such as power supply for the station (solar or isotope generator), environmental control and life support systems, extent of artificial gravity, and experiment priorities. NASA concluded in July 1970 that it could prepare a phase C/D statement of work on the basis of the phase B study results already available, but that was not to happen.

On July 29, 1970, Charles Matthews ordered MSC and MSFC to terminate the ongoing phase B activity and to redefine the effort in a fundamental way. Congressional budget decisions on that day had verified that the

Fig. 17 Space station concept, 1970.

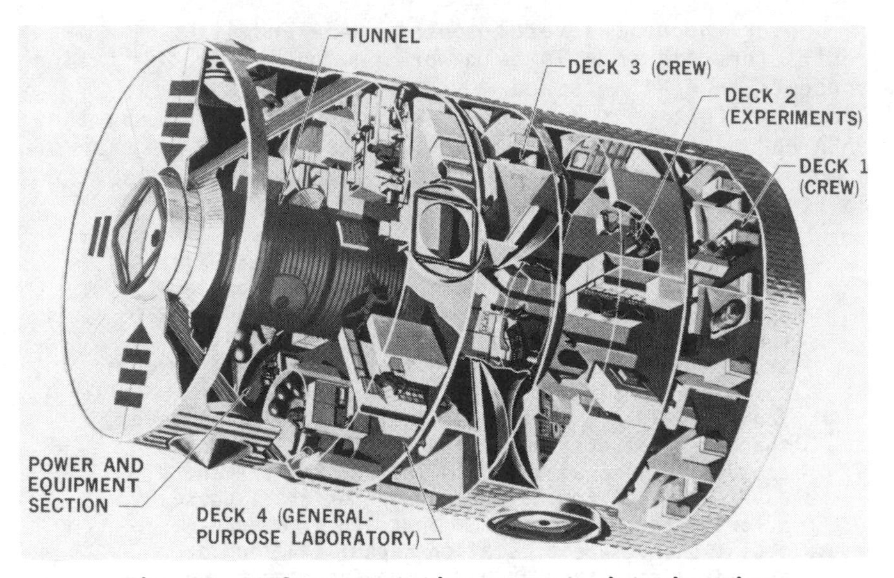

Fig. 18 1970 space station concept, interior view.

Saturn V program, which had been in terminal condition for almost two years, was finally dead, i.e., there would be no booster capable of launching a 33-ft station. The only launch vehicle available for use in putting the space station into orbit would be the Space Shuttle, with its planned 15-ft by 60-ft payload bay. What had started out as the supply vehicle for the station was now its key to survival.

That it might be necessary to go to a Shuttle-sized space station had been becoming more evident in the months preceding July 1970. The feasibility of such an approach had been suggested by studies conducted by the Aerospace Corporation. NASA, while recognizing that it might have to go to a Shuttle-launched station because of the unavailability of Saturn V, wanted to make sure such a station had comparable capability (12-person crew, 10-year lifetime, multidisciplinary R&D facility), since "the development of an in-orbit research facility is the central focus of our current studies and this capability will be rapidly lost with any reduction in crew and vehicle size, vehicle life, and in the general-purpose laboratory and data-processing equipment planned to be on board."[25] On May 4, NASA headquarters told MSC and MSFC to anticipate paying some attention to Shuttle-launched stations, and then, on July 29, directed the centers to reorient their study efforts totally in this direction.

Over the next several months, NASA and its contractors did some interim work on modular space station concepts, and NASA issued a statement of work for a phase B effort on the modular station on Nov. 16. By then NASA had given up on attempting, in the initial version of a modular station, to preserve the same capability as the 33-ft station. The statement of work said:

> The modular station shall consist of individually launched modules assembled in orbit and capable of operating at altitudes of 445 to 500 km (240 to 270 n.mi.) in an inclination of 55 deg. The Initial Space Station shall be sized to accommodate a crew of six with the first module launch scheduled for January 1978. This Initial Station shall also have the potential of growth to a configuration whose capability is equivalent to that of the 12-man, 33-ft-diam station defined in the original phase B effort. The basic program approach of a more evolutionary type of station capability should provide a reduction in early funding requirements. The buildup to this full capability should be complete by 1984 with operations extending through 1989.[26]

It took some doing to skew the study effort toward components of 14-ft diameter; one study contractor commented that "people who were eager to fly in a 33-ft station found the prospect of long stays in the 14-ft station not very attractive." But NASA did issue phase B extension contracts for a modular space station study effort to extend through most of 1971, and North American Rockwell and McDonnell Douglas went to work on the new concept. They developed an "Initial Space Station" concept consisting of a power/subsystems module, a crew/operations module, a general-purpose laboratory module, and a logistics module. As mission requirements demanded, additional research applications modules (RAMs) could be supported by the initial space station (Fig. 19). To achieve a "growth space station" capability, two additional space station modules (power/subsystems and crew/operations) could be added to the initial cluster. This would double the capability of the space station and enable it to perform all the functions required by the growth version. As illustrated in Fig. 20, the growth configuration would have been capable of accommodating six attached research and applications modules or a mix of five attached research and applications modules and several free-flyers cycled through a single docking port. Figure 21 illustrates various features of the

Fig. 19 Initial modular space station.

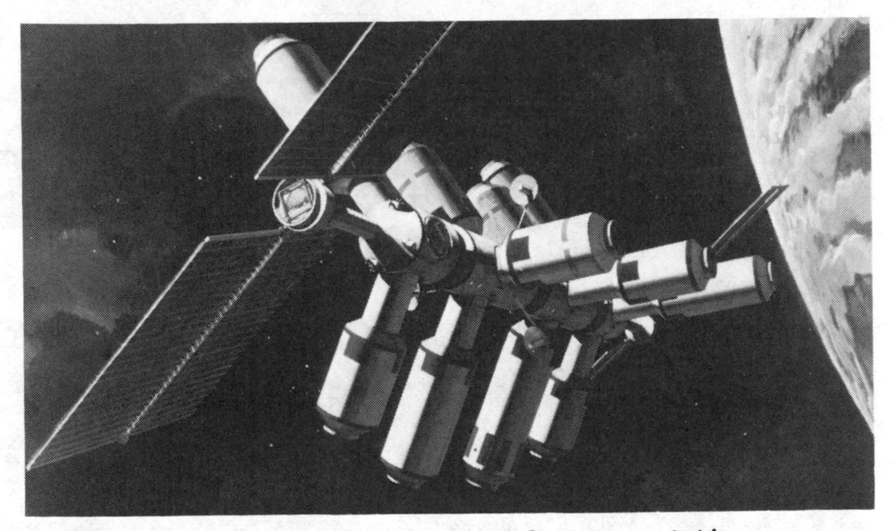

Fig. 20 Growth version, modular space station.

general-purpose laboratory as defined in full-scale mockups built by McDonnell Douglas. The upper left portion of Fig. 21 shows the experiment control console, and the other views show the location of various equipment, such as the automatic film scanner, the color film processor, and the multiformat viewer/editor stations. The large cylindrical structure in the lower right-hand corner of Fig. 21 is an isolated test chamber.

Phase B Studies "Fade Away"

By the time their studies began, however, the likelihood that they would lead to an early commitment to

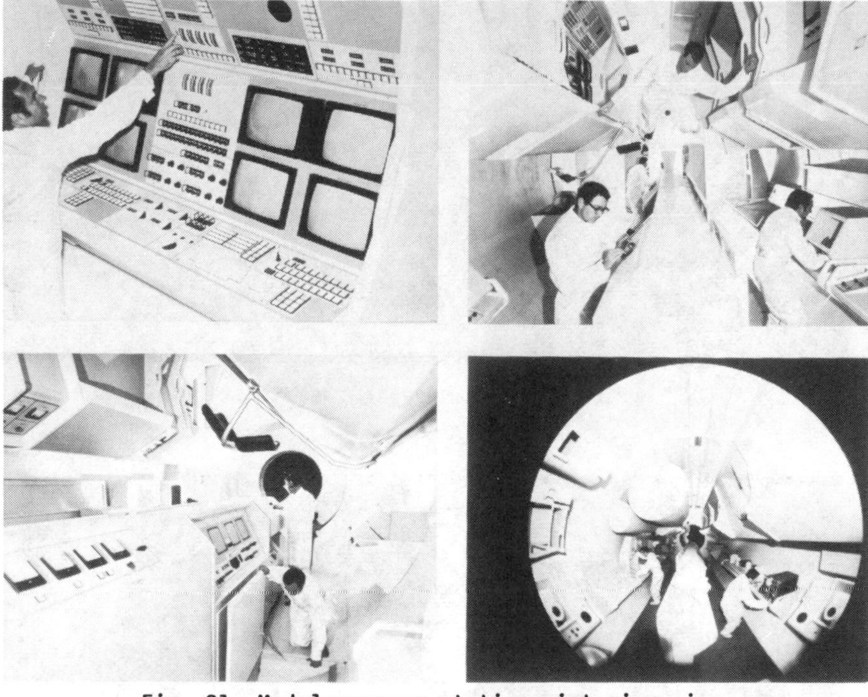

Fig. 21 Modular space station, interior view.

Fig. 22 Manned orbital systems concept, 1975.

station development was already vanishingly small. NASA had been unsuccessful during 1969 and 1970 in its attempts to get an ambitious post-Apollo program approved, and by the summer of 1970, it was becoming quite clear to NASA leaders that only one big program had any chance of Presidential and Congressional approval, and that it was not the space station program. From its start as the "advanced logistics system" for the station and space base, the Space Shuttle had garnered the interest of the Air Force, the White House science office, and many within NASA, and the agency leadership decided to make the Shuttle its top priority program.

Station studies continued through 1970, 1971, and 1972, the final in-house studies being focused on a single research applications module carried into orbit by a Shuttle. This concept led to the Spacelab, developed by the European Space Agency. This was all that remained of what, only a few years earlier, had been plans for truly large facilities in Earth orbit. As a final indication of this reality, on Nov. 29, 1972, the Space Station Task Force was abolished, then immediately reincarnated as the Sortie Lab Task Force. NASA was able to gain approval for Shuttle development in early 1972, and that task occupied the agency's energies throughout the decade. Until the Shuttle was ready, the dream of permanent human facilities in space would have to wait.

Space Station Studies During the 1970's

During 1973, NASA carried out a "quasi-space station" effort, Skylab, which was discussed in a previous chapter. Skylab was not a true space station because the modified Saturn upper stage used to house the orbital facility was not equipped for resupply of key expendable items, and thus the facility could not be used over a multiyear period. The orbital workshop was launched in May 1973, and during the year, three three-person crews occupied Skylab for periods of 28, 59, and, finally, 84 days. The results of the Skylab program provided a major stimulus to NASA's once again beginning to examine the space station concept in the mid-1970's.

Other influences in this direction included the need to begin to identify potential "post-Shuttle" programs and new requirements for manned space operations emerging from a number of study efforts being carried out by NASA in the 1974-1975 time frame. In order to build a plausible rationale for once again proposing a space station as an element of NASA's program, it would be necessary to identify some high-priority missions that could not be accomplished using the Space Shuttle, with its 7- to 30-day orbital stay time, its Spacelab facility for manned experimental activities, and its significant capability for lifting large and/or heavy cargoes to low-Earth orbit. Studies that established requirements for large structures in both low-Earth and geosynchronous orbit, structures that could only be constructed in space, seemed to provide the needed rationale, and space construction became a major theme in space station studies during the 1975-1980 period.

Development of the MOSC and "Construction Shack" Concepts

The first NASA foray into station planning during the 1970's was a 1975 study of "Manned Orbital Systems Concepts" (MOSC) carried out by McDonnell Douglas Astronautics under the technical direction of the Marshall Space Flight Center. This study "examined the requirements for . . . a cost-effective orbital facility concept capable of supporting extended manned operation in Earth orbit beyond those visualized for the 7- to 30-day Shuttle/Spacelab system." Study guidelines included use of available hardware developed for the Skylab, Spacelab, and Shuttle programs "insofar as practical", and an initial operational capability in late 1984.

The context for the MOSC study included a growing concern about the Earth's resource limitations, population growth, and environmental stresses, driven by the widely publicized "limits to growth" debate of the early 1970's. The study noted that "the planning and development of future space programs cannot be done in isolation from the many critical problems facing the peoples of the world during the coming decades" and that "there will continue to be many conflicting and competing demands for resources in the years ahead." This context skewed the emphasis in establishing missions for the facility to "the research and applications areas that are directly related to current world needs."

Though oriented more directly than past station concepts to high-priority global problems, the MOSC study still emphasized the "science and applications research facility" rationale; although such activities as assembly of large structures and operating space manufacturing facilities were examined during the study, the emphasis was on a facility that would "enable the scientific community to pursue programs directly related to the improvement of life on Earth." The final MOSC configuration (Fig. 22) called for a four-man modularized facility; the manned module would be based on the Spacelab design, and Spacelab pallets would also be used to support unpressurized payloads. The MOSC core consisted of a subsystem module and a habitability module placed in a 200-mile orbit by a single Shuttle flight. A second flight would deliver a pressurized or unpressurized payload module and a logistics module. Crews would be exchanged every 90 days, at which time logistics modules could be replaced and additional payload modules added or exchanged. Total program costs for development and operation of the initial MOSC facility were estimated to be $1.2 billion (FY '75$).[27]

In the fall of 1975, NASA decided to shift its space station emphasis from research in orbit to space construction. In explaining its study plans, NASA noted:

> Earlier space station studies emphasized the "Laboratory in Orbit" concept. Emphasis is now being placed on a Space Station as an "Operational Base" which not only involves a laboratory but also such uses as: (a) an assembly, maintenance, and logistics base for conducting manned operations involving antennas, mirrors, solar collectors, transmitters; (b) for conducting launch and retrieval operations for orbit-to-orbit and Earth-departure vehicles which may require assembly or propellant transfer in orbit; (c) for conducting retrieval, maintenance and redeployment operations for automated satellites; (d) for managing clusters of spacecraft and space systems as a central base for support for common services . . . .
>
> Orbital location studies will emphasize the possible exploitation of geosynchronous orbit, as well as low-inclination and polar low-Earth orbit. . . Current planning is directed toward a space station new start in FY 1979.[28]

One reason for NASA's switch in space station justification was the increasingly recognized potential of using space operations for dealing with problems on Earth. This potential was identified most explicitly by a study group which was asked by NASA Administrator James Fletcher in 1974 to provide an "outlook for Space"--"to identify and examine the various possibilities for the civil space program over the next 25 years." The study group concluded that "the great challenges facing the physical needs of humanity are principally the results of the continuing struggle to improve the quality of life. Particularly critical is the need to improve food production and distribution, to develop new energy sources, to meet new challenges to the environment, and to predict and deal with natural and man-made disasters. In each of these areas, we found that significant contributions can be made by a carefully developed space program." The NASA report recognized that "future space programs must provide a service to the public." In responding to the Outlook for Space report, James Fletcher set as a primary NASA goal "accelerating the development of economic and efficient space services for society," such as "resources management, environmental understanding, and commercial returns from the unique contributions of space."[29]

The Outlook for Space report was not directly or strongly supportive of the need for a space station. It did conclude, however, that:

> Most of these activities might well be supported by the Shuttle system, together with associate space laboratories and free-flyers.
> There are more far-reaching objectives, however, which will require human activities in space transcending those supportable by current Shuttle flight plans, such as the construction of satellite power stations or the establishment of a permanent lunar base. It is difficult at this time to assert that either of these activities, or others like them--space manufacturing, space colonies--will be undertaken within the next 25 years. Nevertheless as we looked at the future of space, particularly at those more creative programs directed toward major exploitation of the opportunities which space provides, we inevitably found man to be an integral part of the system. If the United States is to be in a position to take advantage of these potential benefits then it would seem necessary that we develop the capability to operate a permanent manned facility in space in which human crews could operate for extended periods of time. The space facility would be constantly available, although crews would, of course, be periodically exchanged.
>
> The creation of a permanent space facility seemed to us to be the most useful way to continue the advancement of manned-flight technology. With the Shuttle system giving us comparatively low-cost access to space on the one hand, and the economics which could be realized from the use of the permanent space facility on the other hand, the construction of a permanent space station appears to be the next logical step for the manned flight program--not as an objective in itself, but rather for its technological support of a number of other objectives which can

> **benefit from our growing knowledge of how humans can work in space and to provide a foundation for the future.[30]**

In addition to the Outlook for Space study, in the mid-1970's a number of even more visionary efforts were identifying challenging future space goals. One notion, which received wide public attention but had a relatively modest influence on NASA's internal planning activities, was the proposal by Princeton professor Gerard O'Neill that work begin on developing very large human habitats in space--space colonies[31] (Fig. 23). A concept that was quite attractive to NASA's engineers was developed by Peter Glaser of Arthur D. Little, Inc.; this was the proposal that large solar arrays in geosynchronous orbit could provide a source of continuous energy on Earth (Fig. 24). The solar power satellite (SPS) idea was given a great deal of technical attention by NASA during 1975 and 1976. Developing a SPS would require extensive on-orbit manned operations and the capability for assembling very large structures in space. Similar construction requirements were derived from less ambitious schemes

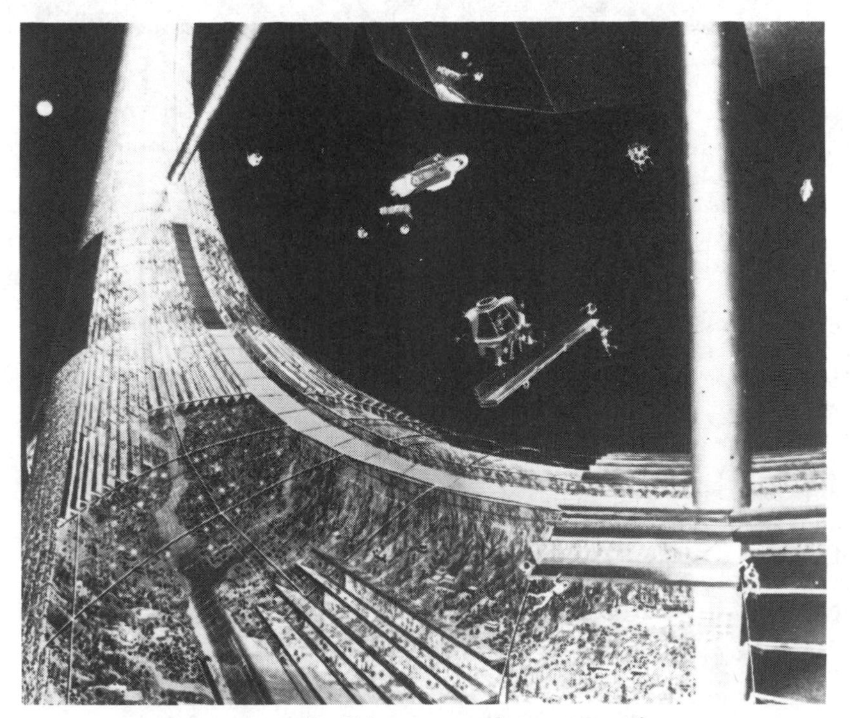

Fig. 23 O'Neill space colony concept.

Fig. 24 Solar power satellite concept.

involving large antennas in space for communications use and scientific investigations (Fig. 25).

By the end of 1975, NASA had developed a plausible argument that space construction might be a major requirement of its programs during the 1980's, and wanted to explore the role of manned orbital facilities in carrying out these construction efforts. In December 1975, the agency issued a request for proposals for a "Space Station Systems Analysis Study"; the study effort was to be focused around the use of a space station to "serve a wide range of operational base and space laboratory activities," such as using the station "as a test facility and construction base to support manufacturing, fabrication and assembly of various sizes of space structures."[32] Aviation Week reported that the station studies reflected NASA's "belief that a large permanent US facility in space is the logical follow-on to development of the Space Shuttle."[33]

Two parallel study contracts for the Space Station Systems Analysis Studies (SSSAS) were let by NASA in April 1976; one went to McDonnell Douglas, which would work

Fig. 25 Antenna construction at a space station.

under the direction of Johnson Space Center (JSC), and the other went to Grumman Aerospace, which would work with Marshall Space Flight Center.

The McDonnell Douglas SSSAS study focused on four objectives, which offered "the promise of: serving important needs of man on Earth; advancing US preeminence in science and technology; and ultimately generating an economic return on investment, essential if permanent Government support is to be avoided." These objectives were:

1) A satellite power system.

2) Providing Earth services for resource management and communications; the major projects here were 30-and 100-m radiometers.

3) Processing materials in space for commercial users.

4) Furthering scientific research.

The study identified a "common point of departure" for meeting these objectives, which was "the basic requirement for construction facilities in orbit." Such facilities would involve crane operations, capabilities for fabrication and assembly of structures, manned extravehicular activity, and long-duration (approximately 90 days per crew) human presence (Fig. 26). These capabilities could best be provided, the study concluded, by a manned "construction shack" as part of the space construction base; this manned element would be the last addition of an evolutionary buildup from sortie flights, through a Shuttle-tended unmanned construction base, to a permanently manned orbital facility. The cost of having a fully capable station in orbit by 1988, and of the associated missions that the construction base would perform, was estimated to be $2.4 billion (FY, 77$).[34]

The Grumman SSSAS reached similar conclusions. It too focused on missions related to solar power satellites, space manufacturing, and scientific experimentation, and also on a large-antenna public service platform, and found that a continuously manned space construction base was an integral element of the capabilities needed to perform these missions; such a base, Grumman estimated, could be in operation by early 1986. The study found that "space solar power emerged as the most compelling mission with the greatest impact on the program. Space manufacturing, public service communications and Earth observation, and the solar terrestrial observation missions fit well within the solar power development needs for large energy power supply, long-duration manned presence in space and an in-space construction facility..."[35] (Fig. 27).

One finding of the SSSAS studies was that scientific efforts could "go along for the ride" on space stations capable of supporting construction, materials processing, and power-generation objectives. One aerospace publication reported, "The space base concept is one whose time seems to be coming rather quickly. Until recently, space stations have been thought of mainly as . . .'the traditional laboratory in the sky.' Some observers were surprised when construction, materials processing and power were given roughly equal status with science. . . . Now, the balance has shifted further . . . space construction work is the 'prime focus' of the studies."[36]

In the spring of 1977, NASA officials, sensing that a start on a space station program was unlikely to come soon, began to stress the idea that "the [Shuttle] orbiter is a significant space station in itself," and were looking toward ways to enhance Shuttle capability to perform many of the missions that the SSSAS studies had assigned to a space station.[36]

The first step recommended for extending Shuttle capabilities was the development of a power module to be left in orbit to service the needs of repeated Shuttle and

Fig. 26 Beam building in orbit.

Fig. 27 Space station concept, 1977.

Shuttle/Spacelab operations. Such a unit could also support free-flying Spacelabs and space construction modules as dictated by requirements of future missions.

Following this recommendation, over the next several years NASA sponsored studies at MSFC and at JSC to investigate in further detail potentially feasible system concepts for providing additional power, thermal control, and attitude control to the Shuttle orbiter and to the near-term free-flying payloads that the STS would deliver to orbit. At MSFC, these efforts led to conceptual studies of a 25-kW power system/space platform for use with both manned and unmanned payloads, and at JSC to the definition of a power extension package (PEP) to provide up to 29 kW to the orbiter. This power extension package (Fig. 28), initially attached to the orbiter, would evolve as mission requirements evolved, to a full-capability orbital support module (OSM). The OSM could operate in a free-flying mode, and provide 35 kW, 28 V of regulated power, symmetric heat rejection, attitude control, and general system support at five payload berthing ports.

Evolution of Space Platform Concepts

In the mid-1970's, the concept of a multipayload platform emerged as a cost-effective approach to flying numerous unmanned payloads in a "car pool" fashion. The growing number, size, and cost of space payloads had grown to the point where NASA budgets could no longer provide a dedicated, custom-built spacecraft to orbit and support individual space payloads. The platform was thus conceived to provide those utilities and carrier vehicle services that were most commonly needed by a variety of payloads. The platform would provide power, data, communications, thermal control, attitude control, and orbit reboost propulsion as centralized functions for the payloads, plus a number of payload berths. It also would

Fig. 28 Power extension package.

provide a Shuttle berthing capability for payload delivery, exchange, replenishment, or servicing.

With a shared platform being used, the investment required to fly any given payload would be substantially less than that required for developing and operating a single-payload-dedicated spacecraft. Moreover, since most payloads are planned for many years of operation, a platform offered the potential of periodic Shuttle revisit services for the benefit of all the platform's payloads. In view of the growing need to conserve costs associated with orbiting and supporting the needs of space payloads, platform concepts began to emerge from NASA and aerospace contractor studies.

Early work on platforms took place at MSFC during the mid-1970's and centered on a vehicle concept called power module. This concept embodied a low-cost approach to the provision of a platform that provided 25 kW for the payload(s). Extensive work was performed at MSFC on this concept, followed by a contracted activity at Lockheed. Figure 29 illustrates this early concept.

In the later 1970's, as the needs for and economic advantage of the platform concept grew more apparent, different concepts were developed and evaluated in an attempt to refine the understanding and potentialities of this new type of payload carrier vehicle (Fig. 30). Figure 31 illustrates two platform concepts that were

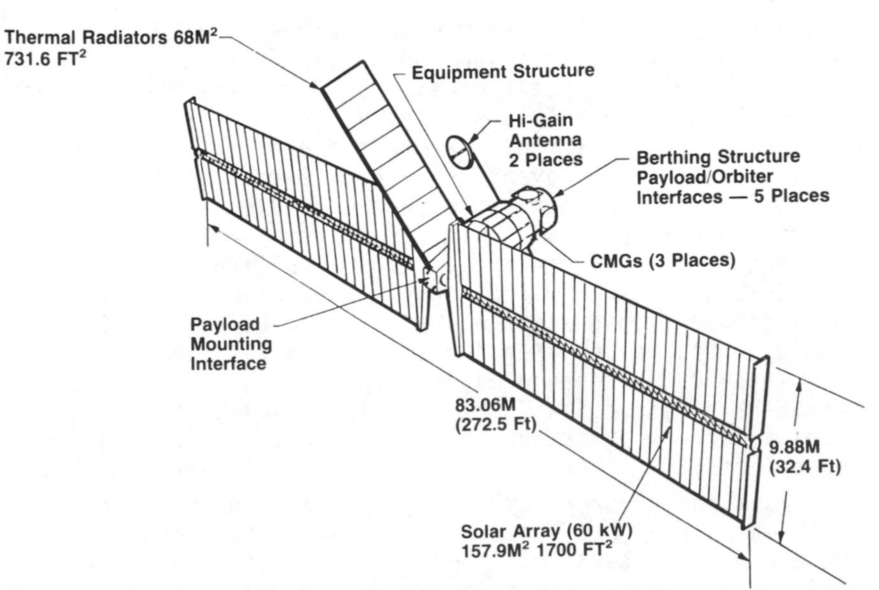

Fig. 29 25-kW platform, early concept.

Fig. 30 25-kW platform, later concept.

developed in 1978-1979 by McDonnell Douglas for NASA's Johnson Space Center. The orbital support module provided five berthing ports for payloads that were mounted in pallets delivered by the Shuttle. The berthing mechanism interfaced with the ends of the pallets and rotated to provide variable viewing directions as required by the payloads. The version illustrated provided 35 kW for payloads.

The deployable platform was to provide 11-22 kW of power, and represented a new concept in payload mounting. Here, the platform itself consisted of a large deployable framework that provided a surrogate Shuttle cargo bay wherein the payloads would be mounted exactly as in the Shuttle. Thus, two advantages were offered: 1) no special berthing mechanisms were required on the payload pallets, and 2) payloads that flew first for five days on the Shuttle could be delivered to this platform for long-duration flight without any appreciable installation and accommodation changes. The payload equipment, of course, would be modified for the longer lifetime.

In 1980, MSFC began major preparations for developing the 25-kW platform. As part of this activity, McDonnell Douglas performed a study called "Science and Applications Space Platform" (SASP), which considered the accommodation of larger and greater numbers of payloads on the platform, as shown in Fig. 32. Late in 1980, McDonnell Douglas and TRW began separate $1 million phase B alternative system design concept studies of the 25-kW power system, which

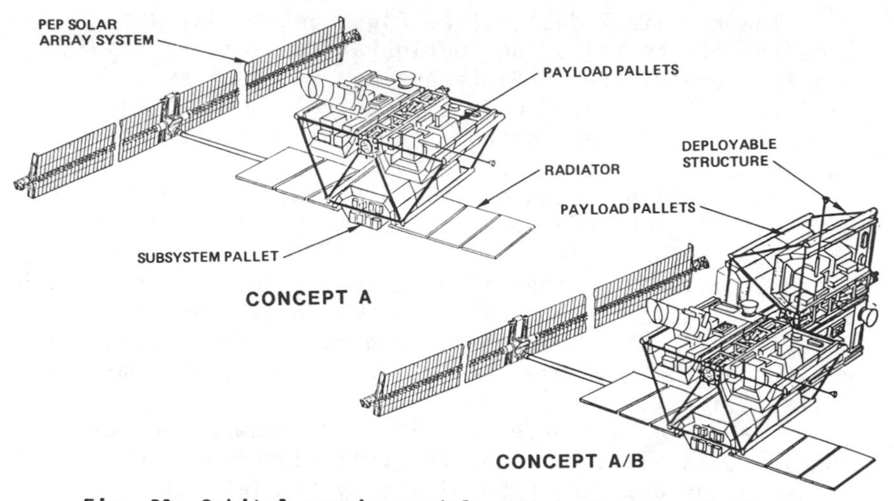

Fig. 31 Orbital service module and deployable platform.

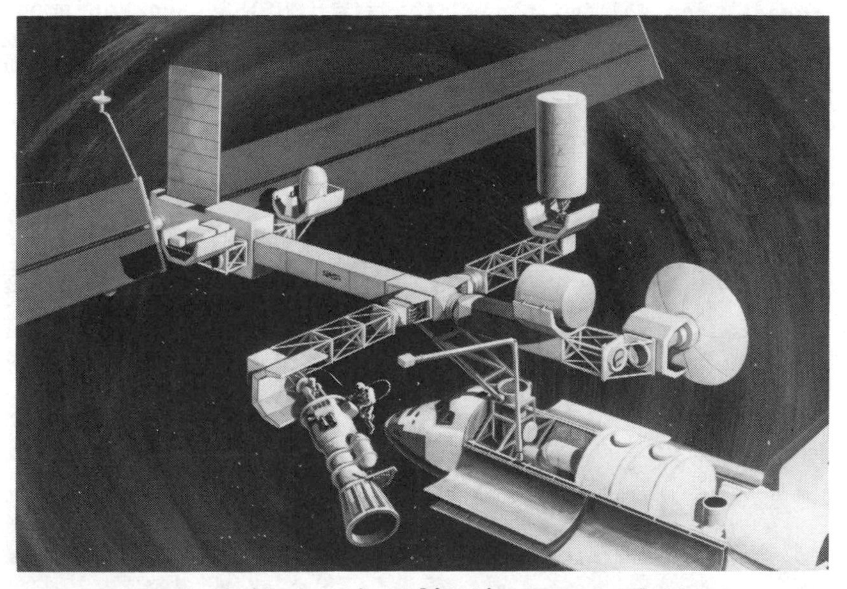

Fig. 32 Science and applications space platform.

developed the detailed information required to embark on hardware development. The McDonnell Douglas platform concept of that time is illustrated in Fig. 33. The key design drivers for the system were 1) maximum payload accommodations, 2) low-cost, low-risk design and operation, 3) minimization of orbiter impacts, 4) facilitation of on-orbit maintenance, and 5) incorporation of modular growth.

The resultant design (see Figs. 34 and 35) was composed of the following configuration elements: solar electric propulsion (SEP) technology solar arrays, radiators, three payload ports, control-moment-gyro (CMG)-based attitude control, separable reboost module, and Ku-band data support.

This configuration maximized payload accommodations by minimizing obscuration from radiator and solar arrays, permitting high payload data rates (230 megabits per second), maximizing thermal heat rejection consistent with electrical power generated (11.5 kW), providing an independent heat exchanger at each port, and incorporating provisions for multiple integral payloads (semipermanent installations).

The design was based on existing hardware wherever possible, as illustrated by the following examples: Tracking and Data Relay Satellite System (TDRSS) propellant tanks, Shuttle orbiter thermal pumps, Landsat Ku-band equipment, Skylab CMGs, Space Telescope attitude sensors, and Multimission Satellite (MMS) S-band equipment.

The length of the platform would permit simultaneous launching with two additional payloads. The system is inert during launch and thus requires no Shuttle functional services. To facilitate on-orbit maintenance,

Fig. 33 Space platform concept, 1980.

key vehicle equipment is accessible for EVA changeout within easy reach of access doors. Provisions for trouble-free maintenance incorporated features being developed for the Space Telescope. The design required no planned maintenance for at least five years of operation.

Provisions were incorporated into all subsystems to permit modular growth, either on the ground or on orbit. The 11.5-kW version was to be built on a 25-kW frame to permit easy conversion. Studies of cost trade-offs supported this approach as the most efficient. Key considerations in these studies were the relatively fixed avionics costs between the two sizes and the low-cost isogrid bolt-together structure.

This design incorporated the following characteristics:

1) Electrical power = heat rejection: 11.5 kW
2) Weight: 11,023 kg
3) Communications: Ku-band and S-band
4) Mission duration: 10 years, with maintenance
5) Low acceleration: $2x10^6$ g
6) Nominal altitude: 235 n.mi., 28- or 57-deg inclination
7) Safehold mode: 72 hours autonomous operation, six months gravity gradient.

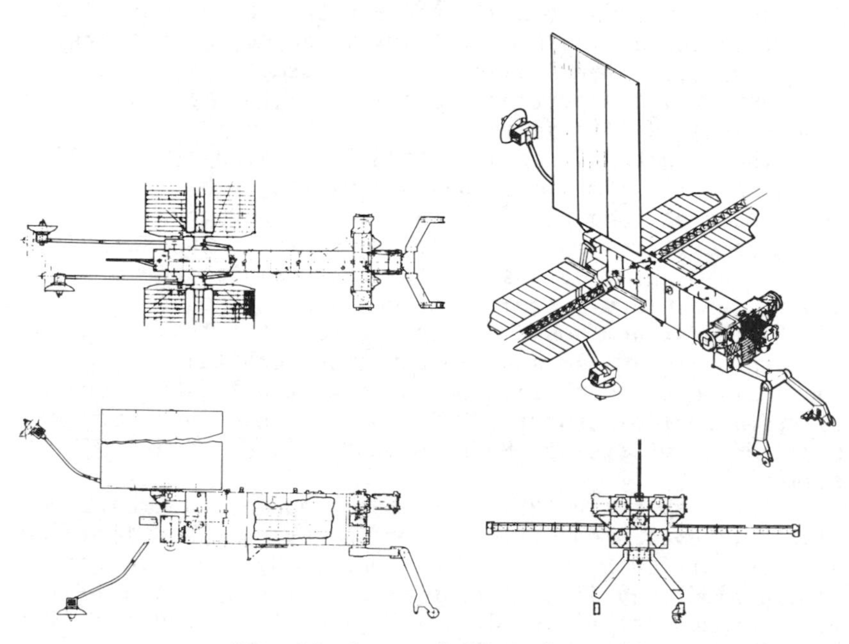

Fig. 34 Space platform features.

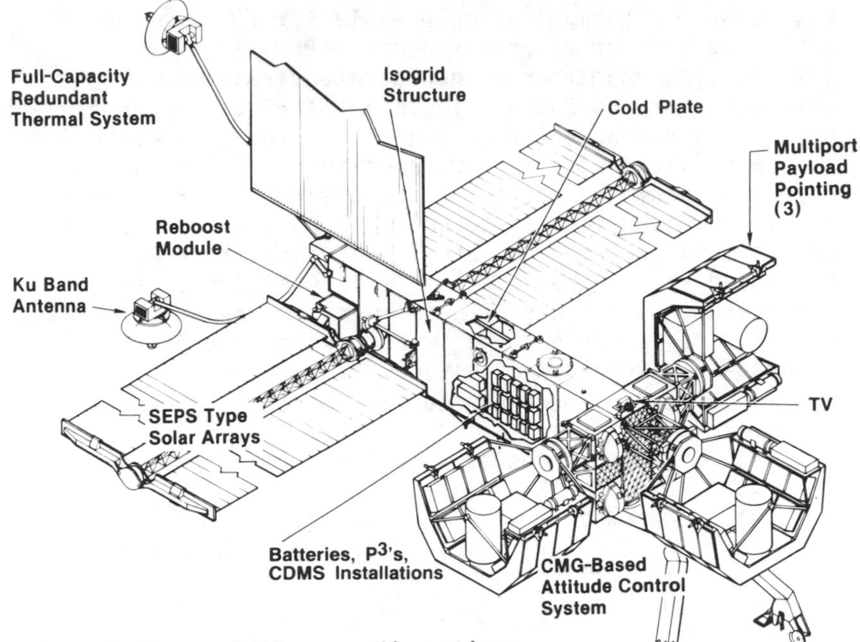

Fig. 35 Space platform configurations.

The payload accommodations of the space platform configuration supported a broad spectrum of potential users, including instruments for astronomy, high-energy astrophysics, solar physics, space plasma physics and environmental observations, and modules for materials processing and life sciences.

The structural configuration was driven by several key demands: 1) maximize payload viewing, which influenced payload port - solar array separation; 2) minimize orbiter impacts, which resulted in the swing-link legs; and 3) provide ease of on-orbit maintenance or growth, which drove equipment location and access.

The platform could fly as a co-orbiter to the space station or in a solo mode in any other orbital inclination. Its mission would be controlled by a space platform control center (SPCC), which would interface with payload scientists and NASA communications and data elements.

Thus, by early 1981, MSFC was prepared to embark on a platform development program for improving the efficiency and cost of flying unmanned payloads. However, in this time period, significant changes were taking place in manned space activities in the United States and the USSR that impacted the unmanned space platform activity.

NASA's major manned system, the Shuttle, became successfully operational, and at the same time the Soviets significantly increased their manned space flight activities on the Salyut space station, which had been flying in slightly varying versions since 1971. Consequently, NASA's attention turned to plans for the next US manned system, namely, the space station.

With the prospect of a space station, the nature of the unmanned platform took a significant change. With a space station in low-inclination orbit, one space platform could fly as a co-orbiter under control of the space station and receiving periodic services from it. Thus, the benefits of an unmanned system would be provided to the payloads, with periodic manned support from the nearby space station via a small maneuvering vehicle, or even occasional docking of the two major vehicles.

Inasmuch as the space station and its manned payloads also had resource and accommodation needs that were common to the unmanned payloads, the prospect of a resource module used by both major systems emerged. Figure 36 illustrates one concept of an unmanned space platform that uses a resource module that could also be used by the space station, but probably with modular additions of more resources. Because many payloads require a polar-type orbit, a second unmanned space platform could be deployed for high-inclination operation.

Very large payloads present a special challenge for accommodations on an unmanned platform. For example, if one or several particularly large payloads (say, 100 m in diameter each) were to be berthed on a platform of the type and size illustrated earlier in Figs. 34 and 35, significant interferences, subsystem impacts, and unwieldy operations would occur. Consequently, in 1980, NASA's Langley Research Center decided that a platform concept should be developed for prospective oversize payloads. McDonnell Douglas was contracted to develop such a concept. Langley's primary interest in it was to define requirements for related advancements in structures technology.

The large payloads that were to be accommodated on this "advanced" science and applications platform included: 1) the 100-m-diam atmospheric gravity wave antenna (AGWA); 2) the 100-m^2 particle beam injector (PBI); 3) the 2-m diam by 18-m astrometric telescope; and 4) the 15-m by 36-m large ambient deployable infrared telescope. After a number of options had been studied in the light of a broad spectrum of criteria, such as assemblability, flight mechanics, attitude control, solar array relationships, adjacency sensitivities, etc., the configuration shown in Fig. 37 was selected.

This T-bar configuration measures 80 by 160 m and dwarfs the Shuttle shown docked at its far side. Four Shuttle payloads would be required to deliver the basic platform, plus at least a half-dozen flights for the several massive payloads. The overall assemblage would weigh 80,000 kg.

The platform capabilities include a 35-kW power demand, including 15 kW for the AGWA payload alone and 7 kW for housekeeping in the spacecraft. S-band and Ku-band data systems were incorporated to accommodate the 10-kilobits-per-second to 7-megabits-per-second payload needs. The propulsion system used 4 tons of propellant per year to keep the giant platform at 500-km altitude because of the approximately 4000 m^2 of drag area. Annual resupply of propellants via Shuttle was planned. Because of the long spans involved, a decentralized heat rejection system was planned, with numerous radiators customized for and co-located with each payload. For the same reason, the low-voltage power distribution cabling will weigh over 2 tons, unless high-voltage equipment is space-qualified in time for such a system. In addition to a power-distribution, weight-reduction technology challenge for this extra-large platform, advancements

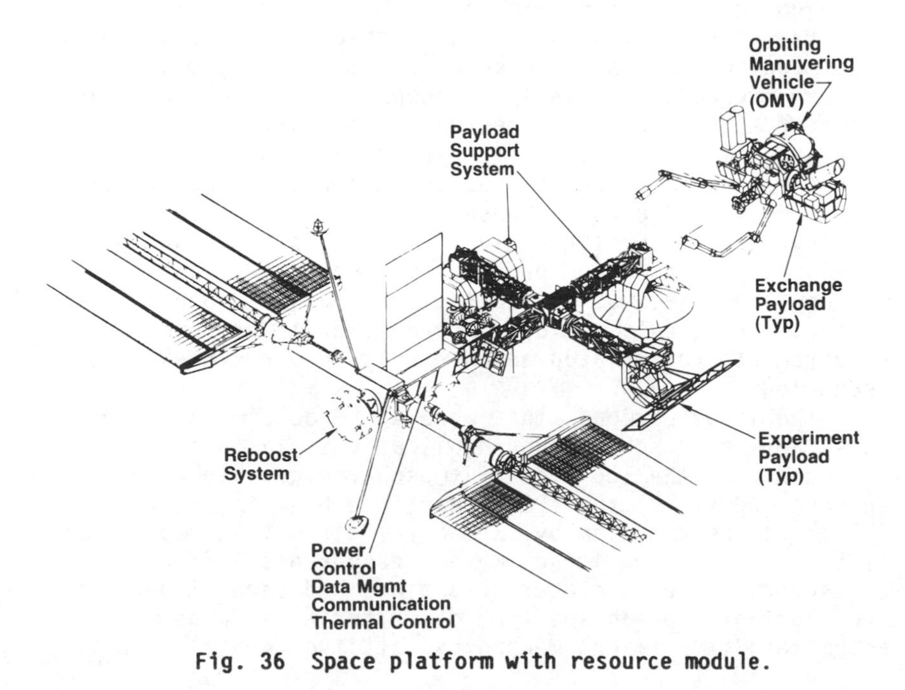

Fig. 36 Space platform with resource module.

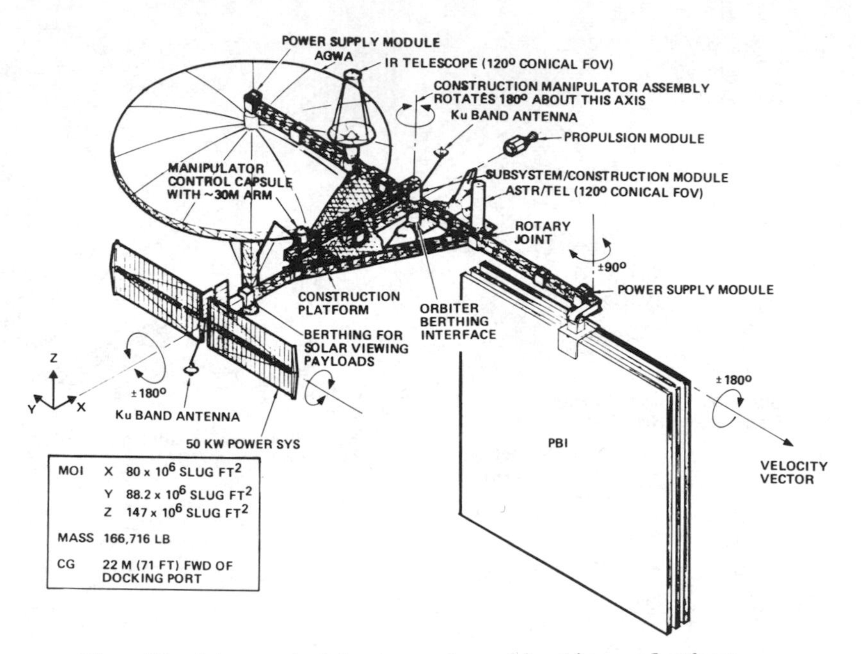

Fig. 37 Advanced science and applications platform.

would also be required in the areas of structural stiffness and dynamics control, and avoidance of contamination from the massive amounts of propulsion by-products expelled during drag makeup. Very large expansion nozzles for the engines, perhaps installed on orbit by astronauts performing EVA, may provide a solution.

In 1982, NASA's Office of Space Science and Applications introduced a new platform concept.(See Fig.38).

This spacecraft, called System Z, was to be a permanent unmanned facility capable of major configuration and instrument changes on orbit during Shuttle visits. Although the spacecraft would operate primarily in an unmanned, automated mode, crews from the Shuttle or space station could perform periodic installation and testing, repair and resupply, and retrieval for upgrading and calibration.

The payloads for System Z were to include global climate, ice budget, and hydrologic cycles, bio-geo-chemical cycles, and biomass dynamics. Moreover, the platform was to provide a facility for the performance of interdisciplinary research through the use of instruments that could contribute to several different science areas of interest.

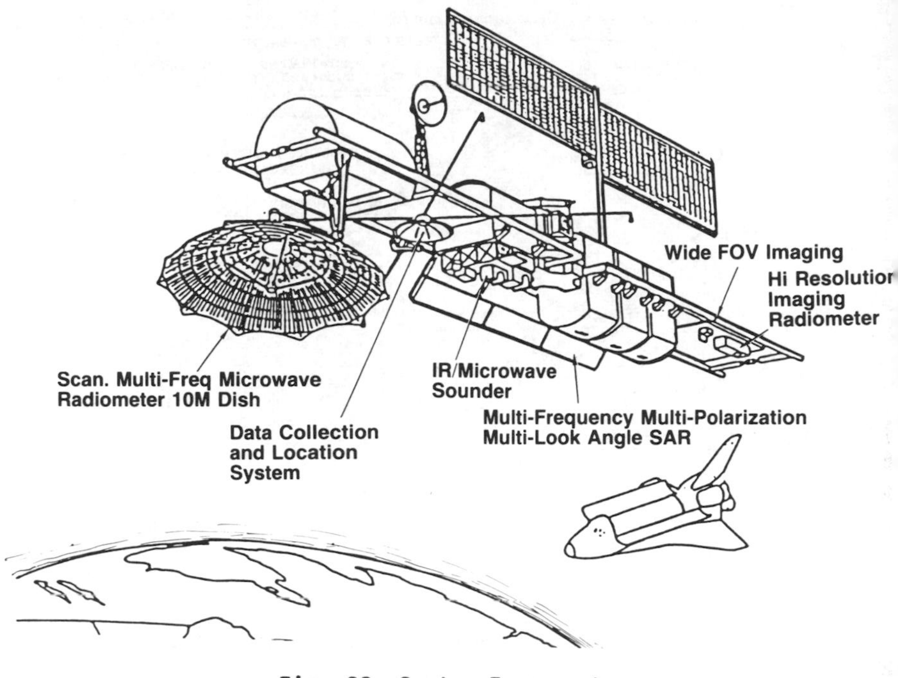

Fig. 38 System Z concept.

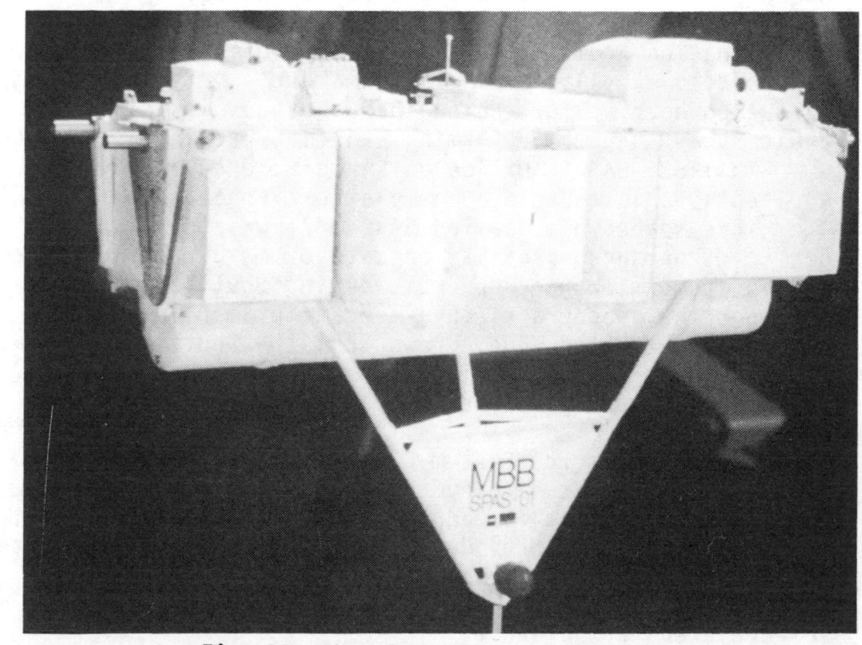

Fig. 39 Shuttle Pallet Satellite.

Representative instruments planned for this system include the following:

1)	Visible IR moderate and wide FOV images	1)	Visible IR high-resolution images
2)	Laser rangers and actimeters	2)	Multifrequency lidar facility
3)	Passive chemical species sensors	3)	Microwave radiometers
4)	IR and microwave sounders	4)	Scatterometers
5)	Real aperture radars	5)	Synthetic aperture radars
6)	Solar sensors	6)	Data collectors and locators

The data system was to be highly advanced, multidisciplinary, and user-oriented. It would include a georeferenced archive with quick-look and interactive capabilities.

During routine, unmanned operation, the spacecraft would have considerable autonomy through the use of expert systems. Functional capabilities would be maintained without ground intervention by means of an ultra-high-reliability (fault-tolerant) computer. This capability would include onboard generation of observational sequences, diagnosis of anomalies, and adaptive modification of subsequent sequences. Basically, the acquired payload data would be automatically processed, diagnosed, stored, and transmitted to ground or reacquired in a modified manner, based on onboard expert systems or artificial intelligence. The vehicle would fly in polar orbit at an altitude of 500 km to make it accessible to the Shuttle for crew-tending. A second vehicle could also co-orbit with the space station for other payloads.

The platform would be highly modular, with a berthing point for Shuttle attachment during module exchanges using the Shuttle remote manipulator. The initial configuration would involve a power level of less than 10 kW, again, with modular growth potential.

In summary, System Z was conceived to permit major advances in interdisciplinary Earth sciences, plus expert systems, robotics, mantending, and interactive data systems. It is currently an in-house NASA study project targeted at a 1991 launch.

Europe has also been active in developing space platforms. Germany's Shuttle Pallet Satellite (SPAS) was deployed and retrieved on STS-7 in June 1983, and is designed for multiple reuse for science and commercial payloads. It was developed by Messerschmitt-Boelkow-Blohm (MBB) using low-cost and modular design features. SPAS-01 is illustrated in Fig. 39. Future flights will be deployed by one Shuttle mission and retrieved by a later one.

The European Space Agency is developing a Shuttle deliverable and retrievable platform called EURECA, which is planned for six months and longer flights. EURECA is planned for an initial launch late in 1987, with retrieval in mid-1988. Phase B studies for EURECA are currently under way, as is the experiment selection for the first mission. Up to six multiuser facilities will be on board, including an automatic mirror furnace, a solution growth facility, a protein crystallization facility, a multifurnace facility, an automatic gradient-heating facility, and a botany facility. Figure 40 illustrates the EURECA concept.

The commercial world is also planning to provide multiuser platform services. Fairchild Industries is developing a platform called Leasecraft, based on the multimission spacecraft built for NASA earlier (Fig. 41). Payloads can be changed on orbit, and it is fully compatible with the Space Shuttle for in-orbit maintenance and repair. The first launch is planned for 1986. Leasecraft contains a propulsion system that permits boost up to 600 n.mi. from the Shuttle parking orbit of 160 n.mi.

The platform concept has come a long way since the mid-1970's, mostly based on the cost advantage of pooling groups of payloads, and the highly advantageous manned delivery and revisit services of the Shuttle. The space

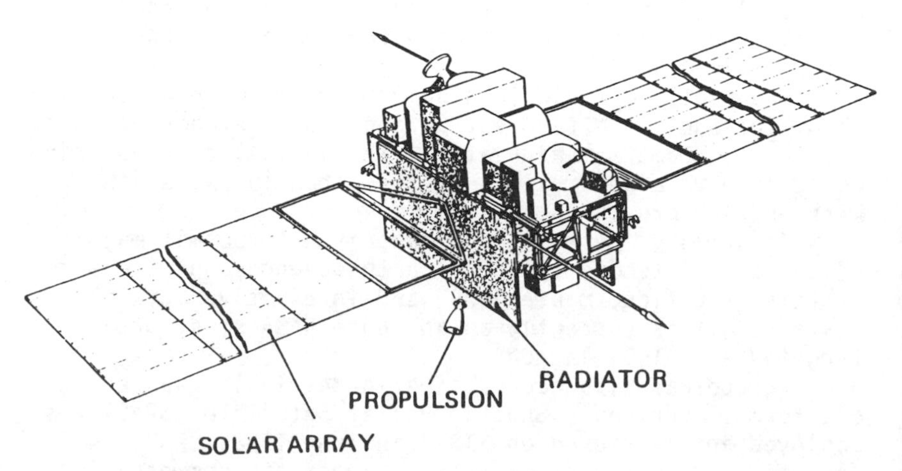

Fig. 40 EURECA.

station program includes plans for one co-orbiting platform in low-inclination orbit and another platform in high-inclination orbit. For the co-orbiter, the space station will serve as an operating base for monitoring, control, selective intermediate data processing, payload replacement, refurbishment or upgrade, and possibly even interim storage. These services will be provided most often remotely via an orbiting maneuvering vehicle, and on lesser occasions by berthing the platforms on the station for major repairs or services. In the scenarios planned for the space station and a co-orbiting platform, a synergism will be achieved that far exceeds their individual potentials.

An Evolutionary Space Station Concept

In early 1981, Marshall Space Flight Center began to examine a low-cost, manned use for the space platform planned originally for unmanned payloads. Studies at that time indicated that the emerging space station concepts required much the same in-orbit facilities and flight control capabilities as the large, unmanned payloads being planned for the space platform. Thus, the concept of a dual-use, manned-unmanned platform or resource module began to take shape. With it, also, emerged the concept of an unmanned platform co-orbiting with and being monitored and serviced by a space station. The space

Fig. 41 Leasecraft.

Fig. 42 Evolutionary space station, 1981.

platform was also viewed as being a highly modularized vehicle that could be configured differently for different missions, but with considerable cost savings through commonalities. Also inherent in that concept was an approach of adding modular capabilities on orbit to accommodate increments of growth; for example, additional power modules could be added to allow growth in the space station through the years. Thus developed a concept with considerable potential programmatic and technical flexibility and cost effectiveness.

McDonnell Douglas was funded to develop this concept in a study entitled "Evolutionary Space Station."[37] The resulting configuration is illustrated in Fig. 42. This rendering was also replicated in a spectacular, wall-size mural at the 1982 World's Fair in Knoxville, Tenn.

The capabilities inherent in this concept were to be introduced in two steps. Initially, the space station would have a crew of three to four, some interior experimentation, and some not-too-complex exterior experiments, primarily sensing instruments. The Shuttle would visit every 90 days for overall logistic support.

The initial capability configuration consisted of six elements: 1) the space platform or resource module; 2) the central docking/safe-haven module; 3) the habitat; 4) a dedicated pressurized payload module; 5) exterior

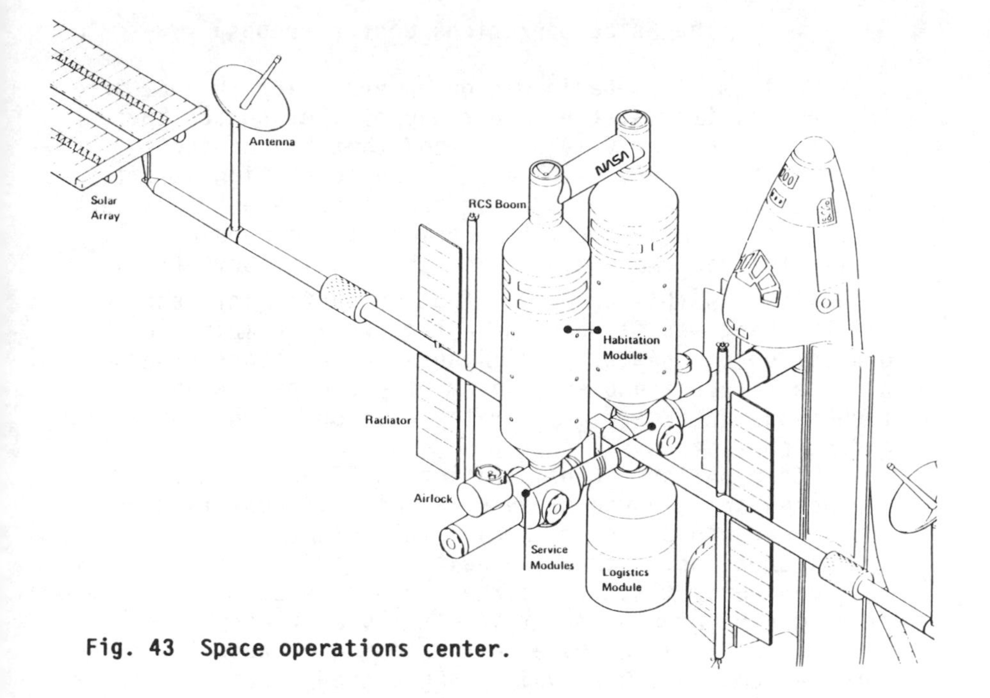

Fig. 43 Space operations center.

payload pallets; and 6) a logistics module that would be exchanged every 90 days by the Shuttle. The growth capability envisioned for the evolutionary space station would be incorporated after three to four years perhaps. At that time operation of the basic station and interior payload operations would have matured to the point where major exterior operations could be accommodated.

In that growth period, the technology base for large, assemblable payloads, station-based launch vehicles, and spacecraft servicing would have reached a point where prototype flight tests would require a major test/service/construction facility on the station. Such a facility would include docking ports for short-excursion orbital maneuvering vehicles, high-orbit transfer vehicles, a long-reach manipulator, propellant storage tanks, exterior stowage and work pallets, assembly fixtures, and, conceivably, an enclosure or hangar for environmental protection of sensitive payloads. In summary, the highly modular nature of this particular space platform/space station concept was seen as providing the programmatic and operational flexibility needed for a facility destined to serve a great number of customers whose exact physical characteristics, support needs, and, most importantly, schedule demands would not always be known far in advance.

The Space Operations Center Proposal

While MSFC emphasis was on an evolutionary approach to space platforms, the leadership of the Johnson Space Center had, by early 1979, decided that the center's efforts should refocus on a major space station effort. Aviation Week reported JSC was "concerned about this lack of continuing assessment for permanently manned US facilities" and was "mindful of the growing Soviet capability in this area."[38] Another factor influencing JSC thinking was "a need for a real goal to maintain the dedication of present participants in the space program and the interest and enthusiasm of young people in space technology in order to motivate their pursuing engineering and science careers."[39]

Based on these considerations, JSC during 1979 conducted an in-house study of a concept identified as a space operations center (SOC). This study was based on two assumptions: "that the next 10 to 20 years will include requirements for large, complex space systems" and "that geosynchronous orbit is clearly a primary operational area in space in the coming decades." If these assumptions were valid, JSC argued, then "the space construction and servicing of these future systems will be more effective with a permanent, manned operations center in space."

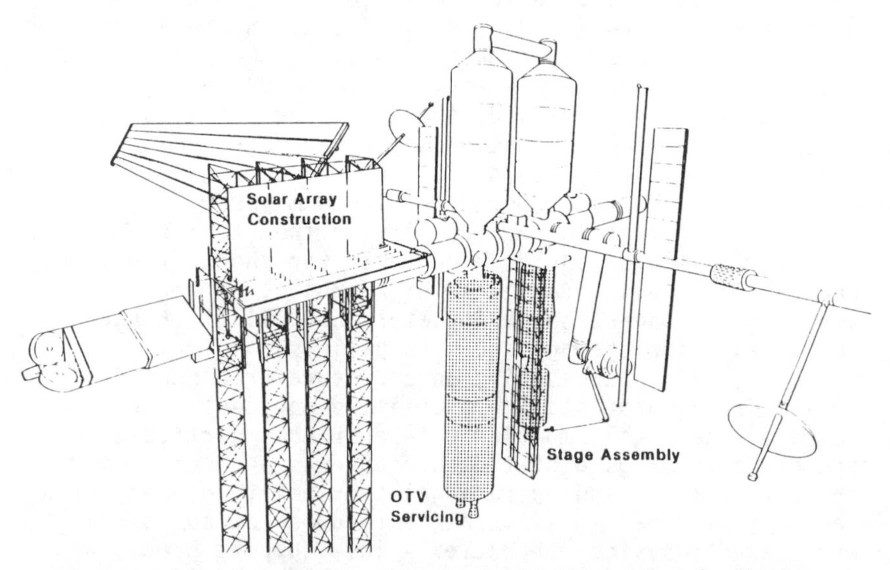

Fig. 44 SOC with construction and flight support facilities.

The primary objectives of the SOC were identified as:

1) The construction, checkout, and transfer to operational orbit of large, complex space systems.

2) On-orbit assembly, launch, recovery, and servicing of manned and unmanned spacecraft.

3) Further development of the capability for permanent manned operation in space, with reduced dependence on Earth for control and resupply.

The SOC study noted that this least of objectives "noticeably does not include onboard science and applications objectives, although the free-flying satellites which would be serviced would include mostly those of this genre. The primary implication of this omission is that experiment and applications requirements will not be design drivers; the SOC will be 'optimized' to support the operational functions of the objectives. However, experiments or applications which can tolerate the operational parameters of the SOC can be operated onboard, or an entire dedicated module could be attached to an available berthing port."

The study developed a concept of a self-contained, continuously manned orbital facility built from several Shuttle-launched modules (Fig. 43). The initial SOC crew would be four to eight people. In addition to a core facility, the full-capability SOC would require a construction facility and a flight support facility (Fig. 44). The costs of this fully capable SOC were estimated at $2.7 billion (FY 78$), with the total facility in place 9 to 10 years after the program initiation.[40]

In July 1980, Boeing won a 12-month SOC systems analysis study; the JSC reference configuration was used as a point of departure to determine system requirements, develop mission models, identify technology requirements, and refine the configuration based on the results. Rockwell International was also given a contract to analyze SOC-Shuttle interactions. In 1981, the SOC study was reoriented by JSC to increase the modularity of the concept to allow a more evolutionary buildup toward full SOC capability; this also permitted a more even funding profile across the duration of the project. Even with its revised approach, however, the SOC concept advocated by Houston focused on supporting space operations rather than on providing a base for orbital experimentation, with the ultimate primary objective of supporting activities in geosynchronous orbit. The cost of the SOC program was estimated at $9 billion.[41]

Conclusion

When NASA Administrator James Beggs created, in May 1982, an agency-wide Space Station Task Force and charged it with developing a station concept and rationale for NASA to offer to the nation, he recognized that that planning effort would clearly not be starting from scratch. Perhaps no other major Government program has had as much thought and study devoted to it before policy approval as has the space station. As this chapter has suggested, the concept of a permanent human outpost in space has been receiving serious engineering attention for over 60 years. Since the inception of space station in 1958, NASA has always had the station on its study agenda, and twice before, in 1959-1960 and 1969-1970, a space station program was considered for implementation. The experience of both the United States and the Soviet Union has demonstrated the values of human presence in space. As the United States moves toward making that presence permanent, it can build on the efforts described herein, efforts that have helped to create a solid foundation for success.

References

[1]David, L., "Space Stations of the Imagination," AIAA Student Journal, Vol. 20, Winter 1982/1983, pp. 6-13.

[2]Ordway, F.I., III, "The History, Evolution, and Benefits of the Space Station Concept," paper presented at XIII International Congress of the History of Science, Moscow, USSR, August 1971.

[3]Ross, H.E., "Orbital Bases," Journal of the British Interplanetary Society, Vol. 8, January 1949, p. 11.

[4]von Braun, W., "Crossing the Last Frontier," Collier's, March 22, 1952, pp. 25-29, 72-74.

[5]Logsdon, J.M., The Decision to go to the Moon: Project Apollo and the National Interest, M.I.T. Press, Cambridge, Mass., 1970, Chap. 2, and Compton, W.D. and Benson, C.D., Living and Working in Space: The History of Skylab, NASA SP-4208, Government Printing Office, Washington, D.C., 1984, Chap. 1.

[6]Bruce Loftin, in "Minutes of the Research Steering Committee on Manned Space Flight," May 25-26, 1959, NASA Historical Archives, NASA Headquarters, Washington, D.C.

[7]Low, G., "Manned Space Flight," in NASA-Industry Program Plans Conference, Washington, D.C., July 1960, p.80.

[8]Butler, G.V., "Space Stations, 1959-To?" in Bluth, B.J. and McNeal, S.R., Update on Space, Vol. I, National Behavior Systems, Granada Hills, Calif., 1981, p. 8.

[9]Hook, W.R., "Historical Review and Current Plans," unpublished and undated paper.

[10]Langley LARC Research Staff, A Report on the Research and Technological Problems of Manned Rotating Spacecraft, Langley Research Center, Hampton, Va., NASA Technical Note D-1504, August 1962.

[11]"Report on the Development of the Manned Orbital Research Laboratory (MORL) System Utilization Potential," Douglas Missile and Space Systems Division, Douglas Aircraft Co., Report SM-48822, January 1966.

[12]Hook, W.R., "Historical Review and Current Plans," unpublished and undated paper, pp. 5-6.

[13]Faget, M. and Olling, E., "A Summary of NASA Manned Spacecraft Center Advanced Earth Orbital Mission Space Station Activity from 1962 to 1969," in Compilation of Papers Presented at the Space Station Technology Symposium, Langley Research Center, Hampton, Va., Feb. 11-13, 1969.

[14]Becker, H.S., "A Historical Sketch of MSFC Space Stations Studies and Considerations for Future Systems," in Compilation of Papers Presented at the Space Station Technology Symposium, Langley Research Center, Hampton, Va., Feb. 11-13, 1969.

[15]Memorandum from NASA Associate Administrator, "Special Task Team for Manned Earth Orbital Laboratory Study," March 28, 1963, NASA Historical Archives, NASA Headquarters, Washington, D.C.

[16]"Summary Report, Future Programs Task Group," reprinted in US Senate, Committee on Aeronautical and Space Sciences, NASA Authorization for FY 1966, hearings, Mar. 25-30, 1965, Part III, pp. 1027-1102.

[17]Memorandum from Deputy Administrator, NASA, "Preliminary Study of a Manned Space Station," July 19, 1966, NASA Historical Archives, NASA Headquarters, Washington, D.C.

[18]The results of the study were presented at briefings given by Charles Donlan on Sept. 29, 1966, and several times in November 1966, NASA Historical Archives, NASA Headquarters, Washington, D.C.

[19]Normyle, W., "Alternatives Open on Post-Apollo," <u>Aviation Week and Space Technology</u>, Jan. 13, 1969, p. 16.

[20]Letter from Charles Matthews to Robert Gilruth, Jan. 18, 1969, NASA Historical Archives, NASA Headquarters, Washington, D.C.

[21]Normyle, W., "NASA Aims at 100-Man Station," <u>Aviation Week and Space Technology</u>, Feb. 24, 1969, p. 107.

[22]Memorandum from Deputy Associate Administrator for Manned Space Flight, "Design Characteristics of the 1975 Space Station Module," Feb. 19, 1969, NASA Historical Archives, NASA Headquarters, Washington, D.C.

[23]Normyle, W., "Large Station May Emerge as 'Unwritten' U.S. Goal," <u>Aviation Week and Space Technology</u>, March 10, 1969, p. 104.

[24]"Bidders Conference: Space Station Program Definition Study," NASA, Washington, D.C., May 8, 1969.

[25]Memorandum from Director, Program Integration, to Deputy Director, Space Station Task Force, "Shuttle-Sized Station Modules," April 4, 1970, NASA Historical Archives, NASA Headquarters, Washington, D.C.

[26]<u>Statement of Work for Phase B Extension: Modular Space Station Program Definition</u>, NASA, Manned Spacecraft Center, Houston, Tex., Nov. 16, 1970, p. 2.

[27]<u>Manned Orbital Systems Concept Study</u>, Book 1, Executive Summary, McDonnell Douglas Astronautics Company, Huntington Beach, Calif., Sept. 30, 1975, pp. iii, 36, 1-2, 30.

[28]US Senate, Committee on Aeronautical and Space Sciences, <u>NASA Authorization for FY 1977</u>, hearings, p. 1046.

[29]<u>Outlook for Space: A Synopsis</u>, NASA, Washington, D.C., January 1976, pp. iv, v, 5-7.

[30]<u>Outlook for Space: A Synopsis</u>, NASA, Washington, D.C., January, 1976, pp. 55-56.

[31]O'Neill, G., <u>The High Frontier</u>, Doubleday, Garden City, N.Y., 1976.

[32]"NASA Seeks Proposals for Space Station Studies," press release, NASA, Dec. 11, 1975.

[33]Couvalt, C., "Space Station Design Studies Planned," Aviation Week and Space Technology, Dec. 1, 1975, p. 48.

[34]Space Station System Analysis Studies, Vol. 1, Executive Summary, McDonnell Douglas Astronautics Company, Huntington Beach, Calif., July 1977, pp. 3, 4, 13, 15, 38.

[35]Space Station System Analysis Studies, Final Report, Executive Summary, Grumman Aerospace, Bethpage, N.Y., June 1977, p. 701.

[36]"Operational Base Concepts Gain in Space Station Studies," Aerospace Daily, Sept. 13, 1976, p. 54.

[37]Evolutionary Space Platform (Station) Concept Study, Vol. I., Executive Summary, McDonnell Douglas Astronautics Company, Huntington Beach, Calif., May 1982, pp. 1-2.

[38]Aviation Week and Space Technology, March 26, 1979, p. 49.

[39]Space Operations Center: A Concept Analysis, NASA, Johnson Space Center, Houston, Texas, Nov. 29, 1979, pp. 1-1, 1-2.

[40]Space Operations Center: A Concept Analysis, NASA, Johnson Space Center, Houston, Texas, Nov. 29, 1979, pp. 1-8, 1-12, 1-13, 1-19, 1-24.

[41]Couvalt, C., "Consensus Nearing on Orbital Facilities," Aviation Week and Space Technology, Feb. 15, 1982, pp. 122-123.

Chapter V. A Summary of Potential Designs of Space Stations and Platforms

A Summary of Potential Designs of Space Stations and Platforms

Richard Kline,* Ronald McCaffrey,† and Donald B. Stein†
Grumman Aerospace Corporation, Bethpage, New York

Abstract

This chapter discusses the potential designs and characteristics of space stations and space platforms, and describes the ability to fulfill currently defined objectives through the year 2000. A definition of mission requirements leads to the functional requirements of the unmanned platform and manned space station. The chapter also describes in detail the major design issues that drive the base configuration and its respective systems for crew habitation; berthing/docking and logistics; satellite and upper-stage vehicle servicing; command, control, and communications (C^3); base resources; and platform commonality.

Introduction

The primary objective of the Space Station Program (SSP) is to provide the benefits of a permanently manned space-based facility that can accommodate specific user missions and support itself for extended periods. The next SSP will rely on the Space Shuttle for its initial launch and subsequent logistic support; it will, therefore, be placed in low-Earth orbit (LEO). As currently conceived, the SSP will consist of manned and unmanned elements of modular design to facilitate evolutionary growth in capability, size, and technology advancements. It is envisioned that the manned orbiting base will:

1) Provide long-term multifunctional facilities for crew and equipment.

---

*Director, Space Station Programs.

Program Manager, Space Station Programs.

2) Possess a high degree of onboard autonomy to minimize ground dependence.

3) Operate with other co-orbiting elements.

Future mission operations will blend human and robotic roles to maximize human productivity inside as well as outside the pressurized crew compartments. As shown in Fig. 1, establishment of this base as a transportation node to assemble and service vehicles destined for higher orbits could expand our capability to efficiently operate in useful higher-energy orbits such as geosynchronous orbit (GEO).

The initial capabilities to be developed for this program will

1) Govern the scope of early mission operations.

2) Develop capability for subsequent mission growth.

3) Influence long-range mission planning.

A facility that will probably attract increased mission usage will be rich in available resources such as volume, power, cooling, data processing, communications and crew.

Many configuration concepts have been considered by NASA and its contractors. Of these the "Power Tower," pioneered by Grumman, was selected as the reference configuration for further study because it maximizes the accommodations of current and future users while demonstrating acceptable design and operations characteristics. The power tower reference configuration is shown in Fig. 2. It flies in a local vertical-local horizontal orientation with its keel along the local vertical and its solar array boom perpendicular to the orbit plane. The Earth-pointed end of the space reference configuration contains Earth-looking payloads. The zenith-pointed end contains solar, stellar, and anti Earth viewing payloads and communication antennas. Non viewing payloads are located at various places on the space station, and the pressurized modules are located near the bottom of the keel. Servicing equipment is located along the keel on either side. The servicing and refueling facilities, OMV and OTV technology demonstration equipment, and satellite storage and equipment areas are located at various places along the structure.

Gimbaled solar array wings provide full power at any relative alignment of the space station and sunline. Heat rejection is provided by a combination of body-mounted radiators on the modules, deployed non-rotating radiators on the traverse boom, and deployed rotating radiators near the bottom of the keel.

Final definition of the initial SSP capabilities, however, must be derived from further detailed studies of

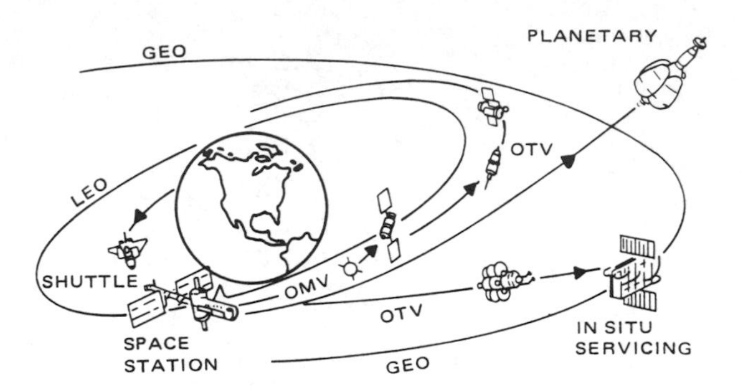

Fig. 1 The space station as a transportation node.

mission operation requirements, alternate design concepts, and technological and programmatic factors.

Mission Requirements and Desired Attributes

Space station requirements are based on missions that either require a space station or could benefit by the availability of a space station and its ancilliary services. Potential missions from domestic and international sources can be categorized as 1) science and applications, 2) space commercialization, and 3) technology development.

Space operations such as on-orbit assembly, servicing, or repair and launch are not considered mission categories, but they do provide service to the three primary categories.

Science and application missions encompass basic scientific research in and of the space environment, and the application of scientific technology to further our understanding of Earth, other planets, and our celestial surroundings. Commercial missions include use of the space environment to support commercial operations in the areas of communication, Earth observations, and material processing. Technology development missions involve the use of a space station to develop technology that enhances the evolution of a space station and contributes to future space programs.

With the assistance of the domestic as well as the international industrial and scientific communities, NASA's Space Station Mission Requirements Working Group developed time phase mission sets for 1991-2000. For each mission category, the orbit (altitude and inclination), location (attached or free-flyer), environment (pressurized or unpressurized), servicing, and transportation requirements were identified (see Fig. 3).

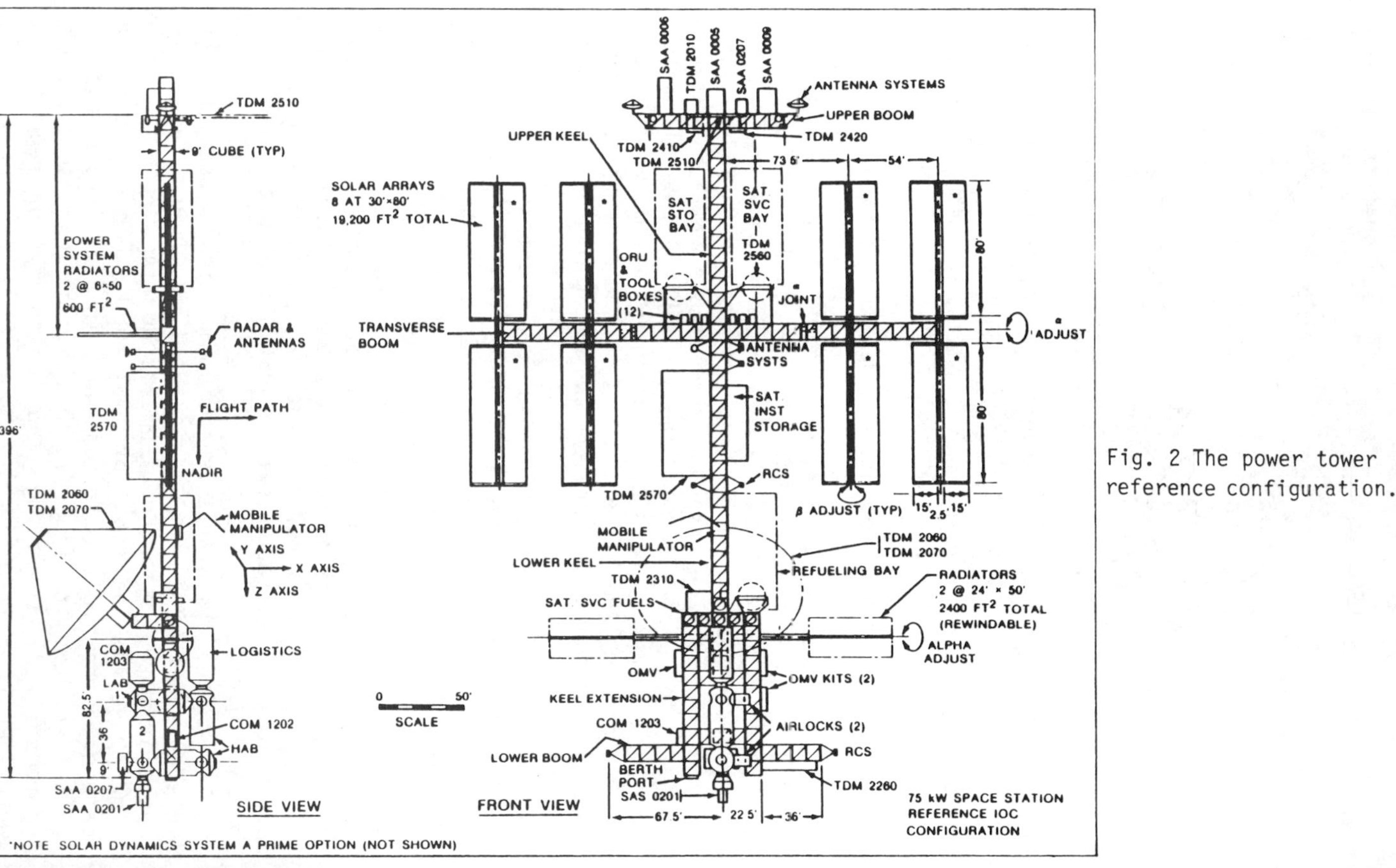

Fig. 2 The power tower reference configuration.

TECHNOLOGY DEVELOPMENT MISSIONS

COMMERCIAL MISSIONS

SCIENCE & APPLICATIONS MISSIONS — ASTROPHYSICS

CODE	NAME	'91 – 2000
SAA0001	COSMIC RAY NUCLEI	ATT 28.5°
SAA0002	SPACE PLASMA LAB	ATT 28.5°
SAA0006	STARLAB	ATT, 28.5°
SAA0003	SOLAR OPTICAL TELESCOPE	ATT. 28.5°
SAA0009	PINHOLE/ OCCULTER FACILITY	ATT 28.5°
SAA0011	ADV SOLAR OB-SERVATORY	28.5° ATT
SAA0004	STRTF	ATT ATT 28.5°
SAA0005	TRANSITION RAC & ION CAL	ATT 28.5°
SAA0007	HIGH THROUGH-PUT MISSION	ATT. 28.5°
SAA0008	HIGH ENERGY ISOTOPE	ATT 28.5°
SAA0012	SPACE TELE-SCOPE	FREE FLYER 28.5°
SAA0013	GAMMA RAY OB-SERVATORY	FF 28.5°
SAA0014	X-RAY TIMING (XTE)	FF 28.5°
SAA0019	FAR UV SPECTRO-SCOPY	FF 28.5°
SAA0010	CORONAL DIAG-NOSTIC MISSION	FF 28.5°
SAA0015	OPEN	FF- 0° & 90°
SAA0016	SOLAR MAX M	FF 28.5°
SAA0017	AXAF	FREE FLYER 28.5°
SAA0018	OVLBI	FF 57°
SAA0020	LARGE DEPLOY-ABLE REFLECTOR	FF 28.5°

▼ - OMV SERVICE △ - OTV TRANSPORTATION

Fig. 3 Integrated time phased mission set.

Science and Applications Missions

Space station roles and the appropriate orbit inclinations for science and applications missions are summarized in Fig. 4. The lowest launch costs for delivery to orbit are for a 28.5-deg inclination, which is the logical site for life sciences, long duration R&D, a transport node for scientific payloads to GEO and beyond, and for celestial and low-lattitude terrestrial viewing.

Earth resources and meteorological payloads (such as System Z) that require a view of the entire Earth's surface fit naturally on a high-inclination (polar) platform. This platform would also support long-duration and ecliptic plane solar and celestial viewing.

The benefits of using a space station for science and application missions are a result of the capability to support instruments for long duration and the ability to permit manned intervention. Service, maintenance, and refurbishment of instruments that may be attached to the space station, platforms, or free-flyers can be performed more economically and efficiently compared to performing these operations from the Shuttle.

Material science activities in internal laboratories will begin in 1991, when the first furnace becomes available for experimentation. Operations will continue throughout the decade as furnaces are added and higher power is provided. Life science missions begin by monitoring and physiological measurement of the onboard crew and are followed by laboratory experiments using animals and plants.

Most externally mounted payloads consist of celestial pointing telescopes. The Space Telescope and Gamma Ray Observatory are among those co-orbiting satellites that could benefit from periodic servicing and maintenance at or from the space station.

Commercial Missions

Prior to the Shuttle, nearly all commercial space missions consisted of communications satellites in geosta-

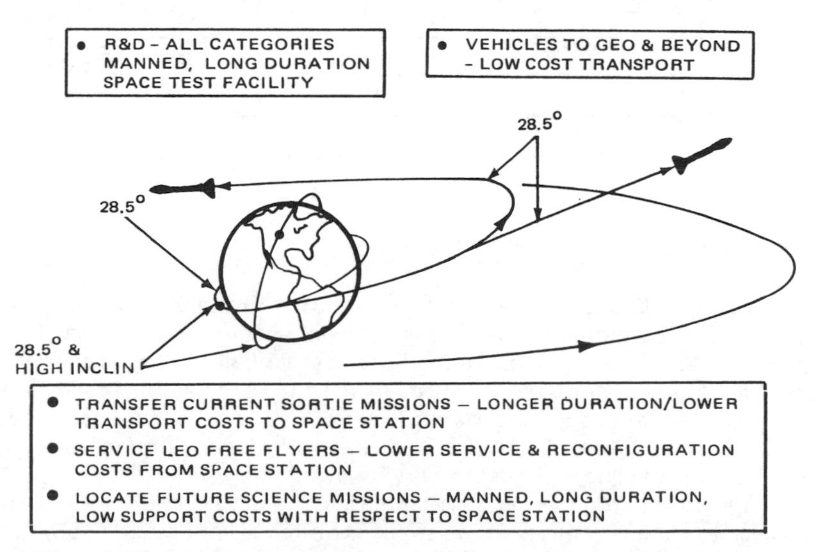

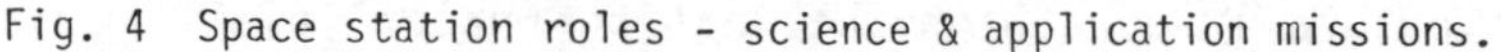
Fig. 4 Space station roles - science & application missions.

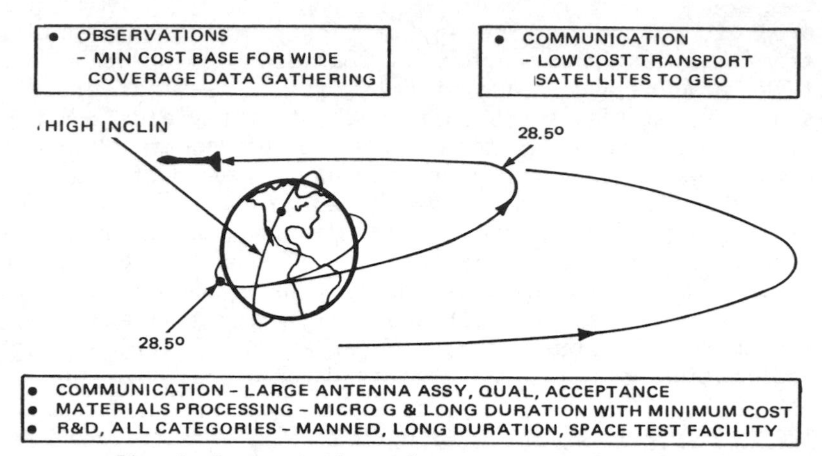

Fig. 5 Space station roles - commercial missions.

tionary orbit. Very few material processing missions have been flown, although a few will be flown during early Shuttle operations. The demand for communications satellites will most likely continue, and space materials processing will also expand. The commercial market for Earth observations, however, is still in its infancy. All of these missions (except Earth observations) could benefit from a low-inclination Space Station (see Fig. 5). Earth observation requires a high-inclination orbit for complete global coverage and would benefit by sharing transport costs and space facilities with other similar missions (compared with dedicated platforms).

Communication missions benefit by lower transportation costs using an orbital transfer vehicle (OTV) based on the space station. In addition, communications satellites, particularly those with large antennas, may be assembled and checked out on the space station before committment to GEO transfer. Communications activities include 1) component R&D, 2) qualification of large antenna satellites, and 3) deployment of satellites to GEO.

Deployment of GEO satellites from the space station could start in 1994, when a reusable, space-based OTV will most likely be available. Before then, it is assumed that traffic to GEO will go direct (using expendables) and bypass the station. Satellites suitable for remote servicing in GEO are expected to be launched in the mid-1990's.

Materials processing missions that require micro-g levels and R&D projects will benefit from human interaction and periodic intervention over long durations. This combination is not available from present or planned spacecraft. Since materials processing and the majority of R&D

projects do not require any particular orbit inclination, the lower-cost transport to a 28.5-deg station is, therefore, advantageous. Additional transport savings are possible by sharing launch costs with other payloads. Materials processing activities consists of continuous R&D efforts to develop new processes and new products, as well as production of already developed products. Commercial R&D is accomplished onboard the station whenever possible; production takes place onboard, or on free-flying platforms that can be periodically tended from the space station.

The stereoscopic multispectral imager is identified as a commercially viable remote observation mission. A high-inclination polar orbit is required for global coverage.

Technology Development Missions

As mentioned for other mission types, human interaction over a long duration is an obvious contribution to R&D activities, and a space station can fulfill these requirements. Generally, on-orbit technology development missions (see Fig. 6) do not require any particular orbit inclination; however, the 28.5-deg station is preferred because of lower space transportation system (STS) costs.

Enabling technology for the initial station will be completed in the mid-1980's, but station upgrading and station operations technology will continue through the

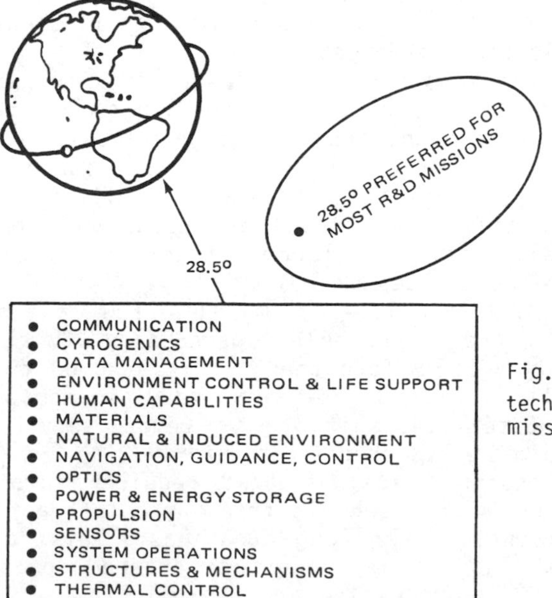

Fig. 6 Space station roles - technology development missions.

1990's. Some of these tasks can be accomplished on the space station itself. Technology development missions generally fall into four broad categories that benefit from on-orbit conditions: 1) large flexible structures, 2) manned interactions, 3) long-duration space tests, and 4) micro-g environment.

Antennas and optics involve large flexible structures, long-duration tests, manned interactions, and use generic and specialized test gear that requires a true micro-g environment. Requirements for the advance propulsion, advanced energetics, and long-duration exposure categories differ, but each makes an equally strong case for the services of a space station.

Integrated Requirements

The preferred space station and space platform orbital inclinations to support the operational objectives of various missions/payloads are summarized in Fig. 7. All missions/payloads the either do not have preferred orbital characteristics, or the destination of which is GEO or beyond, can be placed in a 28.5-deg inclination orbit, since this results in the lowest transportation cost to orbit. (Weight-to-orbit costs are based on $84.3M per flight and projected STS lift capability.)

High-inclination polar orbit provides the best solar and terrestrial coverage for most civil missions. Some science and application missions could perform satisfactorily in a 57-deg inclination orbit, however, which represents the highest achievable inclination from Eastern Test Range (ETR) due to Space Shuttle launch constraints.

ORBIT INCLIN	28.5	57	POLAR
COST TO LEO ($/kg)	2600	3380	6750
EARTH COVERAGE (%)	48	86	100
LEO MISSIONS	o * R&D o * MAIL PROCESS o ASTRONOMY o LIFE SCIENCES	* NAVIGATION * SURVEILLANCE o EUROPEAN S/C	* WEATHER o RESOURCE OBS * SURVEILLANCE o SOLAR OBS
GEO MISSIONS & BEYOND	o * COMMUNICATION o * WEATHER * SURVEILLANCE o PLANETARY		

* MILITARY USERS
o NON MILITARY USERS

INITIAL SPACE STATION AT 28.5°
• LOWEST COST TO GEO & BEYOND

Fig. 7 Activities related to LEO space station at 28.5°, 57°, and polar inclinations.

	SPACE BASE	SPACE PLATFORM
• SCIENCE & APPLICATIONS		
— LABORATORY EXPERIMENTS	✓	—
— OBSERVATIONS - EARTH & ASTRO	✓	✓
— LAUNCH SOLAR EXPLORATIONS	✓	—
— LAUNCH GEO PLATFORMS	✓	—
• COMMERCIAL		
— MATL PROCESS DEVEL, & PRODUCTION	✓	✓
— COMMUNCIATION DEVEL TEST	✓	✓
— LAUNCH GEO COMMUNICATION SATELLITES	✓	—
— OBSERVATION DEVEL & OPERATIONS	✓	✓
• TECHNOLOGY DEVELOPMENT		
— DEMO ADV OPERATIONS	✓	—
— ASSEMBLE LARGE SYSTEMS	✓	—
— DEMO ADV PROP & ENERGY SYSTEMS	✓	—
— LONG DURATION EXPOSURE TESTS	✓	✓
• SPACE OPERATIONS		
— PROXIMITY OPERATIONS	✓	—
— SERVICE SATELLITES & PLATFORMS	✓	✓
— RECOVER & SERVICE ORBITAL STAGES	✓	—
— STORAGE DEPOT		—

Fig. 8 Space station/platform missions.

The European community favors a 47-deg inclination orbit for ease of direct communication with their ground stations.

Generic missions that can be allocated to a manned space base and a free-flying space platform are listed in Fig. 8. An unmanned space platform is generally preferred for missions that require very accurate pointing with low jitter rates, contamination-free environment, or very low micro-g levels that a manned space station may not be able to provide. Space platforms also offer the ability to share payload resources (e.g., power, thermal, communications, etc.) among compatible missions when their separate projected operations for Shuttle launch and logistic support cannot justify a permanent dedicated presence in orbit.

Time phased integrated requirements to support these missions from a manned, 28.5-deg LEO base or a co-orbiting/ polar platform for their net demands on electrical power, pressurized volume crew support, data handling, and space-based vehicle operations are shown in Figs. 9 and 10.

The average electric power required for the 28.5-deg base increases dramatically over time (approximately 60 to 120 kW). This is due mainly to the projected growth of commercial materials that will be processed in space. This trend in space commercialization is also reflected by the

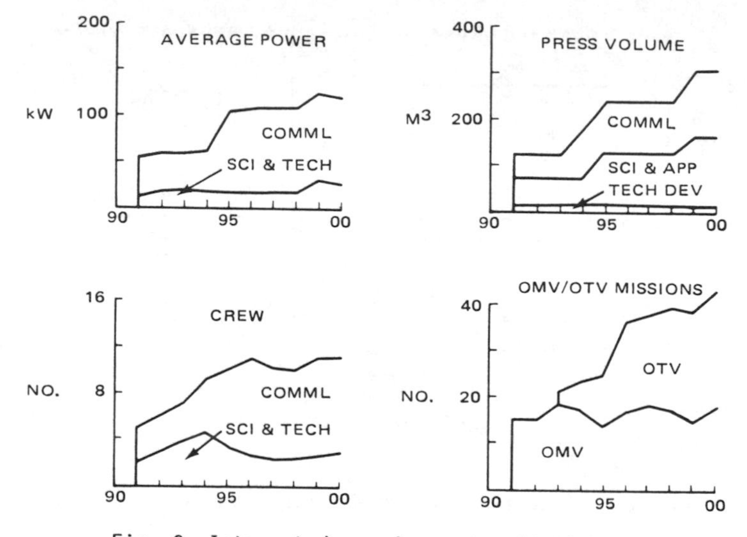

Fig. 9 Integrated requirements, 28.5° base.

need for more pressurized laboratories and added crew support. Space-based missions for orbital maneuvering vehicles (OMVs), however, are needed for modest orbit transfer and servicing of free-flying scientific satellites and space platforms. When reusable orbital transfer vehicles (OTVs) are introduced in 1993, they will support delivery of scientific and commercial satellites to high-energy GEO orbits.

The average electrical power needed for the missions on the polar and 28.5-deg space platforms is less than that required for initial space station operations. Multispectral instruments grouped on the polar platform typically require 12 kW, whereas the 28.5-deg astrophysics free-flyer requires 3 kW, and a co-orbiting epitaxial crystal growth production facility needs almost 24 kW.

Desired Attributes

The initial space station should be manned, placed in 28.5-deg orbit and provide substantial economic, performance, and social benefits. It is clear from the work accomplished by NASA and industry that there are five major roles in which an SSP has high potential payoff for the U.S. Beneficial attributes (see Fig. 11) include a: 1) space test facility, 2) transportation harbor, 3) satellite servicing and assembly facility, 4) observatory, and 5) space industrial park. The gross benefit that could be

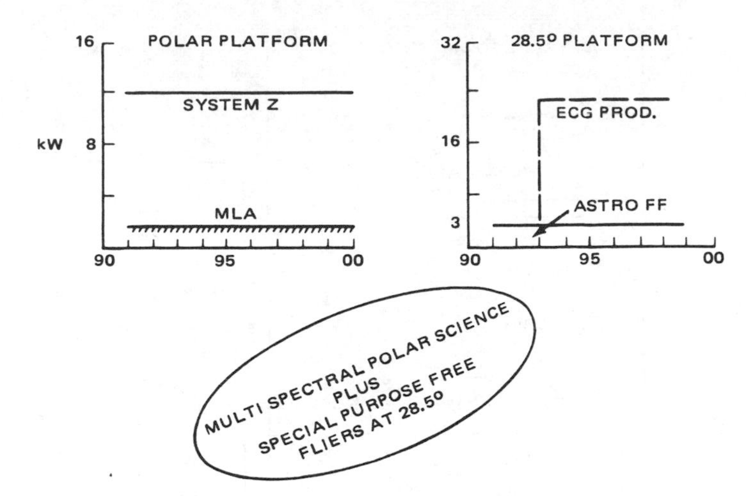

Fig. 10 Integrated power requirements, space platforms.

derived from these roles is estimated to be nearly $10 billion (constant 1984 dollars) by the end of the century.

A space test facility and range could provide the unique opportunity to conduct technology development and proof-of-concept in space with manned interaction in the development process. A "shirt sleeve" environment could provide ideal conditions for an Industrial Research Facility to conduct materials research and development relevant to commercial products and services, as well as a life sciences laboratory. Large structures/antennas/optics deployment and testing as well as heat rejection/radiator development could be performed in external bays. A test range might, for example, permit advanced state-of-the-art antennas to be tested with signal generators and diagnostic subsatellites that are strategically placed down range of the space station so that far-field experiment results would be obtained.

A transport harbor would have great utility and an ever-expanding role. Initially, the harbor would provide Shuttle support and repair in case of Shuttle malfunction. For upper-stage operations, the transport harbor could first support proximity operations using OMVs to fly to and from the space station for satellite or platform deployment/retriveal, inspection, or maintenance. Finally, a reusable OTV could become operational and provide lower-cost transportation to GEO.

Satellite servicing and assembly is another potential area of high payoff for the space station. Assembly/integration, storage, and manifesting of payloads close to the

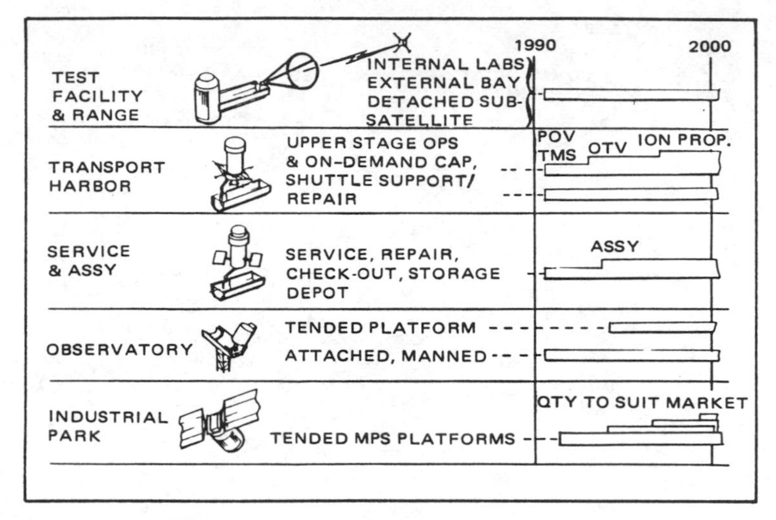

Fig. 11 Space station attributes.

space station represent a significant amount of activity. Satellite servicing could also pay off particularly well for large observatory satellites, such as the Advanced X-Ray Astrophysics Facility (AXAF) that NASA plans to operate in the 1990's. The potential exists for replacement of equipment to maintain the AXAF and to upgrade its capability at appropriate times. A $200 million savings is estimated for such a satellite over a ten-year period.

Another potential area for high payoff is an observatory attached to the space station with appropriate vibration isolation so that delicate instrumentation and telescopes could be pointed with reasonably high accuracy. The observatory function could also be performed by separate platforms/satellites flying in formation with a 28.5-deg space station, or by an independently tended platform at higher inclination.

Materials processing in space will more than likely offer technical and economic advantages over Earth-based manufacturing procedures. Higher-quality products can be produced in the space environment and a significantly higher product yield will most likely result when processing materials in a near-zero-gravity field. Both of these factors and an expected increase in market demand for the many products identified suggest development of an industrial park as part of the space station complex. The industrial park function could be performed as part of the manned base, or by a co-orbiting space platform if very low g levels (10^{-5}) are needed.

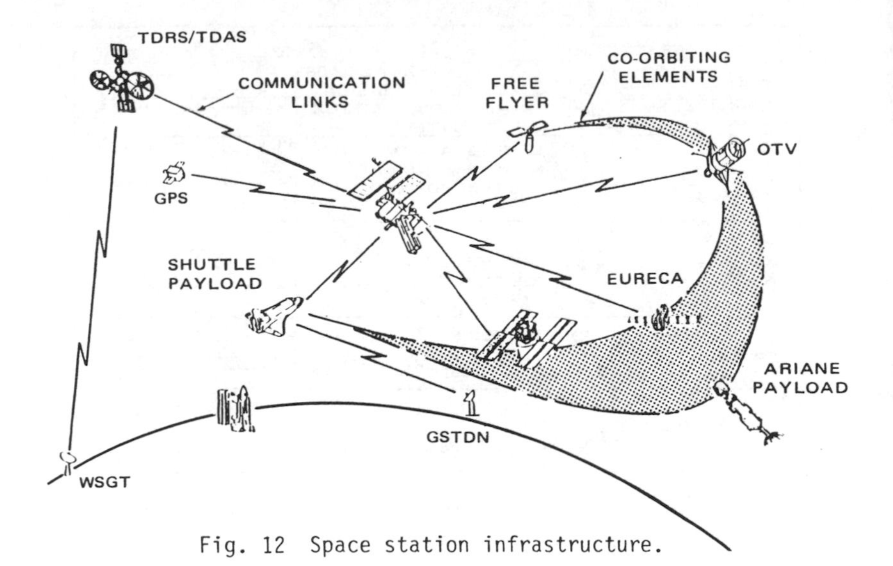

Fig. 12 Space station infrastructure.

Functional Requirements

The initial space station will depend on the Shuttle for launch to orbit, crew delivery/rotation, and logistic support. This permanent orbiting facility will provide essential functions for each user (e.g., volume, power, crew support), allow periodic maintenance/reconditioning of its subsystems for continued operations, and possess inherent capabilities for subsequent system growth. The US space station will be designed for joint operations with international payloads and their interdependent systems. As a combined transportation/communications node, for example, the space station will support nearby operations as well as those in other orbits. Space station functional requirements must, therefore, be derived from analysis of all operations related to 1) mission payloads within the pressurized shell, 2) Attached payloads, 3) extravehicular activity (EVA), 4) close proximity activities, and 5) remote mission activities.

Space Station Infrastructure

Basic elements of this orbit infrastructure are shown in Fig. 12. The primary operational interface of the space station to these elements is via communication links to ground- and space-based terminals. Ground communications will be routed primarily through the Tracking and Data Relay Satellite (TDRS) and subsequent Tracking Data Acquisition Satellite (TDAS) systems. Space station orbital control will be based on navigation with respect to the Global Positioning Satellite (GPS). While mission payloads will be primarily delivered by the Shuttle, some future mission payloads may be delivered in proximity to the space station by upgraded versions of Ariane. For the space station to be effective in the exploration and productive use of space, it will have to perform a variety of functions and accommodate many different spacecraft.

Operational services to other elements of the infrastructure will include such activities as satellite servicing, payload transport to higher orbits, on-orbit testing, in-orbit formation flying, communications (command,

FUNCTIONS / PARAMETERS	MANNED SPACE STATION			PLATFORM SUN SYNCH	
	PRESS. LAB 1	ATTACHED PAYLOADS 2	COMMAND CONTROL COMM SUPPORT 3	TERMI-NATOR	DAY/NIGHT
1 LAUNCH MASS	X	X		X	X
2 RETURN MASS	X	X		X	X
3 LOGISTICS MASS – UP	X	X		X	X
4 LOGISTICS MASS – DOWN	X	X		X	X
5 LAUNCH VOLUME	X	X		X	X
6 RETURN VOLUME					
7 VOLUME	X	X		X	X
8 AREA		X	X	X	X
9 PORTS		X		X	X
10 ORIENTATION		X		X	X
11 MANHOURS IVA-AVG	X	X		X	X
12 MANHOURS IVA-HIGH	X	X			
13 MANHOURS EVA		X			
14 EVENTS – EVA		X			
15 EVENTS – OMV				X	X
16 EVENTS – OTV					
17 POWER – AVG	X	X	X	X	X
18 POWER – HIGH	X	X	X	X	X
19 POWER – PEAK	X	X	X	X	X
20 HEAT REJECT – AVG	X	X	X	X	X
21 HEAT REJECT – HIGH	X	X	X	X	X
22 HEAT REJECT – PEAK	X	X	X	X	X
23 DATA RATE – AVG	X	X	X	X	X
24 DATA RATE – HIGH	X	X	X	X	X
25 DATA RATE – PEAK	X	X	X	X	X
26 DATA STORAGE	X	X	X	X	X
27 OTV REQUIREMENTS				X	X

Fig. 13 Functional requirement parameters.

data reception), stationkeeping, and mission operation control. These services will be on a continuous, periodic, or intermittent basis depending on each spacecraft's requirements. Formation flying platforms and attached payloads (scientific, commercial, and foreign) can be tended, as needed, to meet mission requirements. Tethered constellations of space platforms or instruments can provide ready access to a space station and also reduce the amount of power, data processing, control, and communication systems that they would have to carry as free-flyers. Space-based OTVs will deliver satellites to higher orbits, perform remote servicing operations, and be clustered for unique space exploration missions.

Space Station/Platform Functional Requirements

To provide the desired attributes of a space test facility, transportation harbor, satellite servicing and assembly facility, observatory, and industrial park, the manned space station element must encompass the following eight functional characteristics,

1) Pressurized laboratories: Pressurized work stations to provide payload power, low gravity, and long-duration crew support.

2) Attached payloads: External payload ports with supporting resources plus periodic crew tending and servicing.

3) Command, control, and communications (C^3): Provisions within the space station system to operate the base; its internal/external payloads, and to remotely operate.

SPACE STATION ELEMENT	SPACE STATION SUPPORT FUNCTION	DIRECT SUPPORT TO USERS
• PRESSURE VESSELS	• HABITATION	• LABORATORY • WORK SHOP
• EXTERNAL ACTIVITY ZONES	• BERTHING	• SERVICE • ASSEMBLY • STORAGE
• POWER SUPPLY & RADIATORS	• UTILITIES	
• TRUSS STRUCTURE	• INTERCON-NECTION	

Fig. 14 Larger external elements.

monitor, throughput, and preprocess data for free-flyers and platforms.

4) Deployment, assembly, and construction: Remote manipulators, manned maneuvering units (MMUs), as well as other devices and supports to perform this function.

5) Proximity operations: Maintain, service, and checkout maneuvering payloads that operate within a reasonable distance, nominally 2 km.

6) Remote maintenance, servicing, checkout, and retrieval: Provide remotely operating service vehicle to maintain/service remote payloads in situ, or retrieve payloads for servicing and redeployment.

7) Payload integration and launch: Provide and maintain expendable or reusable transfer stages for delivering integrated payloads/satellites to other orbits.

8) Payload staging for Earth return: Provide for demating, preparing, and storing payloads as well as experiments/captured samples for return to Earth.

In addition, the SSP can include two or more unmanned space platforms in complementary orbits (LEO and POLAR). The space station element can provide exterior payload ports, required resources (power, thermal, attitude con-

CONFIGURATION CHARACTERISTICS	OBJECTIVES					SAFETY
	PERF	MODERATE COSTS				
		DEV	PROC	LAUNCH	OPS	
• PERFORMANCE						
– INTERNAL MODULES	●					
– EXT ATTACHED ZONES	●					
– REMOTE	●					
• CONTROLLABILITY	●					
• ASSEMBLAGE				●	●	
• GROWTH	●					●
• COMMONALITY		●	●			●
• REDUNDANCY					●	●
• MAINTAINABILITY	●				●	●
• SEPARATE WORK STORAGE ZONES-SIMPLE CIRCULATION PATHS	●					●
• CONTAMINATION AVOIDANCE	●				●	●

M84-1086-014PP

Fig. 15 Configuration characteristics and program objectives.

trol, and data management), and operational needs (orbit cleanliness, microgravity, etc.) for allocated missions.

A sample of the mission parameters that have a major impact in space station functional design is shown in Fig. 13. For each parameter (launch mass, return mass, volume, power, etc.), time phased functional requirements can be derived from the missions allocated to the manned/unmanned space station functions.

Space Station Configuration Development

Generic Configuration Issues

In general, a full capability manned space base will have four types of elements (see Fig. 14). Two of these elements, pressure vessels and external activity zones, will provide direct service to the user in the form of laboratories, workshops, service, assembly and storage capabilities. All four elements will have space station support functions that are vital to the existence of the station by sustaining the crew and providing power, cooling, and data handling.

This section discusses overall arrangement of the elements, as well as how they are joined and articulated. The allocation of different external zones to specialized functions (berthing, service, assembly, storage), and the inertia and drag properties of the base and the control of its attitude with respect to its flight path and the Earth, Sun, and stars, are also discussed. Element size and modularization to minimize development cost, fit within the

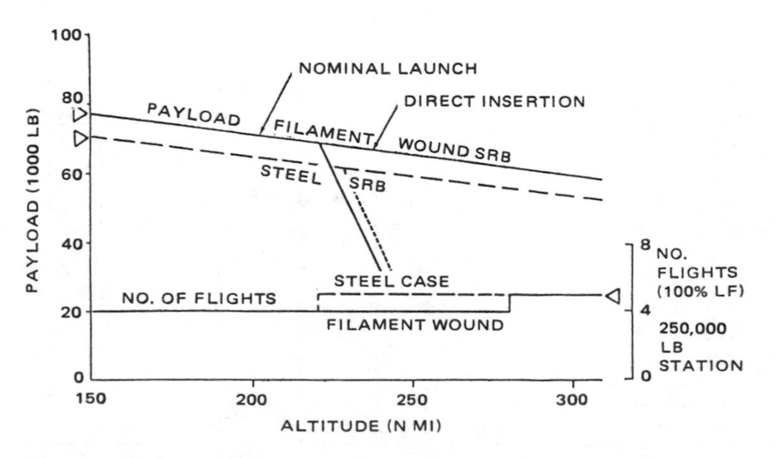

Fig. 16 Direct insertion capability (STS).

launch vehicle, increase inherent reliability, and facilitate replacement and growth are also discussed herein.

The space station must meet three objectives: 1) performance, 2) affordability, and 3) safety. Which configuration is adopted will strongly influence how these objectives are met; however, the main objectives appear to be at odds (for example, most performance increases involve higher cost). As shown in Fig. 15, some individual configuration features can serve more than one purpose. This matrix starts with the three objectives ranged horizontally (costs are broken into subheadings: development, procurement, launch, and operations) and with nine configuration characteristics ranged vertically. Symbols in the matrix field show the objectives that each configuration characteristic promotes. For example, the seventh characteristic, easy maintainability, improves performance by minimizing crew effort for maintenance, reduces operations cost for the same reason, and supports safety by encouraging a high level of readiness. The following paragraphs discuss these configuration characteristics and features.

Performance. To fully serve its many users, a space station will have to provide capabilities in three locations: 1) internal (pressurized laboratory and work modules), 2) external (attached to station, work, and storage zones), and 3) remote (near and more distant orbits). Station user requirements are summarized below.

Internal laboratory and work zones should provide shirt-sleeve working conditions for scientific experiments, instrument development, and (possibly) material processing at industrial levels. Some of the internal zones will have to allow for the attainment of very low acceleration levels (approximately 10^{-5} g), and these should be sited within 10 m of the station's center of gravity. All zones should provide power, cooling, and data handling; and some will have to be capable of being biologically isolated. Usable volume requirements approach 100 m^3 and could grow to two or three times this figure within ten years.

External, attached-to-station zones should provide facilities for handling, servicing, assembly, testing, storing, and pointing a wide variety of payloads. These include: large development articles to be assembled, calibrated, and tested in a true space environment; satellites and orbit maneuvering and transfer vehicles to be serviced, modified, and replenished; material processing facilities to be operated, modified, and replenished; and astronomical instruments to be provided with a steady observation vantage point. The external attached zone should be stable;

provide manipulators, handling aids, power, lighting, and data management; and be capable of storing and transferring fluids and gasses. Work in the external zone will be performed remotely by crew both inside the station and outside in suits. A command and control center, from which internal crew members can oversee the external activity, should be located to give a clear view of the work zone or berthing corridors. External attached work zones should have an initial length of approximately 20 m and a method of growing when mission activity and storage demands increase.

For some users, the space station will have to project its capabilities into other orbits. These remote capabilities may include: co-orbiting diagnostic satellites to aid the development of test articles on the base (e.g., large antennas); servicing of nearby scientific satellites (either remotely in their own orbits or by returning them to the station); deployment of payloads to GEO or beyond (satellites, or spacecraft that service satellites); and the support of co-orbiting industrial platforms (servicing and material change-out). All these extensions of the station's capabilities to other orbits will depend on the presence of mobile units that work out of the base: initially, the orbital maneuvering vehicle (OMV) with modest propulsion and, later, the orbital transfer vehicle (OTV) with greater long-range performance. Servicing, replenishment, and payload integration of these mobile units constitutes a significant part of the activity in the external, attached-to-station work zone described above.

Controllability. This factor will be a key driver of any base configuration. Selected control mode(s) will set the rules by which solar array and radiator articulation versus size trades are made. Controllability will also influence other rules that guide the siting and articulation of astral and Earth observation instruments. Furthermore, imposing an inertia pattern strategy will impact the overall configuration at all stages of operation and growth.

The choice of control modes will most likely lie between the "inertial hold" and the "local vertical." The inertial hold mode is usually restricted to a solar inertial mode with solar arrays aligned to the Sun. This will reduce the complexity of both the solar arrays and the thermal radiators, and provide some advantage for solar or stellar pointing instruments. An alignment of the minimum inertia axis that is perpendicular to the orbital plane will minimize the cyclic momentum inertia axis in the orbital plane. Although this may increase control requirements, it does provide a relatively efficient means for pointing station fixed elements (such as solar arrays) without a separate gimbal mechanism.

The local vertical mode can be gravity gradient stable, with the minimum inertia axis aligned to the local vertical and the maximum inertia axis perpendicular to the orbital plane; or neutrally stable with any other combination of principal axes aligned to the orbital plane. Misalignment between principal axes and orbital plane is an unstable equilibrium condition that generates angular momentum. When aerodynamic moments are considered, the stable position of the axes are skewed to a new torque equilibrium position, depending on the difference between center of mass and center of pressure. Four desirable configuration characteristics have modularity as a strong common feature.

Assemblage. On-orbit assembly of the initial space station will most likely require several Shuttle launches. The station is too heavy and too large for one Shuttle flight, therefore, modularity will be used to divide and package the station to fit inside the Shuttle. Another assemblage feature will be presence of mechanical handling aids, which will be derived from the Shuttle remote manipulating system (RMS) and handling and positioning aid (HPA). Brought up on early launches, these will be used to assist in space station assembly and as a permanent part of the station to assist berthing, payload handling, and base maintenance.

Growth Capability. Similar to initial assemblage, growth capability also covers an extended time scale. A space station must be able to enlarge its capability to respond to actual user experience. Configuration modularity is also the key to flexible, inexpensive growth items such as pressure vessels (more laboratory volume, larger crews), solar arrays and radiators (higher power), and truss structure area (larger external work and storage zones). Initial configurations should provide room for a reasonable amount of growth and a limited number of components (i.e., solar array and radiator articulation joints) should be designed from the outset for the heavier duties that growth will bring.

Commonality. A major feature of the SSP, commonality, is aimed at reducing development and procurement costs. At the overall configuration level, it is clearly visible as 1) modularity of the pressure vessel sizes (the basic structure and, perhaps, the interior stuffing), 2) modularity of solar array and radiator elements, and 3) modularity of connecting truss structure.

Commonality between the space station and the tended platforms is a further area for potential cost containment. Functions common to the two include power supply and storage, thermal control, altitude control and altitude reboost, berthing provisions, data handling, communications, and, perhaps, interconnecting structure. Because the initial space station requires about 75 kW and the scientific platform at high inclination requires about 15 kW, power supply, power storage, and thermal control modularity must be modularized in the 3- to 15-kW range to encourage commonality.

Redundancy. One of the bedrocks of safety, redundancy, is apparent in the use of modules (i.e., multiple pressure vessels, multiple solar arrays, multiple radiator panels). However, rigorous replication is not the only way to provide redundancy. The backup airlock, for instance, could be any full-size pressure module with the interconnecting doors closed and 99% of its atmosphere pumped back into storage before the outer door is opened.

Maintainability. Also similar to assemblage, maintainability requires a configuration that offers easy, mechanical access to all functional areas of the space station. Maintainability, however, has a further requirement in that any element of the station should be replaceable, (preferably removable and replaceable) without lengthy interruption to base operations. Apart from mechan-

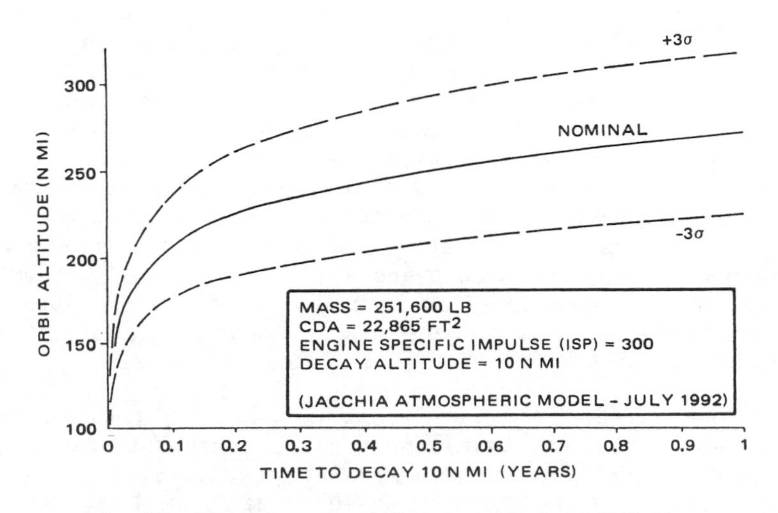

Fig. 17 Atmosphere uncertainty effect on lifetime.

ical aid access, the chief impact on the configuration is in individual mounting of the pressure modules to the base structural skeleton with flexible tunnels between them. This allows for removal and replacement of any pressure module without disassembly of the whole base.

Separate/External Work/Storage Zones and Simple Circulation Patterns. Segregating work zones from storage zones and maintaining distinctly different paths for berthing, payload movement about the base, and EVA crew circulation will be necessary steps toward efficient performance and enhanced safety. This need for segregation will most likely be met by making the configuration big enough; usually by distancing the three active elements (pressure vessels, power and thermal control, and external activities zones) away from each other with an interconnecting structure (which is the fourth element). Spreading out the configuration is strongly constrained by the inertia strategy implicit in the selected control mode.

Contamination Minimization. This characteristic will benefit if the configuration is shaped to achieve three objectives:

1) The main contamination sources (the Shuttle moving along its berthing corridor and any fluid transfer servicing zones) should be positioned as far as possible from contamination sensitive observation instruments and solar arrays.

2) Observation instruments should "ignore" the contamination (where possible).

3) Any combination trapped in the base wake and moving away downstream should not wash through the remainder of the base (this is a combined flight mode/configuration issue).

Altitude

The altitude, or range of altitudes, at which a space station orbits influences the configuration only because possible aerotorques, which impact controllabilities, are more significant at higher densities. However, the choice of altitude has system-wide implications and is governed by 1) STS launch performance, 2) drag effects, 3) radiation effects, and 4) orbital debris considerations.

STS Launch Performance. Figure 16 applies the STS projected launch capabilities for normal and direct insertion modes to a generic space station weight. Assuming that the station is modular by design, the number of launches required to deliver these elements to a given orbit is shown. With direct insertion capability, altitudes from 200 to 310 n.mi. can be attained without a large payload penalty. Also, with this projected STS capability, the complex station could be launched in four or five missions (assuming that it could be appropriately modularized). The results for a filament wound solid rocket booster (SRB) configuration (planned to be operational around 1985) compare within one launch with the steel SRB (not currently programmed).

Drag Effects. Atmospheric drag on orbit lifetime is another factor for altitude selection. Figure 17 shows the time to decay from initial orbital altitude to 10 n.mi. below, including the nominal atmosphere and the associated uncertainty. The advantages of altitudes greater than 250 n.mi. (particularly for the $\pm 3\ \sigma$ (Jacchia atmospheric model for July 1992) are also shown. Altitudes higher than 250 n.mi. would avoid frequent reboosting of the station. For the same atmospheric conditions, Fig. 18 shows the associated yearly propellant required to maintain the orbit of the same generic space station configuration ($W/C_dA = 10$) with a continuously fired, 300-second specific impulse (Isp) propulsion system. Again, altitudes of 250 n.mi. or higher are advantageous.

Radiation Effects. Radiation levels at various atmospheric altitudes and the associated required shielding should also be considered to optimize the altitude of the space station orbit. Figure 19 shows allowable mission stay-time as a function of altitude and shielding levels for a 30-deg inclination orbit and a quarterly eye dose limit of 52 REM. Ninety-day crew rotation periods and shielding effectiveness of 2 gm/cm^2 (or 7-mm-thick aluminum) will limit the maximum altitude to about 260 n.mi. Longer mission durations or higher altitudes will require thicker shielding. However, resultant weight increases become less important when it is considered that Shuttle launches of space station modules are usually length or volume limited, and not weight limited.

Orbital Debris Considerations. The probability of impact with orbital debris varies with orbital altitude (see Fig. 20). As shown, the probability of impact in ten years by different sized objects (orbital debris only) is shown as a function of altitude. Based on the probability of impact (by particles of 0.1 in. diameter), it is desirable to avoid altitudes above 300 n.mi.

Summary

The STS capability via direct insertion and considerations to minimize reboost propellant weight suggest an optimal altitude range of 250 to 300 n.mi. radiation and debris effects tend to favor an altitude closer to 250 n.mi. As a compromise, it appears that 260 to 270 n.mi. would be a good preliminary altitude. However, two more sophisticated approaches use variable altitude. One approach involves letting the station altitude decay between Shuttle visits, with the Shuttle rendezvous occurring at the lowest point of the cycle, followed by station altitude reboost when the Shuttle leaves. The other approach involves changing the mean base altitude as the upper air density varies over the 11-year solar cycle (i.e., flying higher at times of high solar activity, lower at solar quiesence). At the cost of some increase in operational complexity, both approaches could result in a safer average environment without undue launch and resupply cost increases.

Core Systems

Space station core systems are defined as those functional elements (possibly in separate modules) that are pressurized to provide an Earth-like environment to support in-orbit shirt-sleeve operations. For the initial operation of the space station, these functional elements will consist of habitation, laboratory, logistic, and interconnecting modules.

All elements of the space station will be transported to orbit via the STS; therefore, STS capacity imposes physical constraints on the core systems. Commonality of ele-

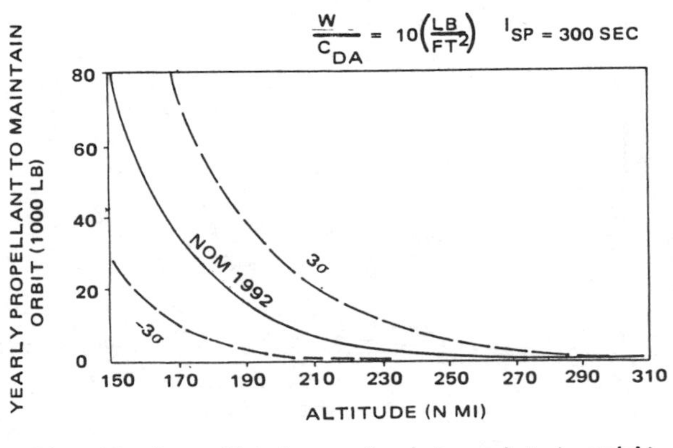

Fig. 18 Propellant required to maintain orbit.

ments can result in significant cost avoidances, which makes STS capacity and core system commonality strong space station design drivers.

The Space Shuttle Orbiter provides structural support attachment points along the length of the cargo bay (see Fig. 21). The payload envelope is constrained to a maximum length of 60 ft between stations 582 and 1302 due to fore and aft bulkheads. Attachment points are located on the main longerons and along the bottom centerline of the bay on bridge fittings between the fuselage frames. To fall

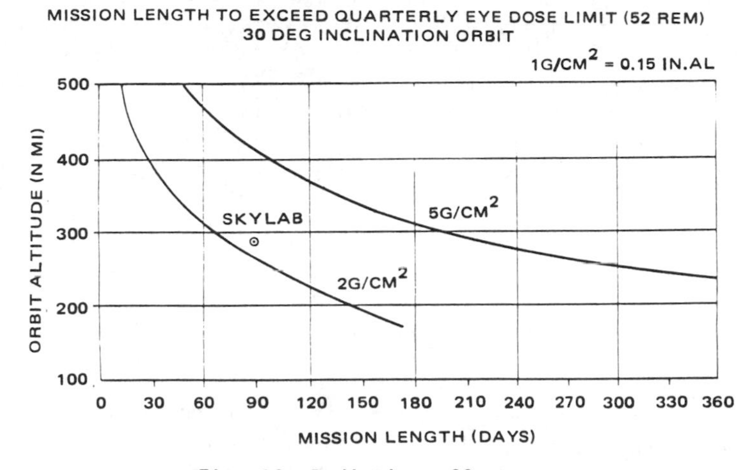

Fig. 19 Radiation effects.

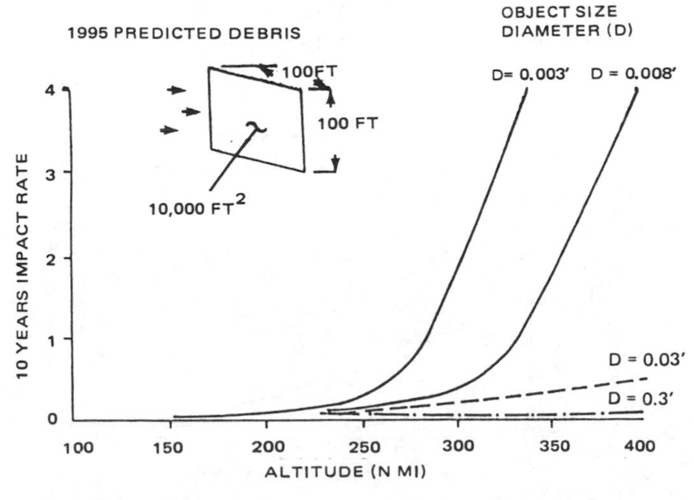

Fig. 20 Collision model (orbital debris).

within the dynamics and thermal envelope, core systems must have a diameter of less than 180 in. To incorporate wrap-around radiators, deployable meteoroid shields, and (in some potential configurations) interconnecting radial berthing rings, the maximum diameter of the pressure shell would be limited to approximately 160 in. For a typical module using a three-point mounting system (consisting of two mounting points on the longerons and one at the keel), the present STS limits the module (cargo) weight between the maximum design launch weight (65,000 lb/29,484 kg) and the maximum landing weight (32,000 lb/14,515 kg) as a function of the center of gravity (c.g.) along the STS X, Y, and Z axis (see Fig. 22). Other alternatives can be considered as the space station evolves, including use of the STS external tank, an aft cargo carrier at the end of the external tank, and modifications to the STS Orbiter.

Commonality

Recent studies have shown that commonality within the space station can significantly reduce design and development costs. Specifically, trade studies indicate that maximizing commonality can yield up to a 20% cost savings.

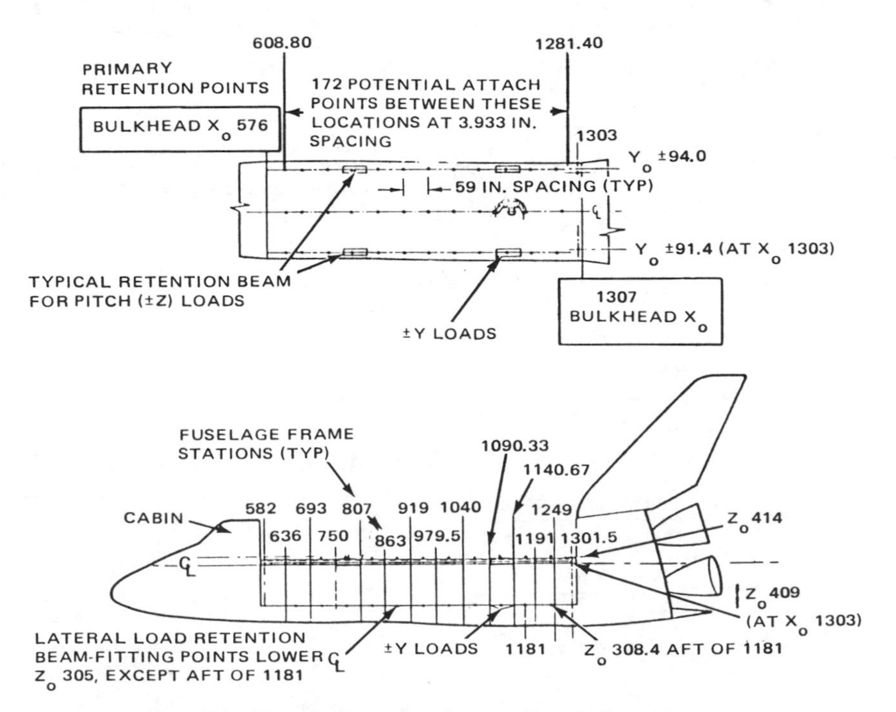

Fig. 21 Standard payload attachment locations.

Commonality also simplifies integration, maintenance, and spares requirements; enhances system evolution; and increases the likelihood of hardware availability throughout the life of the space station.

Commonality goals are applicable to elements, modules, subsystems, and software, (see Fig. 23). Several potential areas of commonality include applicability of space station modules/subsystems and software to space platform, space station design for operation at multiple orbits, and the use of common building blocks to make up space station modules. Building blocks that exhibit the greatest potential for commonality are the cylindrical structure sections (even though there may be some unique penetrations, such as berthing and viewports), end domes, floor and racks, thermal pumps and body-mounted radiators, and lighting fixtures. Additional degrees of commonality can be considered utilizing environmental control/life support system (ECLSS) equipment, controls and displays, and subsystem processors.

The potential for a standardized space station berthing port interface is discussed in the berthing/docking section of this chapter. Figure 24 illustrates a possible architecture for the space station infrastructure and the areas of commonality. Four orbiting facilities are shown: a manned space station and an unmanned platform in 28-deg inclination, an unmanned platform in 57-deg inclination, and another in Sun synchronous (polar) orbit. Sizes and

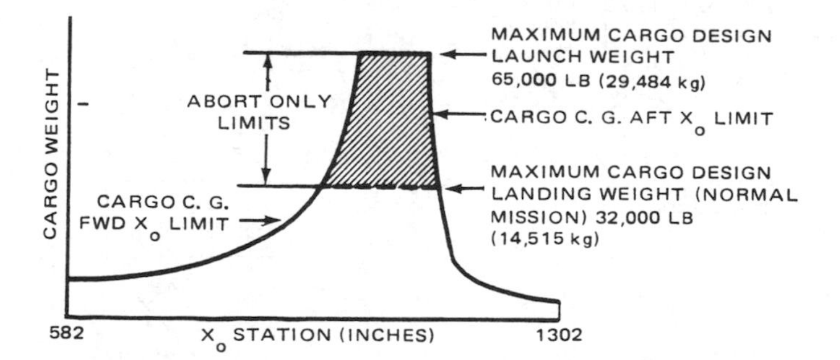

EQUATIONS FOR CALCULATING CARGO X_o (STATION) C. G. LIMITS

$$\text{FWD LIMIT} = \frac{1076.7\,W_C - 3.70 \times 10^6}{W_C}$$

$$\text{AFT LIMIT} = \frac{1108.95\,W_C + 3.4 \times 10^5}{W_C}$$

WHERE W_C = CARGO WEIGHT (LB)

Fig. 22 Allowable cargo c.g. limits (along X axis).

capabilities of the manned space station and the Sun synchronous platform are increased in-orbit by the addition of modules. The types and quantities of the modular elements required for the initial buildup and expansion of the facilities are shown in Fig. 25, which describes a candidate set of space station standard modular elements. The entire architecture could be built using multiples of these elements.

In a typical configuration, the initial space station would consist of a minimum of four pressurized, habitable modules that would house living quarters for the crew, command and control functions, resupply and stores for food and atmosphere, as well as general purpose and specialized laboratories.

Figure 26 illustrates the levels of commonality that must be addressed in a typical design architecture. The commonality emphasis is to subdivide the system functional architecture into modular elements that can be used for multiple purposes. The search for commonality extends within these elements to internal subsystems and to the component level within the subsystems. Figure 27 is a preliminary survey of potential common hardware elements for the structural mechanical subsystem, which shows how each

PRESSURIZED MODULES

- STRUCTURE*
  - PRESSURIZED SHELL
  - FLOOR & RACKS
  - BERTHING MECHANISMS
- THERMAL*
  - PUMPS & RADIATORS
  - INSULATION
- POWER CONDITIONING & DISTRIBUTION*
  - WIRING HARNESS
  - LIGHTING
- ECLSS
  - AIR CIRCULATION
  - HUMIDITY/TEMPERATURE CONTROL
  - CO_2 COLLECTION
  - SAFE HAVEN EQUIPMENT
- C&DH
  - CONTROLS & DISPLAYS
  - PROCESSORS

RESOURCE MODULE & SPACE PLATFORM

- STRUCTURE*
  - TRUSS STRUCTURE
  - MECHANISMS
- THERMAL*
  - RADIATOR (MODULES)
  - PUMPS & LINES
- ELECTRICAL POWER*
  - FUEL CELLS/BATTERIES
  - POWER DIST & COND
- C&DH*
  - S-BAND
  - KuBAND
  - GPS
  - PROCESSORS

NOTE: PRIMARY POTENTIALS - OTHERS HAVE POTENTIAL FOR 'COMMON MODULE' OR 'COMMON SOURCE'

Fig. 23 Potential common hardware.

component can potentially satisfy the needs in several discrete space station elements. The objective of this process is to reduce the total number of elements to be developed and minimize program nonrecurring costs. To achieve this objective, a compromise in performance, weight, and/or other characteristics might be necessary. Slight compromises in these areas are valid tradeoffs, however, since preliminary analysis shows that the cost of the first article can range from two to ten times the cost of a second identical article, and even more in some cases. (The cost of the first article includes development costs.)

The application of commonality at higher levels is represented by the common module concept in which a standard pressure shell, outfitted with rudimentary subsystems, is used as a building block for various space station functions. This is similar to commercial aircraft production in which a standard fuselage is adapted to specific interior needs of various customers.

Figure 28 illustrates one potential solution to a common module approach that is based on a summation of all of space station pressurized module functional requirements. The cylinder is arranged with a floor and ceiling structure parallel to the longitudinal centerline. This arrangement, plus the removable end cones, allows convenient access for ground buildup and checkout. Subsystems are common to all

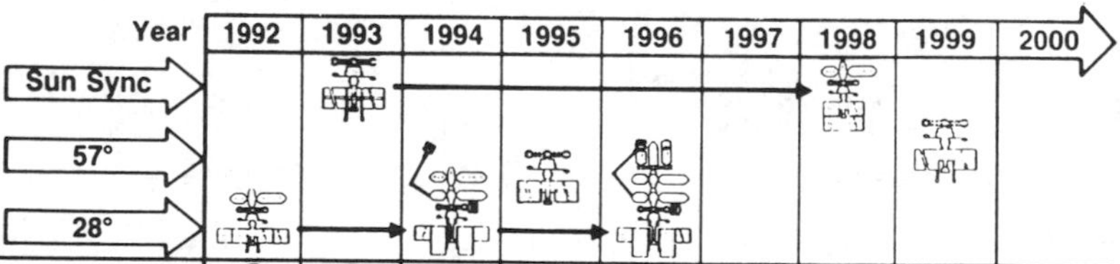

Sun Sync
57°
28°

Year	1992	1993	1994	1995	1996	1997	1998	1999	2000
Modular Element / Facility	(1)	(2)	(1A)	(3)	(1B)		(2B)	(4)	Total Modules
1. Orbiter Docking Module	1	1							2
2. EMU/MMU	2		2				2		6
3. Utility Module	1	1	*	1			*	1	4
4. Propulsion Module	2	2		1				1	5
5. Payload Support Module	1	1		1				1	4
6. Docking Module	1		1				1		3
7. Habitability Module	1		1				1		3
8. Logistics Module	2						2		4
9. Laboratory Module	1		1				1		3
10. ROTV					1				1
11. LH_2 Storage Module					1				1
12. LO_2 Storage Module					1				1
13. Cryo Transfer Module					2				2
14. TMS Prop Module			1				1		2
15. TMS			1						1
16. RMS			1						1
New Elements Total	8	0	3	0	4	0	0	0	

*Electrical Power Subsystem Expansion

Fig. 24 Modular buildup in space station architecture.

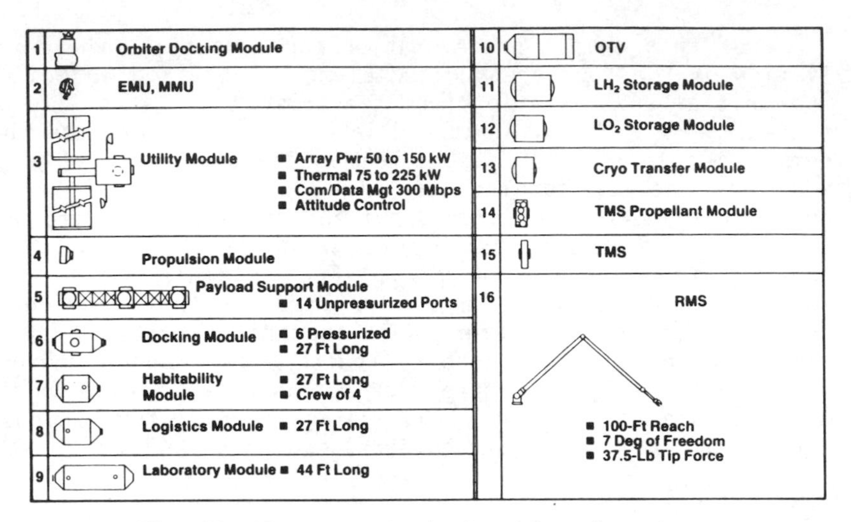

Fig. 25 SS system standard modular elements.

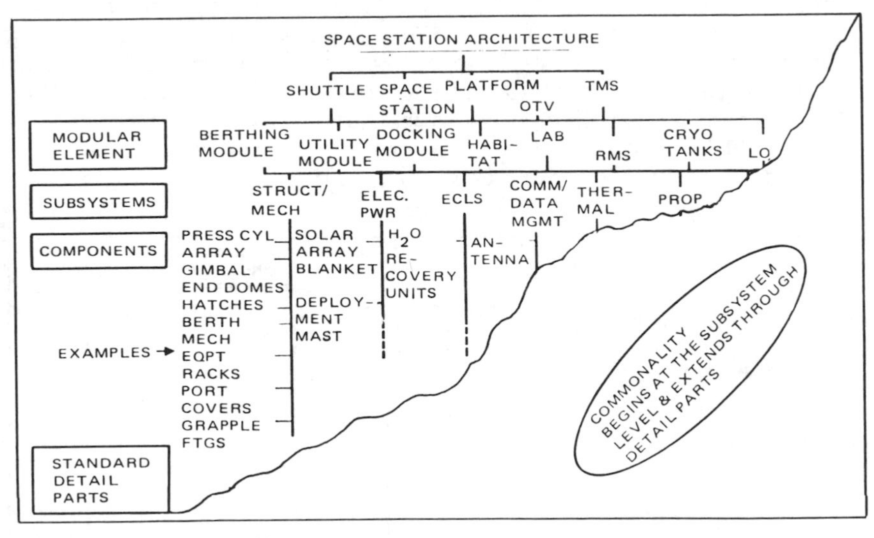

Fig. 26 Levels of commonality.

modules and are installed behind removable floor and ceiling panels to provide a clear interior for the installation of specialized functional equipment. Since the end cones are removable, full diameter sections can be added to the standard cylindrical segment to insert specialized interiors (i.e., without standard floors and ceiling in that section). Shuttle transport support trunions are attached to the removable end cones to provide module end support when full diameter structural sections are inserted to lengthen the module.

For this sample space station configuration, the derivation of the four pressurized elements required to build the initial space station from the common module is illustrated in Fig. 29. Habitat and laboratory elements use the common module without structural modification and simply add the interior furnishings. The logistics module requires the addition of a pressurized tunnel attached to one end cone, and external structure to support tankage for water and atmosphere resupply. The multiple berthing adapter (MBA) requires the insertion of a full diameter section that contains four radial berthing ports, an airlock, and an external support structure for manned maneuvering units (MMUs). These elements can be built using a multipurpose common module and adding interior furnishings, structural extensions, and external equipment and supports.

Habitation Module

Space station mission analyses indicate that a pressurized habitation module will initially be required to support a crew of six to eight. As the demands of the station and crew size increase, replicated habitation modules can be used to house the additional crew. As a minimum, the habitation module should provide the following facilities.

1) Living quarters: Private or semiprivate quarters should be provided for the crew. These quarters should

ELEMENT	RESOURCES MODULE	MBA	HABITATION MODULE	LABORATORY MODULE	LOGISTICS MODULE	PLATFORM
STRUCTURAL/MECHANICAL						
• PRESSURE MODULE CYL STRUCTURE		X	X	X	X	
• ARRAY GIMBAL	X					X
• END DOMES		X	X	X	X	
• HATCHES		X	X	X	X	
• DOCKING/BERTHING MECHANISM	X	X	X	X	X	
• EQUIPMENT RACKS		X	X	X	X	
• PORT COVERS	X	X	X	X	X	
• GRAPPLE FITTINGS	X	X	X	X	X	X
• INTEGRAL RADIATOR/ METEOROID SHIELD		X	X	X	X	
• PAYLOAD PALLET	X			X		X
• UMBILICALS	X	X	X	X	X	X
• DEPLOYABLE RADIATOR	X					X

Fig. 27 Space station standard hardware (preliminary).

provide isolation from other crew members, low noise levels, good ventilation, and proper lighting. An individual sleep restraint will also be required for each person. Other amenities include an audio/video entertainment center, bulletin board, reading/writing provisions, as well as stowage for personal items and clothing. Quarters should be large enough to enable the occupants to dress and undress with a reasonable amount of room for movement.

2) Wardroom: The wardroom/dining area should accommodate all module occupants in one sitting; the area will also serve as a lounge/viewing and recreation area. It should contain audio/video entertainment equipment, game kits, exercise equipment, and internal communication equipment.

3) Galley: The galley should provide equipment to stow and prepare food and cleaning materials. Ambient, perishable, and frozen foods, as well as condiments and snacks, will require tailored equipment for stowage. An oven and water dispenser should be provided for food preparation. Trash bags or a compactor will provide for waste collection.

4) Personal hygiene compartment: A separate compartment will provide for body waste collection/disposal, personal cleanliness, and bathing. Body wastes will be handled with devices for urine and fecal collection and processing, anal cleansing, vomitus collection, as well as processing and specimen collection. Personal care and grooming facilities for full body cleaning, body drying, hair cutting, shaving, and nail care should also be provided.

5) Exercise area: Exercise equipment should be made available to enable the crew to maintain body tone as well as for recreation. The wardroom area can be utilized for this purpose.

6) Storage area: Storage compartments for all new support items can be located near the areas of functional use of the equipment.

7) Housekeeping: Hardware articles should be provided to keep the space station clean.

8) Mobility & restraint: Positive restraint systems should be provided for the crew at all work stations. Handholds and rails can also be included to enhance movement throughout the space station.

9) Subsystems: Life support and space station support systems can be housed in pressurized modules located in floor or ceiling compartments with easy access for maintenance.

In addition to adequate volume, pleasant architectural interior arrangements and colors are desirable, as well as

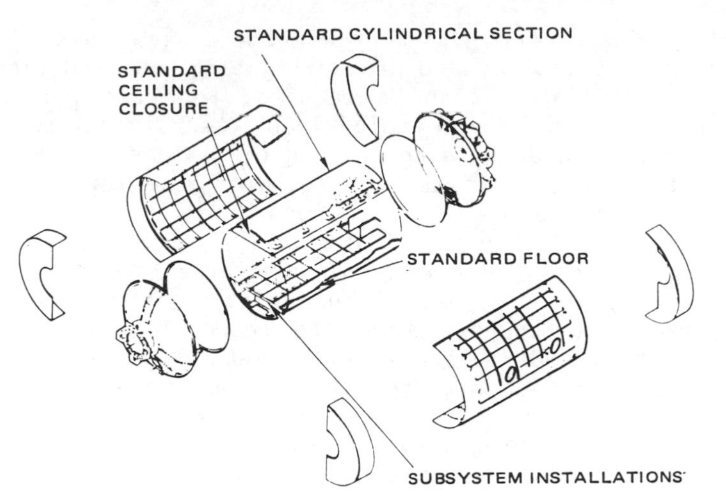

Fig. 28 Space station common module concept.

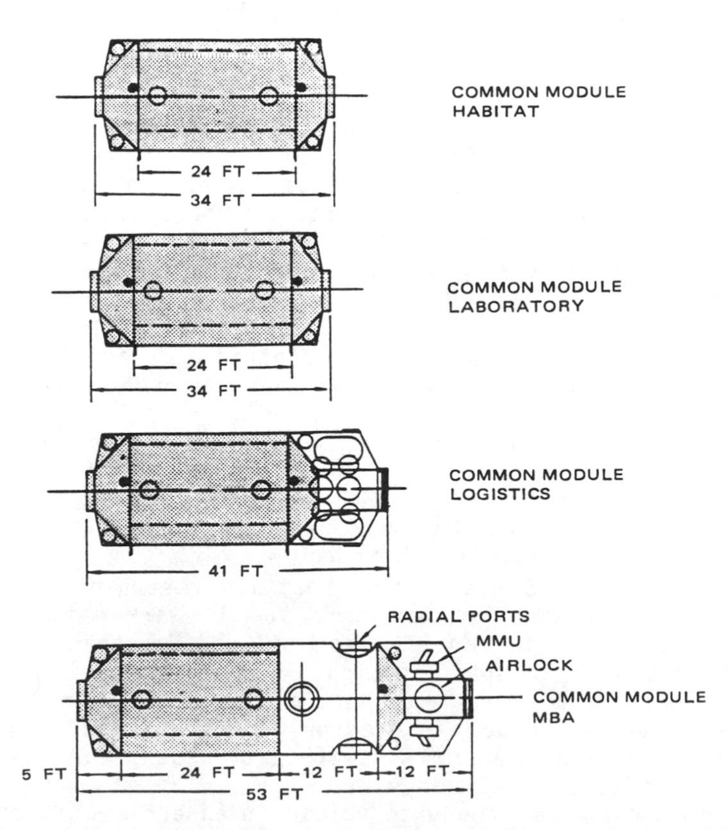

Fig. 29 Basic space station elements derived from a common module.

windows to view the Earth and sky. Accommodations must take the zero-gravity posture of the individuals into consideration. Adequate lighting and ventilation will be required in the living quarters and throughout the station to provide comfort and efficient working accommodations.

Safe operations within the space station will be enhanced by the elimination of sharp corners and edges. This requirement is particularly important to prevent damage to space suits during emergency operations.

Since pressurized volume in the space station is a premium commodity, it must be used efficiently. Crew quarters should be isolated from noisy items including subsystems machinery and equipment, as well as items associated with the hygiene equipment and crew health maintenance equipment. The preferred location for these facilities would be near the crew quarters, but equipment (hygiene pumps, fans, etc.) needed for showers, waste management, and health maintenance (i.e., treadmill and ergometer) can be unacceptably noisy. The unacceptable placement of

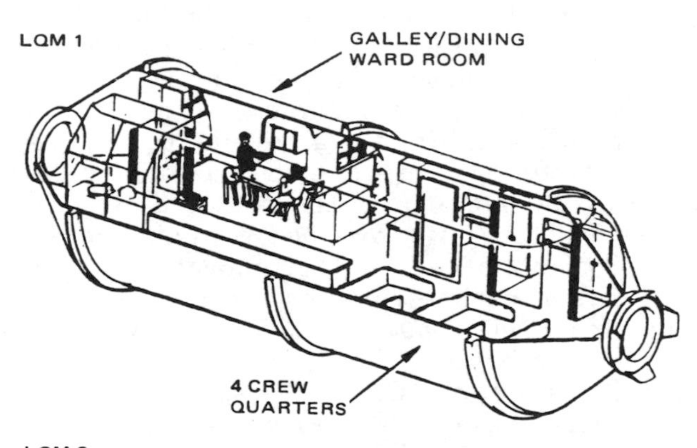

Fig. 30 Living quarters concepts.

galley/dining area/wardroom near hygiene and medical facilities and noisy equipment makes it difficult to achieve efficient utilization of space in a living quarters module. Figure 30 illustrates two concepts for living quarters modules that compromise these requirements by separating the crew quarters with a bulkhead, and placing the galley/dining facility in a separate, adjacent module away from the health maintenance and hygiene facility.

Micro-g Research Module

In addition to having multidiscipline laboratories, dedicated laboratories for micro-g research on materials and life sciences are being planned. The space station will represent man's permanent link to the daily operational use of near space, and the micro-g research lab module will represent industry's link to the development of space resources. Ground-based industries produce the goods by taking raw materials and transforming them into new and unique products. The manufacturing processes usually involve separation, purification, heating, cooling, high pressure, low pressure, and other techniques that will shape or modify the characteristics of a particular material. High gravity is generally used in such areas as separation processes and in the casting of pipe. Low gravity, however, is not available as a manufacturing resource except for short periods. It has been used (e.g., the production of shot) by letting molten metal pellets cool during free fall. Future use of long-term zero-gravity to produce new products offers unlimited potential for products that can not be manufactured on the ground.

The most important requirement for low-gravity manufacturing is microgravity. Secondary requirements include vacuum, power, command and data management, accessibility, mounting fixtures, stowage, and manned access. Placing the lab in orbit may insure a low-gravity environment, but it may not provide a micro-g environment. The acceleration felt at any point for a circular orbit satellite is approximated by the formula:

$$a = \sqrt{5}\,(\omega^2/9.8)d$$

where d is the distance in meters from the satellite center of gravity and ω is the orbital rate in radians per second. This results in a fundamental requirement that a micro-g research lab should be within 10 m or less of the composite c.g. of the entire space station operating in a 200- to 400-n.mi. altitude to achieve a microgravity level of less than 10^{-5} g.

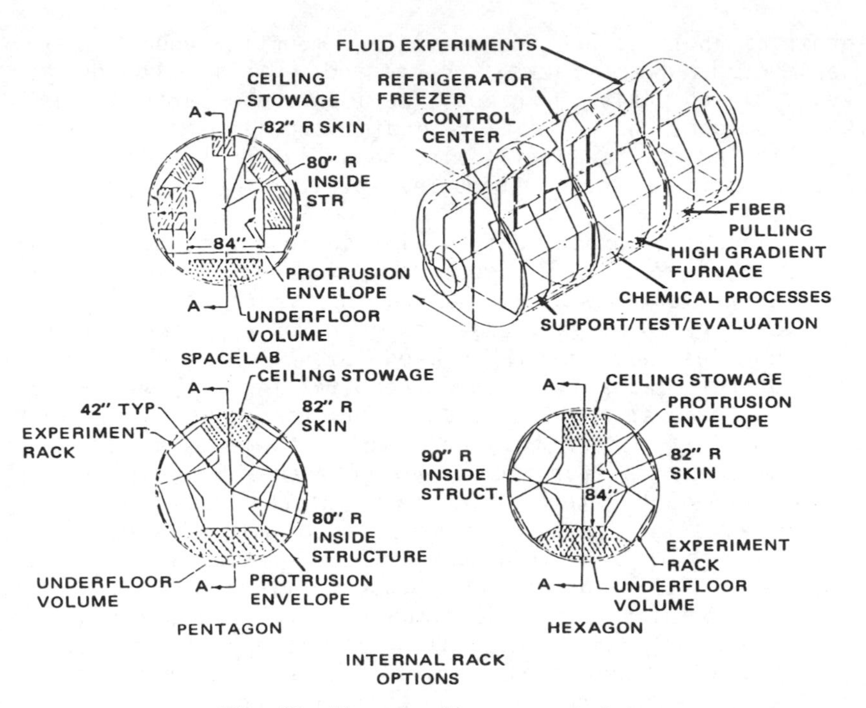

Fig. 31 Microgravity research lab.

Other design requirements for a micro-g research laboratory are derived from a combination of present and projected experiment requirements, reasonable expectations of engineering design capability, the constraints of the launch system, and the necessity of operating in space. Several studies of the resource requirements of individual experiments indicate that most have power requirements in the range of 1 to 5 kW, with some reaching the 10- to 15-kW range.

The time required to perform each experiment can range from one hour to several weeks. A composite analysis of these various requirements indicates that a power level of 16 kW will allow processing of a reasonable set of simultaneous experiments. Peak power should be several kilowatts above this average power level, and the heat rejection capability should be equal to the available power. Biological experiments and the environmental control system will require coolant at a temperature of 40°F.

The configuration of the micro-g research lab module is driven by several considerations. A manned environment will require a pressurized lab, which requires a cylindrical shape and is limited to a diameter of 160 in. by the orbiter payload bay. Length should therefore be on the order of 35 ft to provide sufficient mounting room for experiment hardware racks. A command station should be available to monitor various experiments and perform housekeeping functions. A data recording and transmission system (including video) will also be necessary to record data and inform a ground-based investigator of experiment results.

Since many experiments will require a vacuum, a large-diameter (3 to 5 in.) vacuum line should be provided for some experiment stations. Manned access to the equipment is necessary to perform research, control experiments, analyze samples, and install new equipment. However, operation of this lab as a permanent orbital facility will modify its internal configuration. This is because less floor space will be needed than that which is required for a ground-based configuration. Design trades must therefore be performed to determine optimum internal configuration and rack design. Also, access ports or airlocks must be large enough to pass full racks of equipment, and stowage must be available for unused equipment and samples.

A preliminary configuration for a micro-g research laboratory is shown in Fig. 31. Besides experiment racks, the lab will contain a control center, support/test/evaluation racks, and a refrigerator/freezer to store biological samples. Several possible configurations for internal rack placement are also shown.

Life Sciences Research Module

A life sciences research lab would be used for gravitational biology, radiation biology, exobiology, and human physiology investigations. These disciplines impose various requirements and constraints on research laboratory design and, ultimately, on the overall space station. The equipment required would range from passive measuring devices to centrifuges of various sizes and holding facilities for the test specimens (plants, lower vertebrates, smaller mammalians, and primates, respectively). Equipment such as rotating chairs, sleds, posture platforms, and (eventually) a human centrifuge would be required for human physiological research, and would have a significant impact on design.

Projected mission requirements and STS transportation constraints indicate that module size for a life sciences research laboratory should be approximately 90 to 100 m^3 (3200 to 3500 ft^3). Increasing payload requirements will be met by adding new laboratory modules to the space station.

The increasing use of test animals and longer mission durations is likely to result in a requirement for sufficient space to properly house the animals. The implementation of a separate animal holding module outside of the station's research and habitation areas will probably be needed by the time primates are used, if not sooner. A separate animal holding module docked to the research laboratory could simplify the exchange of an animal population and necessary cleaning. The module could be brought back to Earth while the research laboratory module remains on orbit.

The micro-g environment required for life sciences work is $\sim 10^{-4}$ g and, therefore, is compatible with crew laboratory work. If the station is to be of an operations character with frequent docking and on-orbit assembly, some interference with micro-g experiments could occur. Detailed mission planning, the use of dedicated research modules or platforms, and optimized location of research laboratory modules on the space station is required; the potential of soft docking (berthing) should also be considered.

To determine the influence of frequent shifts between microgravity and the 1-g condition, implementation of a human centrifuge (15 to 20 m diameter) is an explicit ambition of the scientific community. These studies could lead to a determination of whether artificial gravity is necessary for long missions. Whether such a centrifuge could be implemented on a space station in the 1990's requires further investigation. Associated problems would be disturbances in the micro-g environment and inadequate space. The presently proposed centrifuge is a one-man cabin on a rotating arm outside the pressurized modules. An alternative concept could be a tethered centrifuge.

Since an onboard centrifuge (~1 g) is required for plants and animal testing, the location and operation of control centrifuges must be optimized to minimize the influence on the micro-g environment. Furthermore, the centrifuge design (e.g., torodial) must allow the crew to pass through it to reach any one of the two safety exits required on each module. Other required equipment that can influence the micro-g environment, when used, include rotating chairs, sleds, and posture platforms.

The internal configuration of the research laboratory could either be horizontal, as in the spacelab; or vertical, as was used in Skylab. The vertical arrangement might give a better packing ratio, but integration problems and ease of crew training indicate that the horizontal arrangement would be more beneficial.

The research laboratory's environmental control/life support system (ECLSS) must be designed to operate as an integral part of the space station's ECLSS. The use of biological specimens (especially animals) will require adequate protection of the space station crew. This may be best achieved by separate air revitalization and waste management systems for the vivaria, which would isolate the animals from each other as well as from the crew. This would help to guarantee the health of the animals and undistorted test conditions.

The use of animals in the life sciences research laboratory will also require an additional water supply (~100 g/day/kg specimen weight), extra food (~30 g/day/kg specimen weight), and more oxygen (~1.04 g/g food). Storage space and tankage has to be provided either in the research laboratory/animal holding module, or in the general logistics module for the required resupply periods. Waste generated by animals should be collected and stored separately from human waste. The same storage and disposal methods will be used for all waste, which is of particular importance if water recovery is used in the overall space station ECLSS concept. Animal condensate could be recovered separately, purified by the same method as the human waste water, and used for specimen drinking water.

Available crew time and the various skills required to operate a life sciences research laboratory are critical parameters for space station efficiency. The potential for automation is limited, except for animal vivarium functions such as feeding, watering, and cleaning. A dedicated data management system in the life science research laboratory will allow many time consuming calibrations to be automated and achieve simultaneous data collection to reduce valuable crew time.

The life sciences research laboratory will initially be equipped to handle all subdisciplines in one module (see Fig. 32). Expected space station growth and an increase in the number of crew members will eventually require a dedicated medical clinic and health care module. Such a module could handle the cost of continued human physiology research activities, which will separate human and nonhuman activities. The equipment needed for human physiological research and space medicine is very closely interrelated. In fact, most monitoring and measuring apparatus (ultrasound measuring devices, EEG, ECG, blood pressure, biochemical laboratory) is identical, which allows routine use of sophisticated medical equipment in physiological research.

Logistics Module

A logistic module will resupply the space station with all required consumable items. Nominally scheduled to be transported to the space station every 90 days, a new logistics module will be exchanged with the depleted one. Two types of items will be carried aboard the logistics module: short-term consumables and long-term consumables. Short-term consumables include food, water, clothing, hygiene items, chemicals for removing carbon dioxide from the air (e.g., LiOH), and atmospheric gasses (oxygen and nitrogen). Short-term consumables will be sufficient to ensure a 90-day supply for all crew members. Long-term consumables include space station spares, payload spares, and items needed to resupply payloads (raw materials for materials-processing payloads, etc.). Quantity and size of long-term consumables is predicated upon the space station spares philosophy (i.e., maintainability/reliability), payload resupply schedule, and available room aboard the logistics module.

An additional potential function of the logistics module will be to contain crew hygiene facilities including the shower, waste management station, and hygiene console. The logistics module will also be the respository for space station trash.

Assuming an eight-person crew and a 90-day resupply interval, a total volume of 1022 ft^3 is required for the pressurized portion of the logistics module (see Fig. 33). Consumable weight and associated tankage (tanks are stored external to the pressurized portion of the logistic module) requirements are shown in Fig. 34. Consumables include water, oxygen, and nitrogen. Propellant resupply for orbit maintenance and attitude control will depend on the aerodynamic, inertial and orbit maintenance strategy of the selected space station configuration.

All pressurized space station modules should be grouped so that they can be interconnected and passages between all modules should provide for suited and shirt-sleeve conditions. A desirable safety feature would be to have at least two exits from each module.

Two berthing rings on interconnected modules should be available to dock the orbiter to the space station. The orbiter payload bay must be able to clear the space station so that delivered payloads can be transferred to the space station base. Crew rotation and equipment transfer is through these berthing rings. In one concept, a generic type module is used to provide interconnection. The structural shell with berthing rings, radiators, meteorite projection, and secondary structure would be identical for each module. Modules would be outfitted to meet the requirements of a habitation, laboratory, and logistics module, and interconnected in a raft-type configuration (see Fig. 35).

The provision of six adapters in each multiple berthing adapter module is another concept being considered. In this concept, pressurized modules are berthed to the dock-

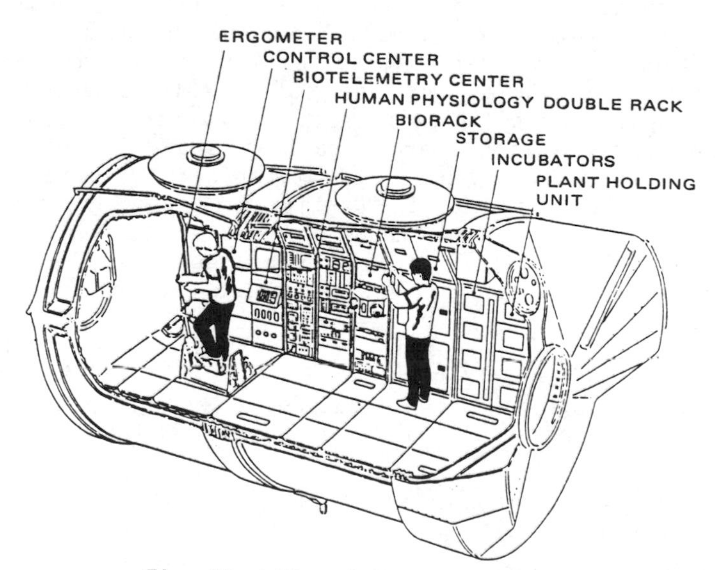

Fig. 32 Life sciences research lab.

ITEM	VOL (FT^3)	WT (LB)
• HYGIENE FACILITY	200	400
• LiOH	170	3780
• FOOD	125	2380
• CLOTHING	150	1800
• PERSONAL EQUIPMENT, STATION HOUSEKEEPING, MISC	257	4100
• SPACE STATION SPARES	60	1800
• PAYLOAD RESUPPLY	45	2650
• PAYLOAD SERVICING	15	375
TOTALS	1022	17,285

Fig. 33 Pressurized volume requirements.

CONSUMABLE	TANKAGE	QTY	WT (LB)
• OXYGEN	35" DIA CRYOGENIC	3	2610
• NITROGEN	35" DIA CRYOGENIC	3	2525
• WATER	68" DIA	2	16,314
	TOTALS	8	21,449

Fig. 34 Unpressurized volume requirements.

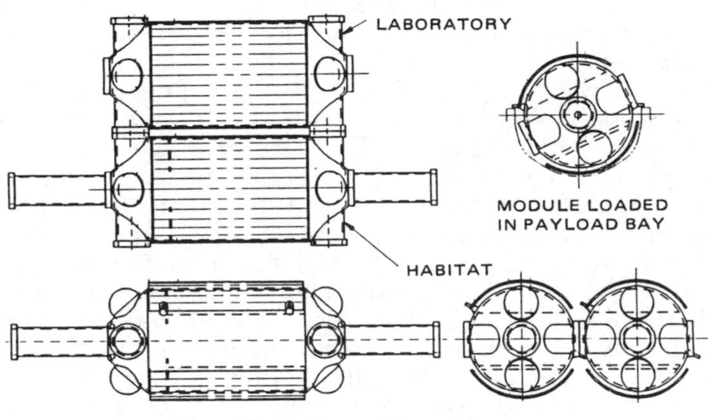

Fig. 35 Raft configuration.

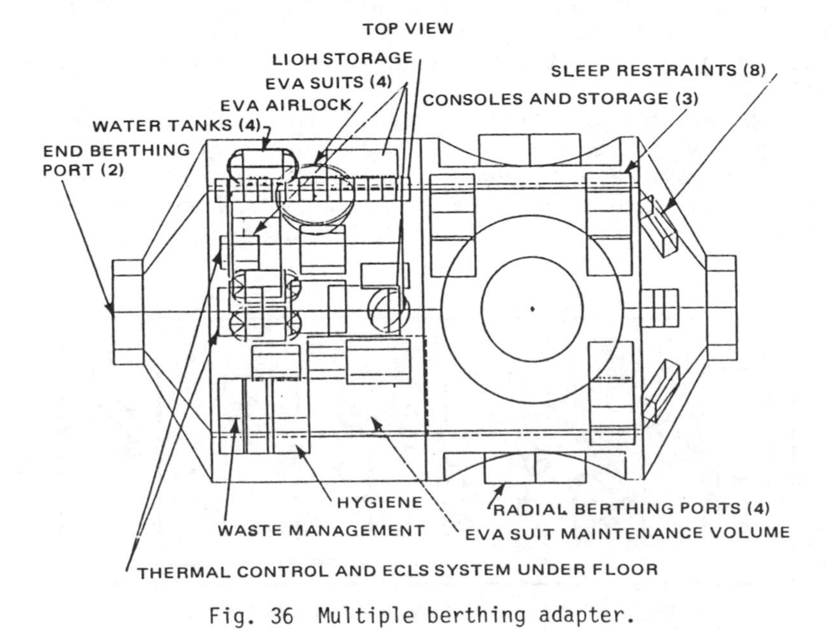

Fig. 36 Multiple berthing adapter.

ing adapter, which provides interconnection between modules (see Fig. 36). Except for radial berthing adapter penetrations, the primary structure could be common to all the other pressurized modules. Additional concepts for interconnecting the modules are shown in Fig. 37.

Berthing/Docking and Logistics

Berthing/docking and logistics have more significance for the space station than for any previous space program because of the 1) variety and quantity of items to be transported to (or serviced at) the station, 2) expense of transferring consumables, spares, payloads, etc., to LEO, and 3) long-term repetitive nature of the operations. The determination of requirements for berthing/docking facilities and development of a suitable logistics system, therefore, must include all aspects of the space station and the associated spacecraft design.

Berthing is defined as the joining of two orbiting elements using a remote manipulator (or similar device) on one of the elements; docking is defined as the joining of two orbiting elements using onboard propulsion to maneuver one of the elements. Either function can be accomplished automatically or manually. Since the space station is to be manned, onboard manual control that uses direct or remote visual cues and other aids appears to be the simplest berthing/docking approach. Manual control with remote visual cues could also be accomplished from the ground. Generally, berthing is favored over docking because velocities and misalignments at contact will most likely be less critical.

The berthing interface concept shown in Fig. 38 allows the transfer of both crew and materials. With this concept, the module and tunnel are installed in the front of the orbiter cargo bay and can be removed when unneeded. Other concepts (i.e., installation of a permanent overhead

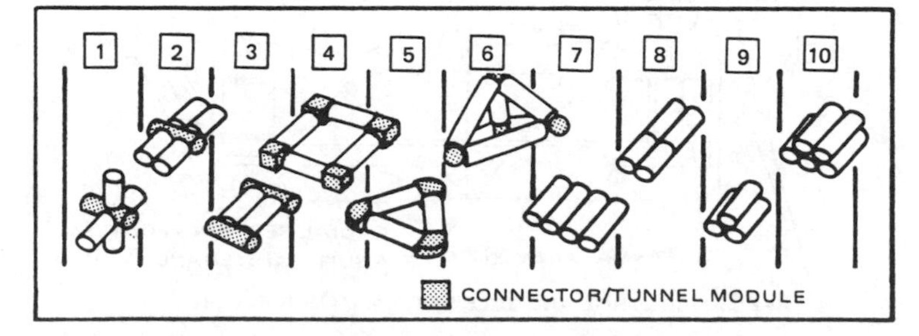

Fig. 37 Module interconnection options.

port) would be advantageous for space station operations, but would require major orbiter modifications. Preliminary docking/berthing design parameters for the orbiter are specified in Fig. 39.

A manned maneuvering unit (MMU) is used to transport an untethered astronaut to areas near an orbiter or space station. Several MMUs can be stored and/or serviced on the station externally, or in an airlock. Final berthing of MMUs to the space station is expected to be performed by EVA and supported by a remote manipulator system (RMS), or a handling and positioning aid (HPA).

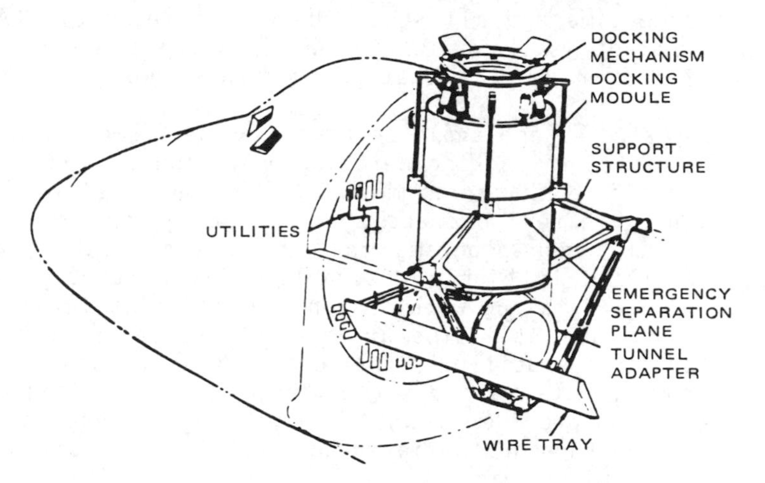

Fig. 38 Docking module concept.

DOCKING DESIGN CONDITIONS

• AXIAL VELOCITY	0.16 – 0.5 FT/SEC
• LATERAL VELOCITY	0.2 FT/SEC
• ANGULAR VELOCITY	0.6 DEG/SEC
• LATERAL MISALIGNMENT	0.75 FT
• ANGULAR MISALIGNMENT	5.0 DEG ROLL 6.0 DEG PITCH/YAW

BERTHING DESIGN MAXIMUM CONDITIONS

• CLOSING VELOCITY	0.05 FT/SEC
• LATERAL VELOCITY	0.05 FT/SEC
• ANGULAR VELOCITY	0.1 DEG/SEC
• LATERAL MISALIGNMENT	0.2 FT
• ANGULAR MISALIGNMENT	3 DEG ROLL 3 DEG PITCH/YAW

Fig. 39 Docking and berthing design conditions.

An orbital maneuvering vehicle (OMV) is used to deploy and retrieve payloads of up to 22,000 kg to or from orbits that are hundreds of kilometers higher than the space station. An orbiting transfer vehicle (OTV) deploys larger payloads from the space station to higher-energy orbits, including GEO. A free-flying platform could be deployed and retrieved by either an MMU or an OMV, or it could have its own propulsion. These platforms would be accessible for servicing while stowed on the space station and, if pressurized, have a hatch for access. Space station servicing could occur at one- to six-month intervals. Several of these spacecraft could be attached to the space station at the same time and all should be within reach of HPAs. Such a concept is shown in Fig. 40 in which a payload is being loaded on an OTV that already has two payloads aboard.

Many studies and analyses have been performed in anticipation of future docking/berthing operations. A standard mating interface that accommodates the space station module mating, orbiter to space station mating, and is compatible with other programs that require the mating of modules/pallets is highly desirable. Such a concept is shown in Fig. 41 along with a standard utilities interface arrangement. The utilities arrangement dedicates specific areas for various utilities that cross the interface: electrical power, data, air distribution, etc. The standard interface also provides a 1-m clear opening to accommodate crew transfer through the interface for either a suited or a shirt-sleeved crewman.

Fig. 40 Evolved space station.

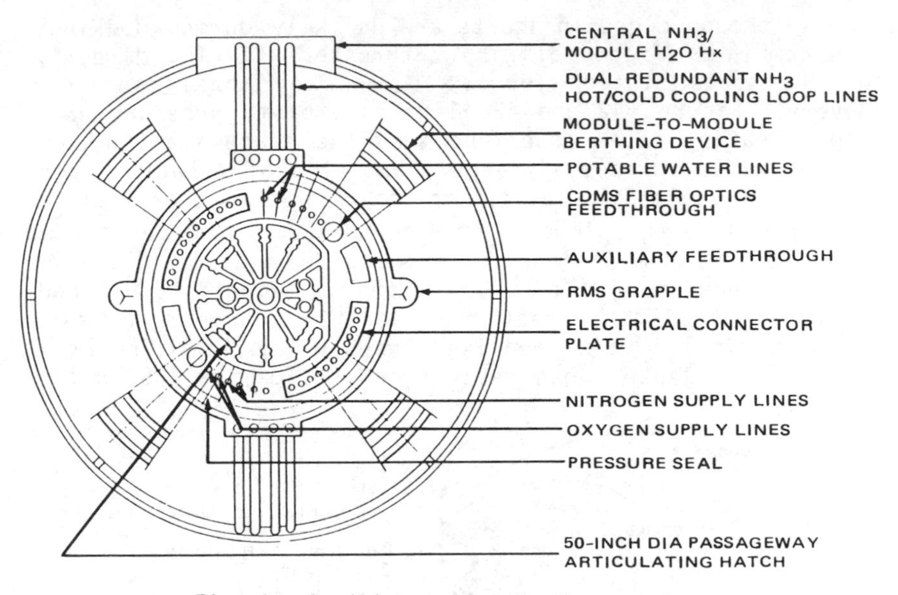

Fig. 41 Candidate common berth concept.

Reaction control system (RCS) plume effects from the Shuttle orbiter will be relatively mild during normal docking operations. Almost all close-in thruster action for normal docking will be minimum impulse adjustments of approximately 80 ms duration. These brief bursts will be mostly single or dual thruster firings and will be aimed away from the space station by nature of the thruster geometry. The little impingement that does exist in these cases will be from the very low flux region of the plume field. The main concern for normal docking is the cumulative effect that mass deposition will have on sensitive space station surfaces. Over the years, mass deposits can build up with many repeated dockings, and contaminants deposited could reach 1000 lb in a 20-year period. These effects require more detailed study.

The space station should be designed so that the orbiter-docked condition will have any single RCS jet in a runaway firing condition. If the runaway jet condition begins before docking contact is made, a more or less normal docking hookup will be accomplished with the jet still firing. If the runaway jet condition begins after this docking commitment point is reached (up to the time the orbiter flight control system is disabled and powered down following normal docking), a docked runaway jet condition is possible. Because as much as one minute could be required to identify and shut down a runaway jet, the space station must be designed for this condition.

Although close-in aborts are unlikely during docking, the combined effects of nine Z-thrusters (firing directly on the space station for two seconds or more) can be severe. Figures 42 and 43 illustrate plume pressure and plume heating rates experienced by the space station and any attached vehicles. Large quantities of exhaust products can be quickly deposited on front surfaces that will be exposed to the plume. Large control disturbances can be induced by plume impingement forces. Shielding will probably be required for OTVs parked on the flight support and servicing facility beneath the service modules. Plume-induced rippling and shearing forces on delicate thermal protection blankets can cause extensive damage. Signifi-

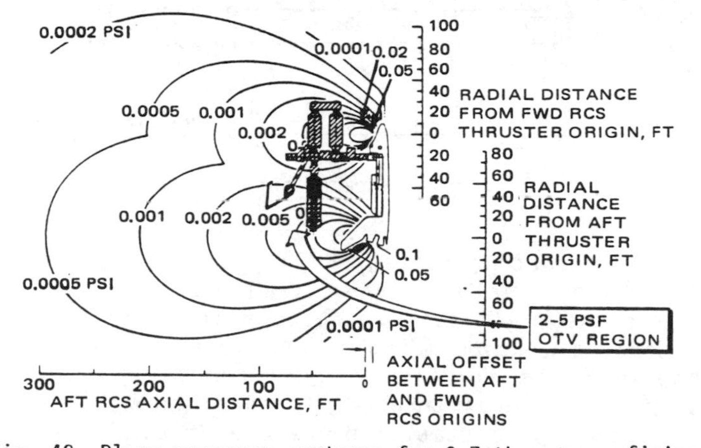

Fig. 42 Plume pressure contours for 9 Z-thrusters firing.

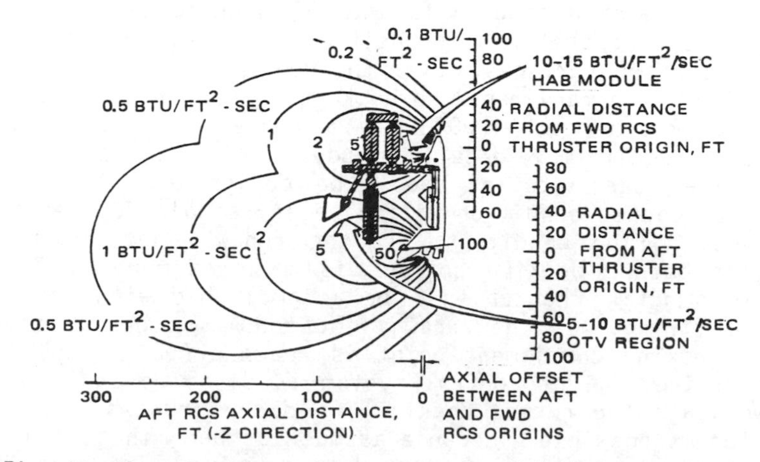

Fig. 43 Plume heating rate contours for 9 Z-thrusters firing.

cant attention must be given in the space station design for protection against these severe plume-induced environments.

A manipulator system that can be used for berthing is shown in Fig. 44. This RMS has been flown on the Shuttle orbiter. Simulation runs have shown that the RMS has the ability to handle large masses and inertias associated with orbiter/space station berthing operations. However, minor software control changes will be required to permit stable control of the arm with large system masses.

The mechanical connection between a manipulator arm (or docking port) and a free-flying spacecraft will usually be some form of probe and drogue that provides the necessary guidance to accommodate misalignments, dissipation of residual energy, and provision of structural and utilities interfaces. A mechanical end effector connection for the orbiter RMS is shown in Fig. 45. The major design task for end effectors and docking hatches will be to accommodate lateral and angular misalignment.

Space station activities and operations will require that supplies, equipment, and personnel be periodically transported to and from Earth in a logistics module. Consumables needed to operate the station include life support items such as food, water, clothes, housekeeping supplies,

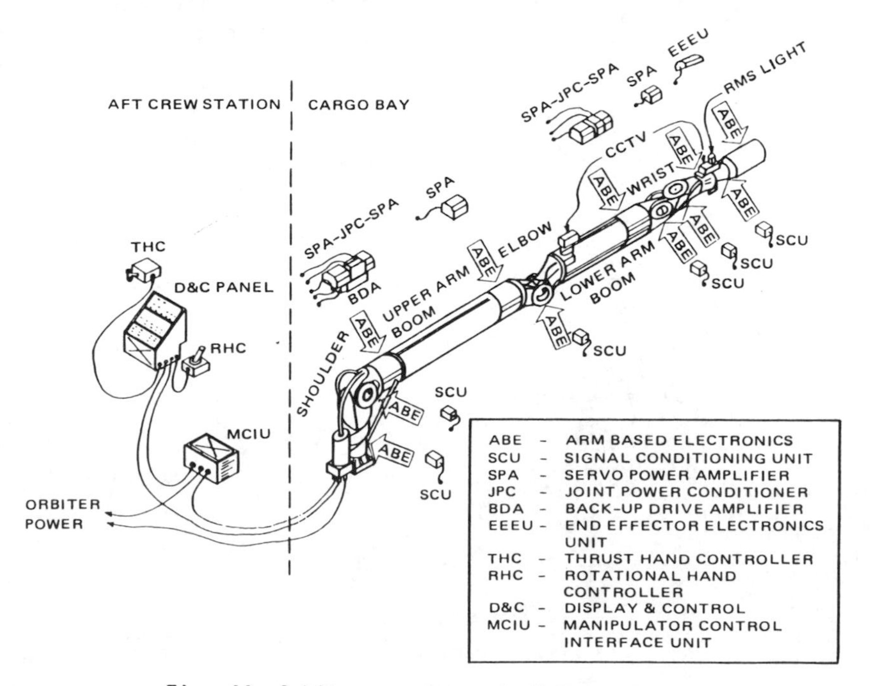

Fig. 44 Orbiter remote manipulator system.

atmospheric gasses, etc. A sufficient propellant resupply is also required to operate mobile ancilliary equipment and spacecraft served from the space station. Replacements and spares for the space station and associated equipment/vehicles must also be accommodated. Materials processing, technology development, and life sciences missions also require replacements, spares, materials, and subjects. Finally, the crew should be periodically exchanged and be provided with a shirt-sleeve environment in which to work.

Station external environment should not be contaminated unless unavoidable (e.g., thruster exhaust). Consequently, most nonpropellant mass transported to the space station should be returned to Earth as waste (or finished products). Another option is to place waste in a "garbage can" that can be destroyed in a reentry orbit or deployed into deep space.

One of the first logistics module design issues to be addressed is how much of it must be pressurized or maintained in a controlled environment. Since repetitive transport of the module and its supplies to LEO is an expensive operation, minimizing the mass and size of the module is a worthwhile objective. In addition, elimination or minimization of the module's pressurized portion is desirable. Liquids and gasses can be transported in their own containers in an unpressurized environment. Basic requirements for pressurization originate from three considera-

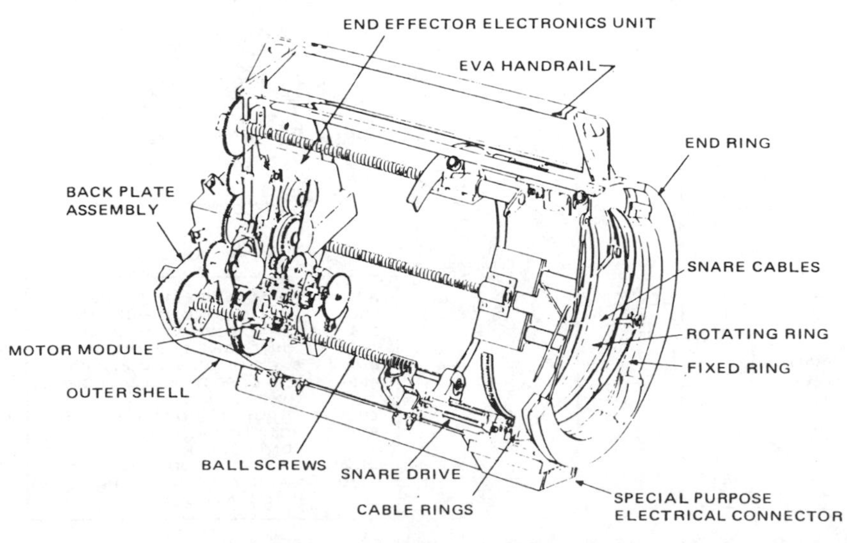

Fig. 45 Standard end effector.

tions:

1) Materials or subjects that cannot survive in a space environment (e.g., plants and animals).

2) Supplies destined for another pressurized portion of the space station (easy manual transfer in same environment).

3) Using the logistics module as a "closet" for unused supplies and items to be returned to Earth.

These considerations suggest the concept of a pressurized module connected to a pressurized station by an open hatch. This will allow free passage by the crew and an aft unpressurized area that is accessible via an RMS. Figure 46 illustrates such a logistics module concept. An integral hygiene facility could be used to collect waste for return to Earth. Such a module would remain connected to the space station until a subsequent orbiter resupply trip, at which time it would be exchanged with another module.

Logistics module design depends on the quantity of supplies to be carried in a controlled environment as well as the quantity to be carried externally. Supplies that are stored in tanks should not require a controlled environment; they can be transported separately in the orbiter cargo bay or attached to the logistics module. Tanked supplies that support the internal station (e.g., oxygen, water, etc) can be carried by external attachment to the module with a transfer interface at the hatch. These tanks then become "station tanks." Tanked supplies for servicing external spacecraft or equipment (e.g., propellants) could be carried as separate cargo in the orbiter and stored externally on the space station.

The size and mass of the resupplies and the logistics module is a function of the resupply interval, which depends on STS operational availability, cost, station operations, and schedules. Although STS availability is expected to limit flight frequency to 30 days or more,

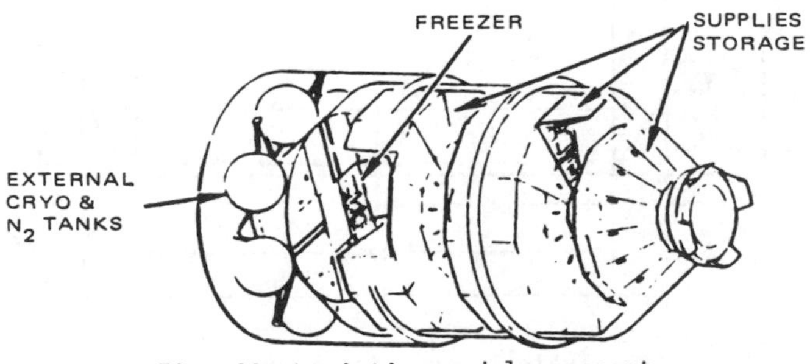

Fig. 46 Logistics module concept.

longer intervals are preferred because supplying the station with lightly loaded orbiters is inefficient. Ideally, any resupply flight should be fully loaded and the crew should also be transferred during resupply flights. Therefore, if a maximum tour of duty of 90 days is assumed, the maximum resupply interval would also be 90 days.

Because the quantity of supplies required at the space station increases as the station system expands with time, the size of the logistics module should be selected. A small module would be adequate for initial operations, but would require more frequent resupply intervals for final operations. One philosophy would be to design for the projected average size and resupply the mature station more frequently.

Present limits of the STS for transfer to orbit are 1) a maximum design cargo mass of 65,000 lb (29,478 kg) and 2) a maximum design cargo length of 60 ft (18.29 km). When manifesting cargo for the orbiter, the docking module, packaging, and mounting fixtures must be included. The orbiter also has a landing cargo mass restriction of 32,000 lb (14,512 kg), which is about half the orbital transfer limit. Since "what goes up must come down" (except for expendables such as propellants), half of the orbiter's load should be propellants (tanks and fixtures included). The other half should be a loaded logistics module (i.e., the maximum mass of the logistics module should not exceed 32,000 lb). Initial sizing results in a pressurized module of 2780 ft^3 (see Fig. 47).

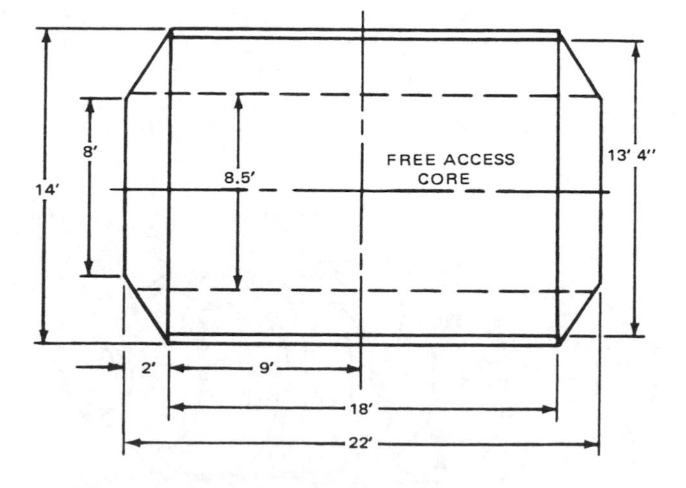

Fig. 47 Typical logistics module dimensions.

Although desires for commonality suggest that the logistics module be common with other station modules (to reduce procurement costs), special requirements are needed to minimize weight and size because of repetitive STS operating costs. The size and weight of station modules are not as critical because they are transported at infrequent intervals.

All items inside or attached to pressurized logistics modules need restraint during boost to orbit and during coast in orbit. Internal racks are expected to accommodate a variety of supplies and equipment packaged in fitted, Styrofoam-like material. Racks must be accessible for manual loading/unloading at the launch site and when berthed at the space station. Oxygen and nitrogen are expected to be carried in external cryogenic tanks, and water may also be carried externally in insulated tanks.

A complete environmental control system in the logistics module may not be required. If the crew is transported in the orbiter, only life sciences specimens will require complete environmental control. These specimens could be transported in the orbiter or have dedicated control units incorporated in their packages. The environment would be controlled by the space station when berthed. Another option is environmental control of a section of the pressurized module. Pressurization is still necessary, however, because the module must be pressurized when berthed and it is easier to fill the module on Earth than in orbit.

Packaging is an important logistics issue. Stores should be restrained in suitable containers, some must be removed and replaced in the logistics module for transfer to or from the space station. Mass and volume of the containers should be added to that of the stores and accommodated by the logistics module. The packaging size of the logistics elements must also be compatible with the size of module hatch openings and associated module aisle widths.

Space Station Servicing System Capabilities

Space station servicing includes activities that implement, activate, or restore the functional capabilities of a system. These activities are distinct from the actual exercise of the capabilities (i.e., operation of the system).

Well-documented flight experience (see Fig. 48) summarizes the value of the flight crew aiding or directly contributing to mission success and/or flight safety. The Apollo program provides a case in point: While enroute to the Moon, the crew literally saved their lives after the command module explosion. Crew versatility on the lunar surface points out the ability to retrieve samples of substantial scientific importance, and to core, erect, and strategically locate scientific instruments. Past flights have also demonstrated man's ability to manually control a spacecraft and "fly" it back to Earth after automated control system malfunctions.

Skylab accomplishments are also well documented. Particular, innovative EVA involvement of the flight crew prevented total mission loss and substantially increased orbital lifetime of the station. The role of the flight crew in unscheduled maintenance events and the variety of subsystems acted upon is especially significant. Most recent Space Shuttle orbiter experiences further indicate the extremely important role that man can play in both IVA and EVA. These missions provide an extremely strong case for the benefit of man in space. The human role on the space station is specified in Fig. 49.

The evolution of space station servicing systems begins with the inherent requirements for servicing the station itself. This servicing can encompass assembly and activation of station modules during the buildup phase,

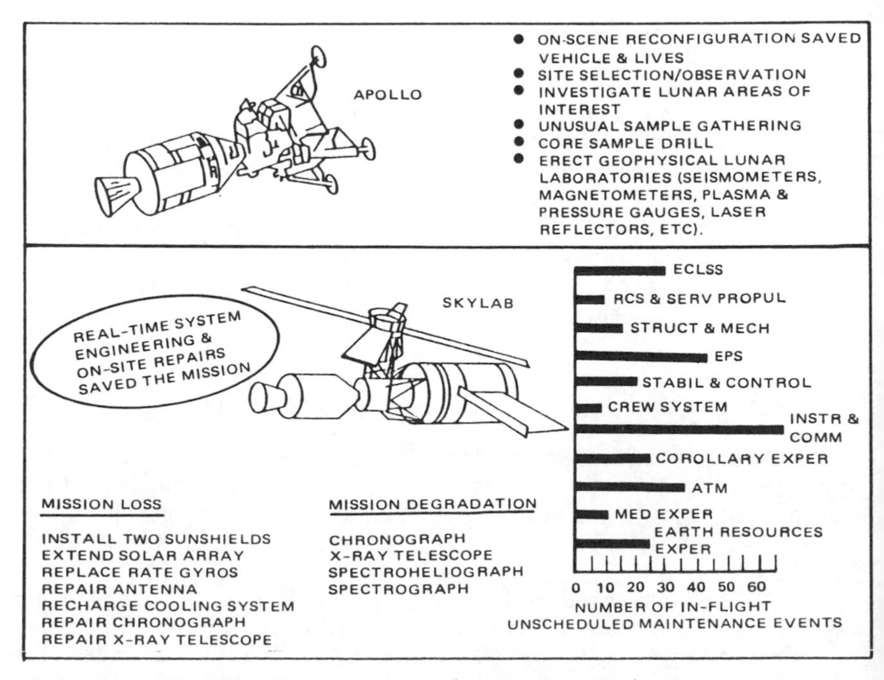

Fig. 48 Previous manned systems experience.

change-out of resupply (logistics) modules, replacement of orbitally replaceable units (ORUs) as required, and work-arounds. A number of design considerations are imposed by space station serviceability; the most immediate will be the need to design crew access to ORUs by EVA or IVA. The maintenance level is a second consideration (i.e., whether ORUs will contain entire subsystems or allow replacement at the component and card levels). An important consideration for critical systems will be the need to continue system operations during servicing. Provisions should be made for onboard system and component conditions monitoring, trend analysis, as well as fault detection and isolation. System descriptions and servicing procedures should be available to the crew from an onboard data base.

Several studies of space station buildup sequences have concluded that manipulator and holding capabilities are needed by the crew early in the buildup. Multiple fixed-position manipulators, mobile tracked manipulators, and self-repositioning manipulator concepts have been identified as options. In addition, an RMS will be developed for the orbiter for use as a berthing aid (see Fig. 50).

Space station servicing extends to laboratory module systems, to externally attached payloads, and to free-flying platforms designed to exploit this capability. Carry-on experiment hardware and control consoles can be installed using ORU change-out capabilities.

Predecessors of space station servicing systems for free-flying spacecraft and platforms are systems that now exist or are under development for near-term spacecraft

HUMAN CAPABILITIES	HUMAN FUNCTIONS ON SPACE STATION	CONSTRAINTS/ LIMITS
• FIVE SENSES • DEXTERITY • SELF-MOBILITY • STRENGTH (VARIABLE) • ADAPTABILITY • LEARNING • REASONING	• INDUSTRIAL USERS' SUPPORT • SCIENTIFIC USERS' SUPPORT • MILITARY USERS' SUPPORT • SPACECRAFT SERVICES • STAGE SERVICES • FACILITY SUPPORT • FACILITY IMPROVEMENT • REMOTE SERVICES	• LIFE SUPPORT • FATIGUE • STRAIN/ANXIETY • HAZARDS/SAFETY • TRAINING

Fig. 49 The human role on the space station.

such as the Solar Max, Space Telescope, and Gamma Ray Observatory. The space station-based OMV is an integral part of the station's architecture. Functions related to servicing include redelivery of STS-delivered spacecraft to their operational orbits and retrieval of free-flying spacecraft and platforms for attached servicing. Major design considerations for OMV, free-flyer, and platform servicing include 1) servicing support structure, 2) docking/berthing interfaces, 3) spacecraft positioning/handling systems (analogous to the flight support system rotation platform), 4) dextrous manipulator end effectors, 5) fluid resupply systems (propellants, pressurants, instrument gases, and cryogens, etc.), and 6) utility support during servicing (power, data, thermal, and checkout communications).

Automation to conserve valuable crew time will be a significant issue. The availability of servicing capabilities at the space station will impact spacecraft design by allowing spacecraft to be delivered to the station in a configuration designed for the launch environment. Spacecraft can then be reconfigured by the crew to suit the operational environment. Ultimately, this will lead to an assembly capability for large spacecraft.

Space station-based servicing is expected to be particularly advantageous to spacecraft flying in formation with the station. The concept of formation flying uses different modal regression rates at different attitudes and different orbital decay rates due to different spacecraft ballistic factors. By exploiting these differences, formation-flying platforms and free-flyers can be brought to the station or visited by the OMV for remote servicing

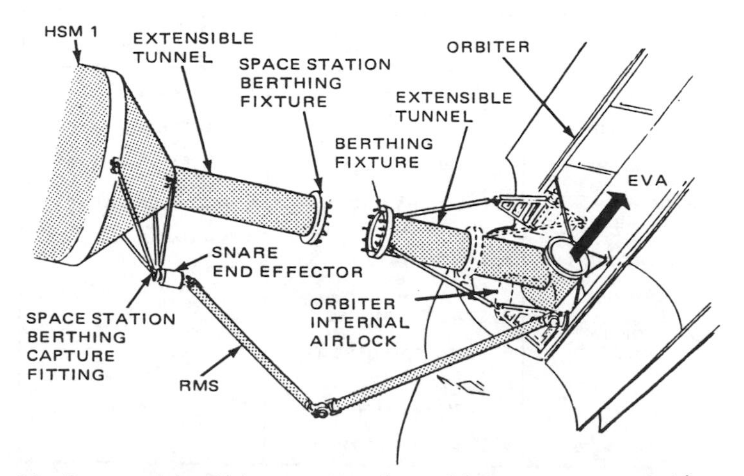

Fig. 50 Proposed berthing system for orbiter to space station mating.

SERVICE OPERATIONS	TEND			SAT. LAUNCH	
	ATT'D PAY LOADS	CO-ORBIT'G SATELLITES	REMOTE ACCESSIBLE SATELLITES	LOW ENERGY ORBIT	HIGH ENERGY ORBIT
EXAMINATION	●	●	●		
RETRIEVAL		●			
MAINTENANCE/REPAIR	●	●	●		
RESUPPLY	●	●	●		
RECONFIGURATION	●	●	●		
ON-ORBIT ASSEMBLY				●	●
MATE UPPER STAGES					●
TEST & CHECKOUT	●	●	●	●	●
ON-ORBIT STORAGE		●		●	●
DEPLOY		●		●	●

Fig. 51 Space station satellite service missions.

(e.g., payload change-out) without incurring severe propellant penalties for plane change maneuvers.

The space station concept extends satellite services beyond its 400-km altitude and its 28.5-deg inclined orbit by using it 1) as a transportation node for assembly and deployment of satellites, 2) for on-orbit support of attached and retrieved payloads, and 3) as a base for in situ servicing of remote satellites in LEO. Because the space station is decoupled from ground launch constraints, it can immediately examine and repair satellite random failure situations. The probability of random failure mission completion or scheduled maintenance for observatory class satellites could be as high as 20%. The space station and its crew could also support the buildup of large systems in orbit, such as an i.r. interferometer in LEO, a cosmic coherent optical system for GEO, or a new, large, interplanetary spacecraft.

Service operation types that can be performed on the space station are shown in Fig. 51, and are keyed to generic service missions. Many of the co-orbiting satellite services are the same as those required for attached payloads. While most of these service operations can be performed with the Shuttle orbiter, the space station offers other services including 1) on-orbit assembly of large systems, 2) mating of large upper stages, and 3) the option for on-orbit storage of satellite hardware if predeployment test and checkout fails.

Satellite maintenance operations will be tailored to the needs of each satellite. Scheduled servicing of observatory class satellites (i.e., AXAF) could entail changing scientific instruments, replacing degraded components, and resupplying consumables. As shown in Fig. 52, similar services can be performed on other satellites if the critical components are serviceable and accessible to the suited astronaut.

Satellite services from the space station can offer the user community benefits that reduce the cost of operating in space and enhance overall mission success. This capability can be provided by the development of key generic equipment that will extend space station capabilities and support user needs for satellite deployment and retrieval through on-orbit support.

Two equipment items that will be widely used in nominal and contingency situations are the manned/unmanned OMVs, and free-flying vehicles that use a cold gas propulsion system for attitude and translational maneuvering. The OMVs can operate manned or unmanned (manned remote) at distances up to one-half mile from the space station. The cold gas propulsion system minimizes potential contamination to payloads (or to the space station) that results from propellant effluents produced by hydrazine and other bipropellant propulsion systems.

Figure 53 identifies mission applications for an unmanned OMV. Initial applications include satellite examination and the examination/retrieval of cooperative satellites. Because of the high usage rate projected for these applications, they are considered key design drivers.

	SCHEDULED SUPPORT	POTENTIAL SUPPORT	
TYPICAL SATELLITES	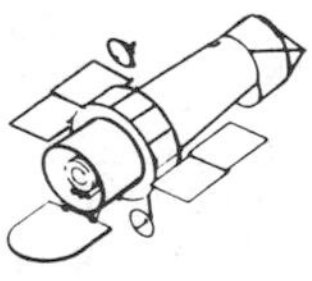 AXAF	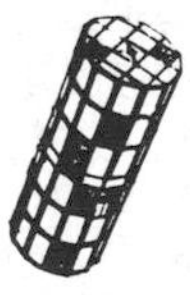 LDEF	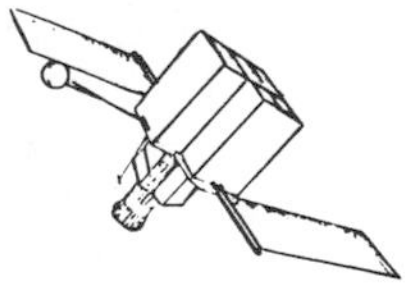 LAMAR
ON-ORBIT SERVICE			
• RECONFIG/REPLACE	SCI INSTR	EXPERIMENT TRAY	
• MAINTAIN/REPAIR	S/S ORUs GYRO SENSOR ANTENNAS RATE GYRO STAR TRACKERS SOLAR ARRAY		COMM/DATA MOD ATT CTL MOD POWER MOD PROPULSION MOD ANTENNAS SOLAR ARRAY
• RESUPPLY	DETECTOR GASSES CRYOGENS		PROPELLANTS
ON-ORBIT REVISIT	2-3 YEARS	~ 1/YR	~2 YR (EXTEND OPS) OR AS NEEDED

Fig. 52 On-orbit satellite maintenance operations.

Still another application for the OMV is a manned sortie close into the space station. A manned OMV, which uses the manned maneuvering unit (MMU) as a propulsion system, offers a number of applications for nominal and contingency operations. These include the following:

1) A portable work station for in situ emergency repair (i.e., satellite appendage or payload release mechanism hangup).

2) Backup to a remote manipulator system for satellite deployment.

3) A vehicle to stabilize/retrieve a satellite that has been deployed with higher than nominal attitude rates, or one that experiences subsystem failure soon after deployment.

4) A free-flyer to transfer replacement modules and assist in situ servicing.

Retrieval of satellites at distances up to 1000 ft from the space station could also be accomplished by a manned OMV.

Figure 54 shows a crewman repairing a satellite appendage hangup. The crewman is released from the MMU and held by the foot restraint to provide him with unobstructed access to the frontal work zone.

Figure 55 shows a manned OMV "flying-in" the end effector to engage the satellite's grapple fixture. As most of the major hardware elements for this concept exist or are in late stages of development, the manned OMV could be a more readily available approach for near-term satellite

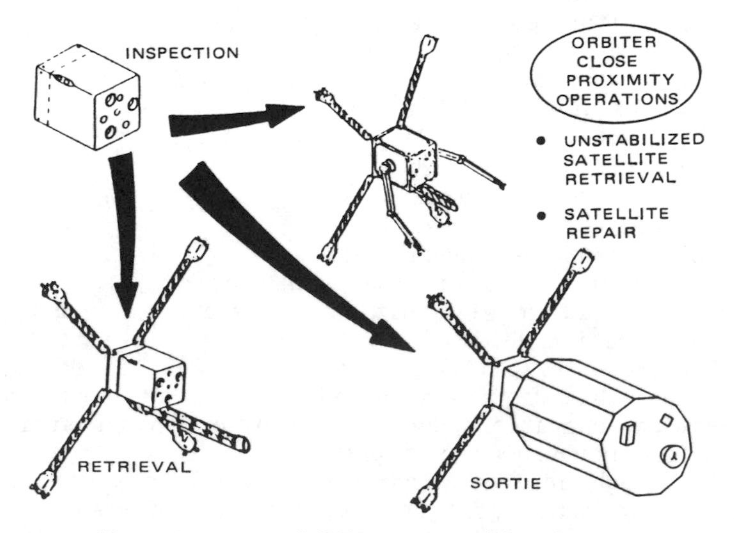

Fig. 53 Unmanned OMV system applications.

retrieval missions. The astronaut would stabilize/position the satellite within the reach distance of the RMS arm using the flight control capabilities of the MMU. The OMV could then detach itself from the satellite's grapple fixture to allow the RMS to capture the satellite. After capture, the RMS would place the satellite on a tilt table or HPA to enable on-orbit servicing.

The next step in the evolution of space station servicing systems is expected to support operations of a station-based reusable OTV. The major servicing system involved is the propellant depot. Depending on the OTV payload capability and propellant (storable or cryo), the propellant depot may be attached to the station tethered to the space station or fly in formation. In addition to propellant services, the scale of OTV turnaround operations between missions (including maintenance operations and payload/OTV integration) has led to concepts of large, unpressurized service hangars. The technical basis for many of the OTV/payload service operations will be established by OMV/payload operations and by near-term support to upper stage missions (e.g., delivering the payload and upper-stage by separate Shuttle launches).

On-orbit assembly and deployment of large spacecraft (e.g., a 60-m-diam GEO communications platform) would begin with delivery of the folded platform to the space station. A dedicated orbiter flight is postulated because the folded platform would probably fill the payload bay. The platform is transferred to the station, where it is supported by an HPA during unfolding operations. Space station crewmen would

1) Monitor deployment of folded appendages.
2) Release hangups, if necessary.
3) Support platform operations ground control during system checkout and calibration.
4) Check out the OTV.
5) Control upper-stage mating.
6) Verify system interfaces.
7) Control vehicle separation.
8) Monitor the launch to GEO.

While these tasks could also be performed by the orbiter, the space station will not be constrained by orbiter mission duration limits. Therefore, the space station offers greater flexibility to deal with satellite deployment situations that require extended calibration operations for the payload operations control center or other system activation contingencies that might arise.

Although an OTV turnaround scenario is sensitive to the traffic model, specific turnaround requirements are insensitive to the traffic (see Fig. 56) Postflight

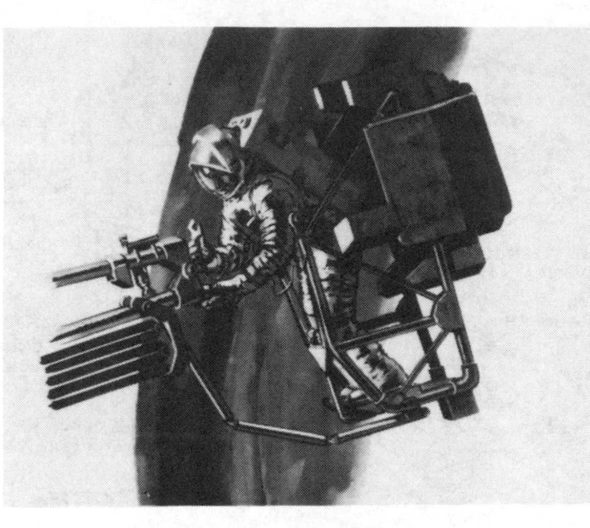

Fig. 54 MMU/work station - backup for satellite appendage hangups.

Fig. 55 Manned proximity operations module - satellite capture.

and/or periodic maintenance is performed at the space station, and major maintenance or overhaul is performed on the ground. Maintenance includes the following procedures.

1) Postflight only: Assumes that operational flight instrumentation subsystem performance, as well as postflight safety and damage inspections, are satisfactory. The vehicle is serviced at the space station, configured for the next mission, fueled, and overall systems checked out for initiation of the next mission. This is the

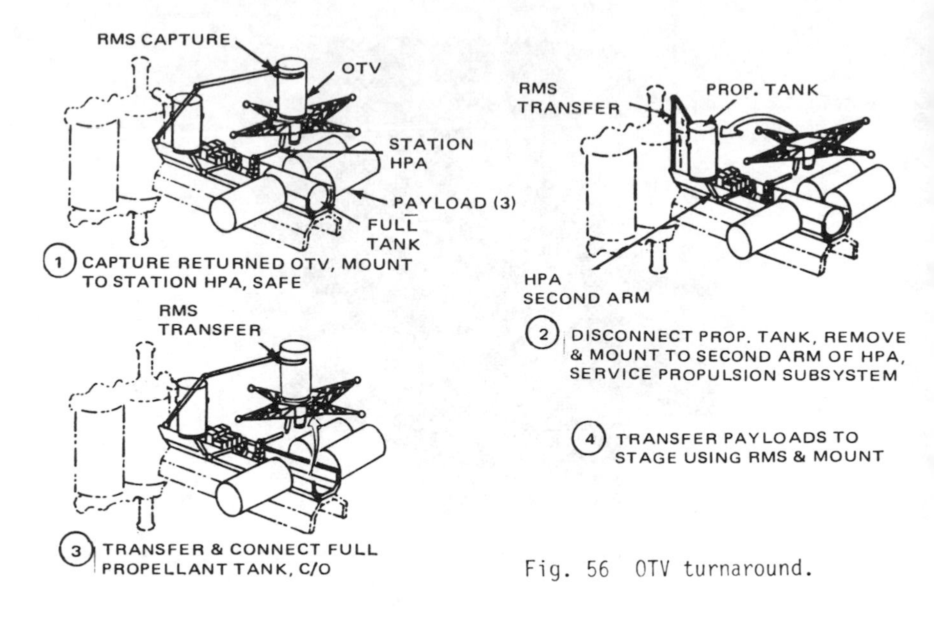

Fig. 56 OTV turnaround.

minimum maintenance mode requiring no refurbishment, which will be the standard mode for every flight except those that require periodic maintenance or overhaul.

2) Periodic maintenance: Assumes limited maintenance at the space station, including on-line calibration or checkout of navigation hardware and/or software (as required) as well as replacement of one or two ORU modules. Activities include those discussed above, plus specific limited maintenance items.

3) Overhaul: Includes a complete ground inspection, performance checks of all subsystems, calibration of all sensors, change-out of limited-life items (including engines) and required vehicle modes, as well as replacement of discrepant or suspicious ORUs. Overhaul is paced by engine-limited lifetime and is conducted on the ground.

Turnaround requirements arise from two basic considerations: subsystem hardware requirements (checkout, calibration, and refurbishment) and vehicle handling and transportation requirements. A periodic requirement to modify one or more subsystems (including structure) will correct marginal design or fabrication conditions, eliminate generic problems, or provide increased capability. In an overall maintenance program, these requirements will generally be integrated into a set of scheduled (routine) and unscheduled (corrective) operations that can be performed in an unpressurized or a pressurized hangar. Studies indicate that a pressurized hangar is a viable operational option.

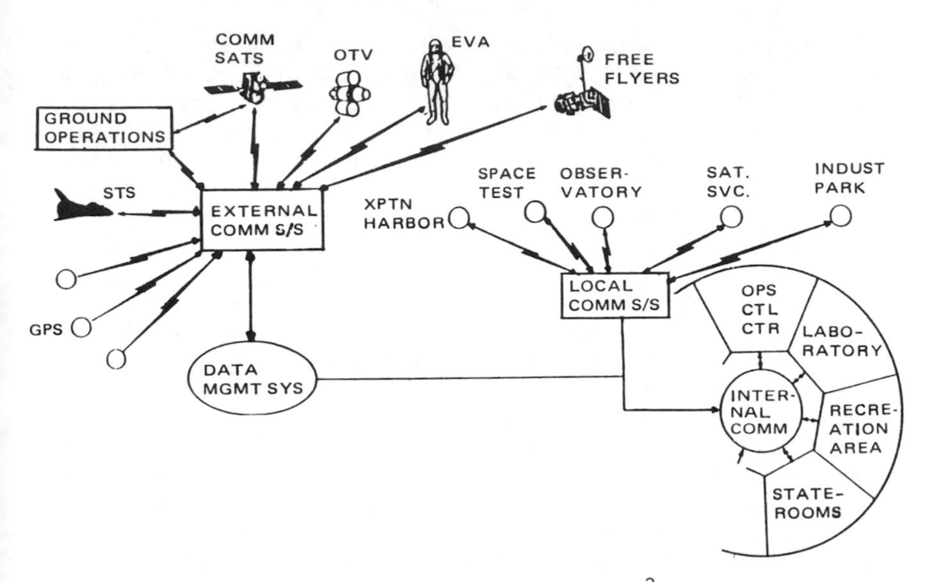

Fig. 57 Command, control and communication (C^3) system elements.

KEY PARAMETERS		HABITAT	MILITARY FACILITY	SPACE TEST FACILITY	SATELLITE SERVICE FACILITY	INDUS-TRIAL PARK	OBSER-VATORY	TRANS-PORT HARBOR
•DATA RATES (KPS)	MISSION	11.6	704.0	1875.0	0.5	1.9	3300.0	–
	OPERATIONS	45.0	23.0	6.0	6.0	6.0	15.0	15.0
•STORAGE (X 10^9 CAPACITY BPD)	MISSION	1.0	61.0	86.0	0.04	0.17	280.0	–
	OPERATIONS	0.6	0.6	0.2	0.2	0.2	0.2	0.2
•PROCESSING SPEED (X 10^6 OPS)	MISSION	0.005	1.4	0.5	0.0003	0.0004	2.1	–
	OPERATIONS	1.3	1.3	0.1	0.1	0.1	1.0	0.2
•COMMUNICATION RATES (X 10^6 BPS)		0.008	4.2	7.5	0.001	0.001	13.0	0.1
•AVERAGED OVER A 24 HOUR PERIOD								

Fig. 58 DMS performance requirements.

For space station turnaround of an OTV at LEO, a typical scenario begins after rendezvous. The returning OTV is captured, berthed, and prepared for maintenance, which may consist of safety and damage inspection, replacement of defective hardware (ORUs), and reconfiguring for flight. For those missions in which the OTV must be returned to Kennedy Space Center (KSC) for ground overhaul, the OTV is captured, berthed, and transferred to the next orbiter for ground maintenance; no maintenance is performed at the space station. Mission preparation would follow maintenance and consist of servicing the required systems, refueling, and final systems check prior to GEO transfer.

Servicing a manned OTV at the space station in an unpressurized hangar will use EVA as required. When the OTV returns from a flight, it will be captured by a manipulator and berthed to a pressurized part of the space station so that the crew can disembark. Assuming that the next mission is also manned, work platforms can be positioned around the vehicle at suitable heights for the EVA service crew to perform their tasks. Logistics pallets can then be positioned within reach of the service crew. After servicing, platforms and pallets are removed, drop tanks installed, and propellant is transferred to the core. The mission crew will board the OTV, check out the systems, and separate the vehicle from the space station using the berthing manipulator.

In a pressurized hangar, the scenario for servicing a manned OTV at the space station could use unsuited crewmen. When the OTV returns from a manned flight, it is captured by a manipulator and berthed to a pressurized part of the space station to load the crew. If the next mission will also be manned, the manipulator transfers the vehicle to a pressurized hangar that can accept the crew capsule and appendages. Servicing will be performed from work platforms in the hangar. The vehicle is raised from the hangar, rotated through 180 deg, and then lowered back into the hangar so that the propulsion subsystem and the subsystems located between the propellant tanks are within the hangar. Subsystems are then serviced; after servicing, the vehicle is raised from the hangar, drop tanks installed, and the propellant is transferred to the core. The OTV crew boards a small capsule that can be transferred by the manipulator to berth with the OTV crew capsule. The crew can then board the OTV; the transfer capsule is removed, the vehicle systems checked out, and the space station separated. A berthing manipulator can be used to avoid possible fouling.

Preparation of an OTV for an unmanned mission would use shirt-sleeved crewmen working in a hangar at the space station. The scenario begins when the OTV returns from a manned mission. After the crew disembarks, propulsion core/crew capsule interfaces are released, and the core is transferred by the manipulator and lowered into the pressurized hangar, engines first. As with the preceding scenario, propulsion subsystems are serviced. The propulsion core can then be raised out of the hangar, drop tanks installed (using the manipulator, possibly, EVA assistance), and the payload is installed in the same manner. After checkout, the vehicle is separated from the space station by the berthing manipulator.

Many satellites and large platforms (or constellations of platforms) in GEO can be enhanced by servicing systems that perform ORU change-out and fluid resupply in situ. Such servicing systems can be automated, remotely controlled, or provided by men working from within a manned OTV. In GEO, environmental concerns include solar flares and auroral phenomenon and their impact on radiation shielding requirements. The practicality of spacesuit EVA operation in these environments versus shirt-sleeve operation from within the pressurized cabin (using internally controlled manipulators) is a design issue under study for servicing in GEO.

Servicing functions in LEO or GEO will most likely be accomplished by a mix of EVA, manual remote control (with and without sensory feedback), and autonomous/automatic (robotic) control. This will depend on economics, risk, safety, repetition, and performance requirements.

Command, Control, and Communications System

The command, control, and communications (C^3) system will be the link between man and machine. The system efficiently helps man to 1) determine which actions to take, 2) perform these actions, and 3) communicate to appropriate elements of the station infrastructure. It significantly impacts the SSP through automation, autonomy, and evolution. The proper mix of manned and automatic operation will increase the productivity of both the space and ground crew. Furthermore, autonomy for the space station and its elements (or modules) will improve cost effectivity of development, integration, test, and operations.

Evolution of the SSP over its extended life will occur through incorporation of new technology and added capability, which will be enhanced by appropriate architectural design of the C^3 system. The system will play a key role in all space station operations, housekeeping, and in the conduct of various missions. It should also allow unmanned operations to occur when a manned presence is not required. The C^3 system must serve as the nerve center for station operations and mission conduct while still permitting growth in all areas.

The C^3 system (see Fig. 57) consists of two major flight elements: the data management system (DMS) and the communication system. The DMS includes all onboard computer-related hardware and software required to assume and exercise control of space station activities. The communication system provides both external and internal com-

	ONBOARD	GROUND	SHARED
• USER/PI INTERFACE	0	3	2
• SYSTEM COMMAND & CONTROL (C&C)	4	2	0
• MISSION SUPPORT	2	1	0
• S/S HARDWARE MAINTENANCE	3	0	3
• S/S SOFTWARE MAINTENANCE	1	0	4
• CREW HEALTH MONITORING/MAINTENANCE	2	0	4
• SPACEBORNE EXPERIMENTATION	2	1	3
• S/S ONBOARD SUPPORT	13	0	1
• S/S SUPPORT SUBSYSTEM C&C	2	0	0
• S/S MISSION SUBSYSTEM C&C	2	1	0
• S/S SUPPORT SUBSYSTEM MONITORING	4	1	0
• S/S MISSION SUBSYSTEM MONITORING	4	1	0
• MISSION DATA DISTRIBUTION	1	3	1
• ENTERTAINMENT	3	0	1
• DATA STORAGE	1	1	1
• PERFORMANCE EVALUATION	2	2	0
• MILITARY SUPPORT	1	0	0
• TRAINING & SIMULATION	1	0	0
TOTALS	48	16	20

Fig. 59 Onboard versus ground functions.

munications. Associated ground equipment is also considered part of the C^3 system.

Requirements

Preliminary design of a C^3 system begins by defining 1) missions to be performed, 2) station operations and functions, 3) technologies available for initial operation (1992), and 4) technologies anticipated through the life of the space station (2000 and beyond).

A total of 81 generic missions (nine basic categories) were used to generate C^3 performance requirements. Definition of these requirements (i.e., acquisition data rates, storage capacities, processing speeds, and communication rates) allowed sizing of C^3 system elements (see Fig. 58).

After performing a mission requirements analysis to determine system functions, a functional partitioning analysis determines hardware and software architecture. This functional partitioning consists of 1) allocation between onboard and ground functions, 2) allocation of functions to onboard subsystems, and 3) identification and evaluation of human participation.

Functional partitioning analysis identifies a set of 18 major C^3 functions that can be further subdivided into 84 lower-level functions. Analysis of each function will determine where it should be performed (i.e., onboard, on the ground, or shared) and assess its criticality. This will serve as a guide to deriving DMS architecture (see Fig. 59). A second analysis is necessary to evaluate the operations, quantity and type of data required to control the subsystem, as well as the quantity of telemetry data anticipated from each subsystem. The data are summarized to size the DMS for housekeeping (see Fig. 60). An operations analysis is then required to determine and assess human involvement with the DMS. This analysis identifies potential tasks to be performed by crew members and determines the type of capabilities that the DMS should possess.

Basic crew operation and interaction can be determined by developing a set of "strawman" activities and deriving basic data management requirements. A crew activities time line can illustrate relative durations of various crew activities including:

1) Monitoring and configuring of space station subsystems.
2) Generic payload mission operations.
3) Housekeeping activities.
4) Crew scheduling activities.
5) Docking operations.
6) Emergency operations.
7) Off-duty operations.

SUBSYSTEM	COMMAND RATES (kBS)	TELEMETRY RATES (kBS)
ELECTRICAL POWER	0.02	0.08
ECLSS	0.2	0.3
GN&C	10.0	90.0
ATTITUDE CONTROL	2.0	6.0
PROPULSION	0.4	0.6
THERMAL	0.4	1.6
RADAR	0.4	0.6
DOCKING	0.4	0.6
REMOTE MANIPULATION	0.4	0.6
STRUCTURAL	0.02	0.08
COMM	0.2	1.0
DMS	0.1	0.1
TOTALS	14.5	101.5

Fig. 60 Preliminary estimates of subsystem command & telemetry rates.

TECHNOLOGY	OPTIONS	CRITERIA	COMMENTS
PROCESSOR-CPU	• MIRCOCOMPUTER (boxed) • MINICOMPUTER • LARGE SCALE	• FAULT TOLERANCE • COST • COMPLEXITY • ARCHITECTURE	• 700-1000 kOPS BY 1995 • RAD HARD/SPACE QUALIFICATION WILL IMPACT SELECTION
MAIN MEMORY	• CMOS (boxed) • MAGNETIC CORE • PLATED WIRE	• WEIGHT/SIZE /POWER • COST • ACCESS TIME • VOLATILITY	• RAD HARDENING IS A MAJOR FACTOR • 128k BITS/CHIP BY 1990
MASS MEMORY-ELECTRONIC	• BUBBLE MEMORY (boxed) • CCD MEMORY	• WEIGHT/SIZE /POWER • CAPACITY • VOLATILITY	• EXPECTED TO ACHIEVE 10^9 BITS/CHIP BY 1990 • 1.5×10^6 BITS/SEC-TRANSFER RATE
MASS MEMORY-MECHANICAL	• OPTICAL DISCS (boxed) • MAGNETIC DISCS • MAGNETIC TAPE	• STORAGE CAPACITY • SPACE QUALIFIED	• 2×10^9 BITS/DISC BY 1990 • 7×10^{10} BITS/TAPE BY 1990
HIGHER ORDER LANGUAGE (HOL)	• ADA (boxed) • FORTRAN • JOVIAL	(SEE HOL TRADE STUDY)	

M84-1086-065PP

Fig. 61 DMS technology trade summary.

Technology Assessment

Based on required C^3 performance parameters (i.e., rates, capacities, and throughputs), data management system and communication technologies are surveyed. The survey is performed based on a "technological transparency" philosophy to allow controlled upgrading throughout the system life cycle. Upgrading must be planned to ensure that the C^3 system can efficiently and economically take advantage of new technology as it becomes available.

A number of trade-offs are made in the selection of DMS hardware for the preliminary design concept. The trade-offs and criteria used for their selection are summarized in Fig. 61. The trade-off study shows that no DMS technology barriers must be overcome to implement and launch a space station in the early 1990's. However, additional trade-offs must be performed for internal and external communications (see Figs. 62 and 63).

Architecture

After defining system requirements, studies will be initiated to develop the architecture for the data management system, the communication system, and the ground seg-

STUDY AREA	OPTIONS	CRITERIA	COMMENTS
COMM FREQ PLAN	• S-BAND • X-BAND • KU-BAND • 20/30 GHz • MILLIMETER • OPTICAL	• TDRS/STS COMPATIBLE • BANDWIDTH/DATA RATE • SPACE QUAL HARDWARE • LIFE CYCLE	• BY 1995, FREQ PLAN TO INCLUDE MILIMETER & LASER COMM WITH TDAS
COMM XMITTERS	• SOLID STATE • TWT	• XMIT PWR TO SATISFY DATA RQMTS • RELIABILITY • TECHNOLOGY TRANSPARENCY • LIFE CYCLE COST	• SPACE QUAL SOLID STATE TECHNOLOGY AVAILABLE AT 10W XMIT PWR LEVEL
COMM ANTENNAS	• OMNI • SECTOR • GIMBAL DISH • PHASED ARRAY	• SPHERICAL ANT. COVERAGE • ANT. GAIN TO SATISFY DATA RATE RQMTS • ANT. TRACKING RQMTS • RELIABILITY • LIFE CYCLE COST	• TRACKING PHASED ARRAY ANT. TECHNOLOGY AVAIL., BUT EXPENSIVE • PHASED ARRAY BEST APPROACH BY 1995

Fig. 62 External communication technology trade summary.

STUDY AREA	OPTIONS	CRITERIA	COMMENTS
• INTERNAL COMM TECH	• FIBER OPTICS • HARDWARE	• BANDWIDTH/DATA RATE • EXPANDABILITY • TECH TRANSPARENCY • COMPLEXITY • RELIABILITY • LIFE CYCLE COST • SIZE, WT, POWER	• FIBER OPTICS IS BEST APPROACH IN ALL CATEGORIES.
• MULTIPLEXING	• FREQ DIV MUX (FDM) • TIME DIV MUX (TDM) • WAVELENGTH DIV MUX (WDM)	• DATA RATE • SNR/BER • EXPANDABILITY • TECH TRANSPARENCY • COMPLEXITY • RELIABILITY • LIFE CYCLE COST	• TDM SELECTED FOR MUXING FUNCTIONS AT COMM INTERFACE UNIT • WDM SELECTED FOR MUXING STATIONS (CIUs)
• ARCHITECTURE	• REDUNDANT COMMON BUS • BIDIR RING • ARPANET • STAR	• RELIABILITY • COMPLEXITY • EXPANDABILITY • COUPLING LOSSES	• COUPLING LOSSES LOW & EQUAL FOR ALL STATIONS • ONLY STAR JUNCTION REQUIRES REDUNDANCY
• LOCAL COMM	• RF • OPTICAL	• BANDWIDTH/DATA RATE • SNR/BER • SIMPLICITY OF INTERNAL COMM INTERFACE • RELIABILITY • LIFE CYCLE COST	• OPTICAL (IR) XMISSION DIRECTLY COUPLES TO FIBER OTPICS

Fig. 63 Internal/local communications technology trade summary.

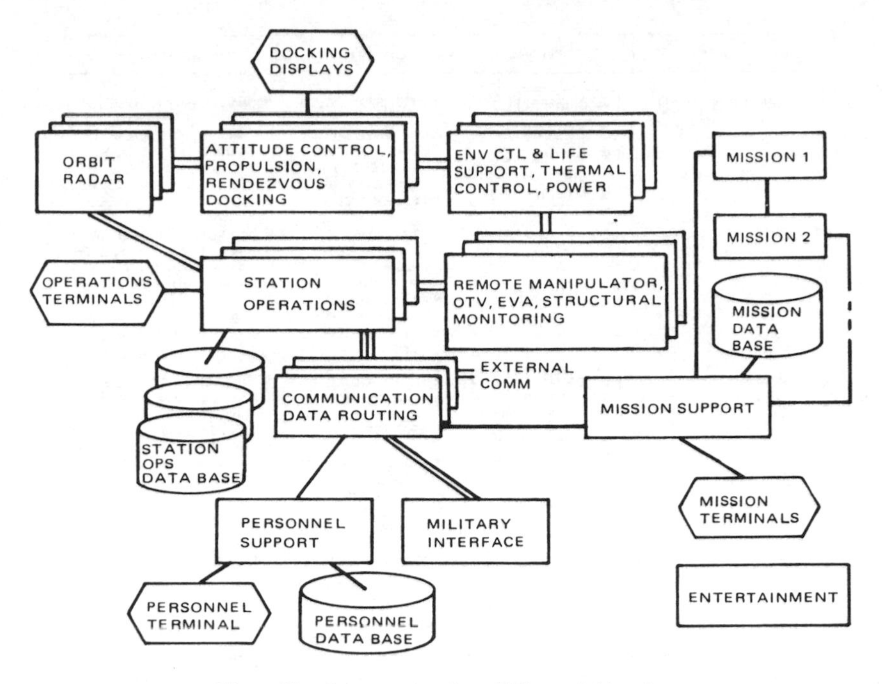

Fig. 64 Space station DMS architecture.

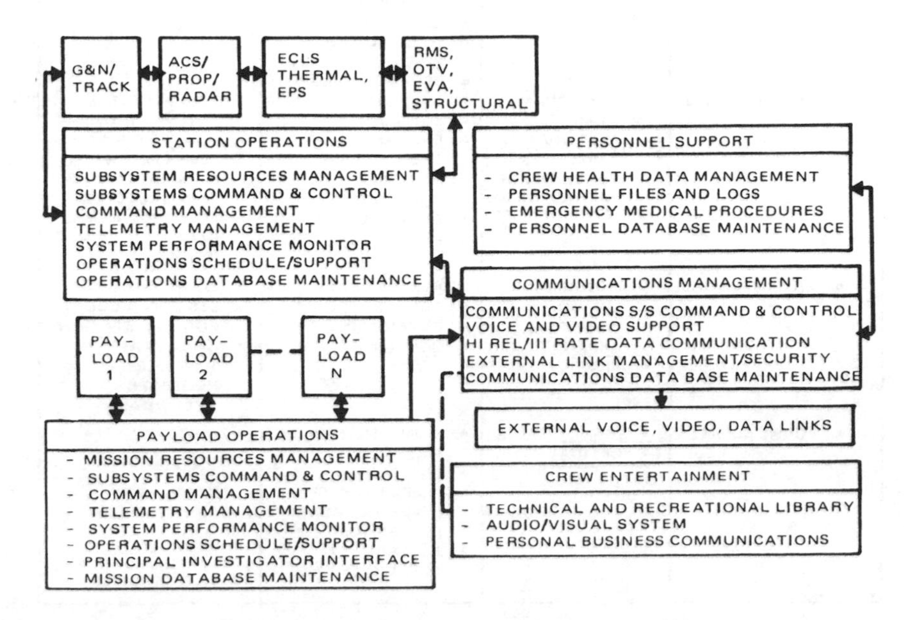

Fig. 65 DMS application software overview.

FUNCTION	LINES OF CODE (k)	MAIN MEMORY (MB)	MASS MEMORY (MB)
STATION OPERATIONS	25	1	150
MISSION OPERATIONS	25	1	150
COMMUNICATION MANAGEMENT	15	0.5	10
PERSONNEL SUPPORT	15	0.5	150
ASTRONAUT PERSONAL DATA	15	0.5	150
STATION SUBSYSTEMS	15	1	--
SUPPORT SOFTWARE	75	(INCLUDED IN ABOVE ESTIMATES)	
TOTALS*	185	4.5	610

* BASED ON BOTTOMS-UP ESTIMATES USING SIMILAR FUNCTIONS PERFORMED BY PAST SPACE PROGRAMS

Fig. 66 Preliminary software estimates.

ment. Development of DMS architecture must consider distributed processing, network topology, and

1) Fault tolerance: Fault tolerance includes automatic fault detection, isolation, and recovery. Since the space station will be manned, highly critical functions involving crew safety and health are required to exhibit "fail-operational/fail-safe" performance; highly critical functions must be performed correctly in the presence of any system failure. If a second failure is detected and recovery is not possible, the system must revert to a fail-safe mode.

2) Flexibility: To ensure an extended life cycle, the space station must have the flexibility to add, modify, and delete mission and station operation functions. Many planned missions will be of short duration and will be phased in and out of the project. Also, additional and potentially different habitation, observatory, and materials processing modules will be added and deleted as the station evolves. These changes to DMS operational requirements will require DMS support with relatively minor system impact and cost.

3) Technology transparency: As the station evolves, obsolete technology will have to be upgraded and DMS architecture must provide for this upgrading with minimal impact on the system.

4) Processing rate: An aggregate processing rate estimated to be in excess of 8 million operations per second (MOPS) and an aggregated communication rate of 25 million bits/s (Mbps) must be supported. The high processing rates

are mission data preprocessing functions; station operational functions require much lower rates.

The sequential methodology used to develop DMS system architecture first involves determining the degree of distributed processing that is appropriate for the DMS. This is accomplished by analyzing functional interfaces, computation requirements, and the critical nature of each function. After a basic distributed DMS architecture is designed, methods for interconnecting the distributed processors are determined through analyses of the characteristics of network topologies and the interfaces to be implemented. Finally, an approach to fault tolerance (for different levels of criticality) is determined and applied to the distributed network. A basic DMS architecture that might result is shown in Fig. 64.

Communications

The space station communications system includes all internal and external subsystems. External subsystems provide communication between the station and external users (i.e., manned and unmanned spacecraft, the ground data system, etc.), as well as navigation, tracking, and surveillance functions. Internal communication subsystems provide intercompartment voice communication, closed-circuit TV, and audio/video/digital data transmission. Local communication is a subset of internal communication in which some free-flying modules of the space station require a radio-frequency (rf) or optical transmission link. The DMS links internal and external communications by configuring the communication interfaces and by performing overall command and control functions.

Ground Systems

The ground portion of the C^3 system must perform its allocated functions according to a functional analysis and several test and simulation operations. The total set of ground system functions can be grouped by commonality and assigned to phases of the space station evolution where they are needed. A functional view of such a ground complex has been derived in which each function is allocated to a specific compartment. The result is a high-level software representation; a hardware configuration can then be developed with primary objectives of expandability, low cost, DMS compatibility, evolution, and flexibility. All hardware and software needed to support communications stations, (i.e., TDRSS ground stations) are not currently envisioned as part of the ground segment; however, such a ground segment will provide a test bed for the onboard DMS.

Software

An assessment of the derived requirements and DMS architecture will enable the definition of onboard software requirements and major software elements, as well as some preliminary sizing and timing parameters. Figure 65 shows a top-level overview of DMS applications software. Five major areas are included: 1) station operations, 2) personnel support, 3) payload operations, 4) communications management, and 5) crew entertainment. Also included are the operations and mission networks. The dashed lines leading to astronaut entertainment indicate that there is no direct link to the rest of the DMS. These six major areas reflect currently derived architecture of the DMS; each area is a separate set of processors connected via a star network with the data routing node (the center node) that contains the communication management software.

At a top level, DMS software is divided into two major categories, system software and applications software. Applications software is derived from space station functional requirements. Support software is generic to all application areas and used by each; it provides all support required by the applications software.

Onboard software size can be estimated based on its functional requirements. This accomplished by comparing space station functions with similar functions performed by other space systems (manned and unmanned). The following algorithm is used.

1) For each function with a subsystem or partition in the system, a number (1 to 10) is generated based on its estimated complexity.

2) Rules are established for size as a function of complexity:

a) In major partitions, the largest subfunction is allotted 100 kbytes, and the smallest is allotted 10 kbytes.

b) Within subsystems, the largest subfunction uses 10 kbytes, and the smallest is allotted 1 kbyte.

3) Subfunction sizes are also estimated according to relative complexities and a total size for the function is derived as shown in the following ACS subsystem example:

Function	Complexity	Size kbytes
Sensor data processing	3	3
Control algorithms	5	5
Electronics configuration	1	1
Mission support	5	5
Self-test/checkout	1	1
Total		15

4) Minimum lines of code are estimated assuming 2 bytes for a machine instruction and five machine instructions for each higher-order language (HOL) statement in a function. Exceptions to this procedure are in the support software areas, where sizes for the executive, data base management system, and network software packages are derived from existing software systems.

Based on previously derived functions and the guidelines imposed by the sizing algorithm, it is possible to estimate the line code, as well as the main and mass memory capacities (see Fig. 66).

As shown by the preceding discussion, the C^3 system is quite extensive. It touches all space station elements and subsystems, and must act in accordance with them. The approach used to develop the requirements, architecture, and preliminary design concept involves the classical systems engineering process.

Resource Module and Platforms

Several concepts have been developed for supplying basic utilities (i.e., power and attitude control) to the space station. Although some of the concepts are based on distributing utilities throughout the station, most are based on use of a separate resource module to provide power generation, energy storage, heat rejection, data manage-

SINGLE MODULE (75 kW)	SPLIT MODULES (37.5 kW EACH)
• ONE MODULE TO PROCURE, CHECKOUT, & LAUNCH • ALLOWS REPLICATION OF MODULE FOR 150 kW STATION • FEWER PHYSICAL INTERFACES BETWEEN SPACE STATION MODULES • SIMPLER REGULATION OF UTILITIES • LESS OVERALL STS LAUNCH COST	• LESS DEVELOPMENT PRESSURE FOR LARGE SOLAR ARRAYS & RADIATORS • CLOSER COMMONALITY TO SPACE PLATFORM • MORE EASILY ADAPTED TO POLAR ORBIT • GREATER POTENTIAL FOR GROUND RETURN • AERODYNAMICALLY BALANCED 75 kW SYSTEM • GROWTH OCCURS NEARER C.G. (MINIMUM UNBALANCE) • RELIABILITY/REDUNDANCY ADVANTAGE • OFFERS POTENTIAL TO OPERATE AT 37.5 kW INITIALLY, THUS SPREADING PROGRAM FUNDING • LESS DESIGN, DEVELOPMENT, & PROCUREMENT COST (FOR A TOTAL OF 75 kW)

Fig. 67 Single module and split module advantages.

ment, attitude control, stabilization, and reboost. These centralization concepts are based on two major arguments. First, the resource module will be the first space station module to be launched and must, therefore, sustain itself. Second, overall program costs could be reduced by imposing a high degree of commonality between the resource module and other space platforms.

Depending on the configuration, one, two, or four resource modules can be used. Because the space station is expected to increase in size and capability from its initial configuration, subsystem capabilities must also increase (or evolve) in an incremental manner to at least twice their capability. This doubling can be accomplished by a replication of the number of resource modules or by a doubling their contents after the station is on orbit.

In addition to a space station, two space platforms are currently envisioned: one in polar orbit and the other in a similar low-inclination orbit. To reduce development cost, both platforms should be as identical as possible. Sizing will be determined by payload requirements and by the orbiter's ability to launch the platform into polar orbit. Each platform must accommodate a variety of changeable payloads. The platform must supply the same type of services to these payloads that the resource modules supply to the space station (and its payloads).

Power requirements are of primary import when sizing the platform or the space station, because they directly affect the sizes of 1) solar array (primary energy source candidate), 2) energy storage, 3) thermal control radiators, 4) drag (reboost) requirements, and 5) torquers (momentum management).

Current studies indicate a space station initial power requirement of 75 kW at the distribution bus, with eventual growth to 150 kW. Space platform power requirements range from 6 to 25 kW at the distribution bus.

Another consideration that has significant design implications on the space station configuration is the number and location of resource modules to be used. A single, end-mounted resource module would yield a space station configuration with an offset between the c.g. and center-of-pressure. This differential would create aerodynamic torques, that would complicate station control; on the other hand, dual, end-mounted resource modules would yield a balanced configuration and simplify space station control. Some key advantages for single and dual resource module options are provided in Fig. 67.

Because the space station is designed to operate for many years, periodic maintenance is essential, and technological breakthroughs are likely to require some initial equipment replacement. Although this maintenance could be performed on unpressurized resource modules and platform modules with ORUs replaceable via EVA, placement of some resource module equipment in a pressurized environment would simplify the process. A completely pressurized environment would, however, add to the initial cost of the program; and since bulky equipment would be less safe in a pressurized environment, a reasonable compromise would be to provide internal access for only a small number of equipment items. Experiments conducted on Shuttle orbiter flights and in the Marshall Space Flight Center immersion facility have shown the feasibility of removing and replacing large ORUs by EVA.

Several approaches to resource module growth have been identified to satisfy the projected increase in utility demands during the program's lifetime. The initial resource module could be replaced by an upgraded version to accommodate more difficult/advanced missions, or additional resource modules could be placed on the station to support growing mission requirements. If the initial station is not designed to accommodate growth, capacity could be added through replication of the initial station in the near vicinity of the first station. A growth capability inherent to most basic concepts is to design the initial station to accommodate anticipated subsystem improvements.

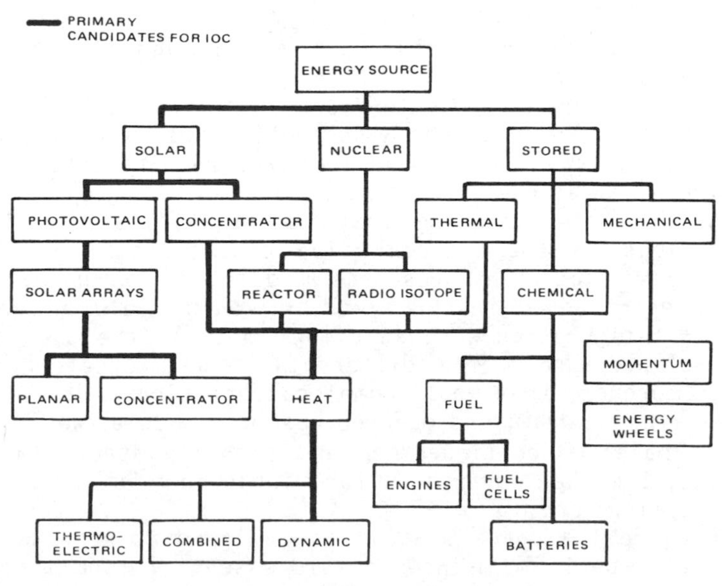

Fig. 68 Energy source choices.

Energy Sources

There are only three sources of energy for a spacecraft: 1) directly from the Sun, 2) stored chemical energy (which comes indirectly from the Sun), and 3) nuclear energy.

Figure 68 illustrates the choices. Stored energy is not practical for a spacecraft that will have a lifetime of more than a few days, because the weight of the system would be prohibitive. The primary energy source candidates for initial operation in 1992 are solar arrays.

Solar arrays convert the Sun's energy directly into electrical energy by means of photovoltaic cells, but the solar array wings must remain pointing at the Sun. To date, most solar arrays have been of the planar type, in which solar cells are applied to a flat substrate. Currently, planar solar arrays use silicon solar cells, which vary in cell thickness, resistivity, and purity; as a result, they provide varying efficiency, cost, and output characteristics. The planar array would be of a fold-up or roll-up configuration, with some lightweight deployment mechanism (see Fig. 69).

A concentrator solar array uses smaller solar cells with mirrored reflectors to provide a concentration of sunlight on the cells. One design for a high concentration ration (100:1) miniature Cassegrain concentrator solar array (see Fig. 70) has several potential advantages over a planar array.

The concentrator array uses high-efficiency gallium aresenide (GaAs) solar cells that can operate at higher temperatures. This will result in a smaller array (less than two-thirds of the planar array) and provide a

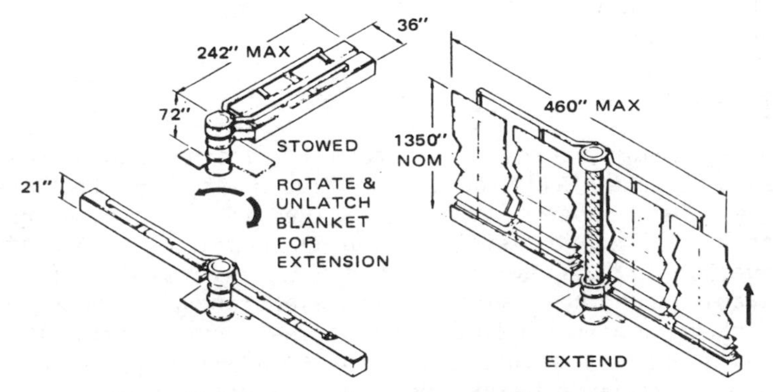

Fig. 69 Planar solar array configuration.

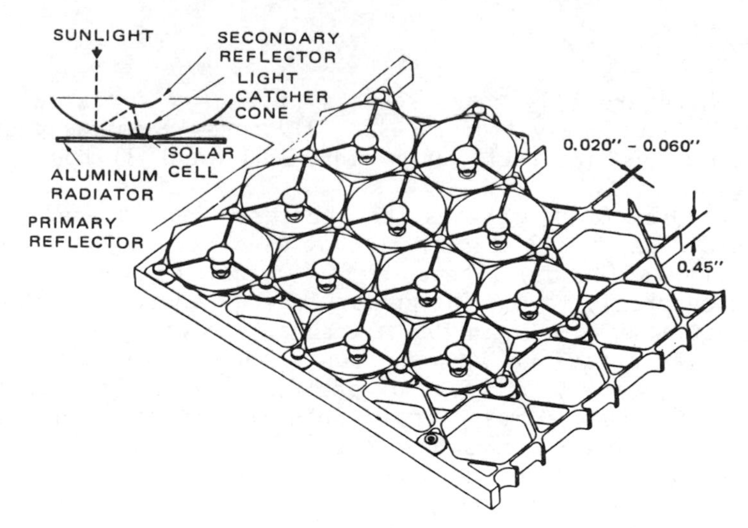

Fig. 70 Concentrator solar array configuration.

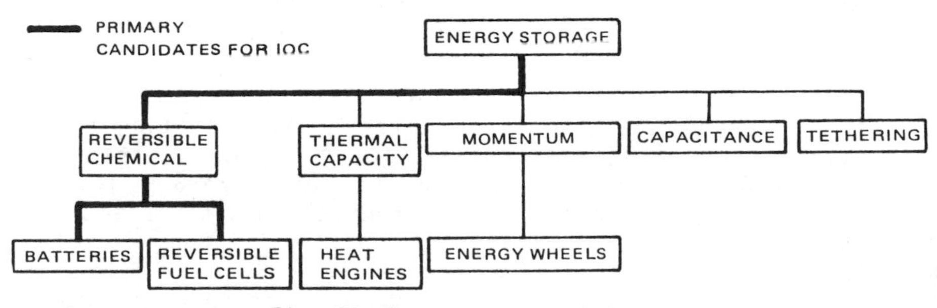

Fig. 71 Energy storage choices.

considerable advantage in lowered drag and mass for altitude maintenance. The concentrator array requires a pointing accuracy of at least 3 deg to prevent a rapid decline of output.

Energy Storage

Using solar energy, some form of energy storage is necessary because all low-inclination Earth orbits have solar eclipses (see Fig. 71). The three promising approaches to energy storage are the various types of rechargeable (secondary) batteries, reversible (or regenarable) fuel cell systems, and electrodynamic tethers.

Nickel-cadmium (NiCd) batteries represent the industry's current standard for spacecraft energy storage. They are relatively heavy, but they can be recharged over many cycles. Their lifetime is a very complicated function of operating temperature, depth of discharge (DOD), charge and discharge rates, overcharge, number of charge/discharge cycles, recycling strategy, and other factors. Lower temperatures and shallow DOD result in more cycle life. Battery sizing is determined by the voltage desired (number of cells in series) and by the energy storage requirements in ampere-hour (A-h) capacity. The A-h requirement is determined by the load, by the maximum eclipse time, and by the chosen DOD. The required number of cycles is determined by the mission time and the orbits per day.

Batteries may be deeply discharged to save battery weight and initial cost, with a planned replacement (and cost of the replacement); or run to shallow DOD, increasing the battery weight and initial cost. Life cycle trades indicate low-temperature (0° C) operation with moderate (20% to 40%) DOD and higher-than-minimum initial battery weight as the best system approach.

Batteries must be charged under controlled conditions so that the charge rate is not too high and so that overcharge does not occur. Charge and discharge results in a net inefficiency and in waste heat. NiCd batteries are available in capacities at several increments up to 100 A-h and can be made as large as 150 A-h.

NiH_2 battery technology is much more recent than NiCd. Serious development started in 1970, and, thus far, only one system has been flown (at synchronous altitude). Current cells are limited to 50 A-h size, but NiH_2 batteries can have slightly deeper DOD than NiCd batteries. The lifetime of these batteries has not yet been proven, and less is known about the effects of operating conditions on that life. The expectation is that the NiH_2 life cycle will be greater than that of the NiCd.

Although several fuel cell systems have been flown as energy sources, on manned spacecraft, reversible fuel cell (RFC) systems have not yet been perfected. In an RFC system, the fuel cell converts the hydrogen fuel and oxidizer (oxygen) into electrical energy (and water) during the eclipse. An electrolysis unit is used to electrolyze the water (using electrical energy from the energy source) back into H_2 and O_2 during sunlight. Water and gases are stored in separate tanks at low pressure.

There is a net inefficiency in this energy conversion cycle, and overall RFC efficiency is generally lower than that for a battery. The efficiency can be varied over a considerable range by design choices. Lower current densities result in higher RFC weight, but in higher efficiency. Increasing the operating temperature improves the voltage regulation (and effectively the efficiency) and lowers the weight, but it reduces the life. Increasing the pressure improves efficiency and raises the weight, but also reduces the life. These interrelationships are very complex and are currently under investigation. RFC lifetime is also determined by the necessary pumps, compressors, valves, and regulators.

At low-inclination orbits, interaction between a gravity gradient stabilized insulated conducting wire tether and Earth's magnetic field can provide a source of electrical energy. Electrons can be collected by a metallized film balloon at the upper end and ejected by an electron gun at the lower end. This causes a current to flow through the wire by virtue of moving at high speed through the Earth's magnetic field. The return current will close by spiralling along the magnetic field lines that intersect the ends of the wire. The magnitude of the power developed depends on tether length and control of the electron gun's current. The system extracts energy from the kinetic energy of motion through a small force that reduces system altitude. As an energy storage system, the altitude could be increased by driving a solar-array-produced current through the wire and placing the electron gun at the upper end instead of the balloon. The interaction of the current in the tether and the Earth's magnetic field produces an electromotive force that can increase the kinetic energy of the space station and, thus, its altitude.

Energy Distribution

Each user wants to have energy available in a convenient form; however, the requirement may occur at a considerable distance for the source. Efficiency of power distribution (or transmission) is important because the source and storage are so expensive that no energy should be wasted. The choice of dc vs ac distribution as well as the choice of voltage (and frequency, if ac is chosen) is very important. The regulation approach also has a great effect on efficiency.

Most spacecraft have been designed with the same type of 28 Vdc power distribution initially derived from aircraft experience and components. However, with power requirements from 10 to 150 kW (and more), and with possible distribution lengths of hundreds of feet, this low level of voltage becomes inadequate.

When current requirements this high, (150 kW at 28 V is more than 5300 A), conductor size and weight, switch and regulator semiconductor losses, and the resultant heat will all increase accordingly. Higher voltages can keep distribution system losses and weight down and the efficiency high. Studies show that the higher the voltage (within limits), the less the life cycle cost.

Three limitations restrict the voltage choice: 1) plasma interactions (in LEO), 2) semiconductor technology (voltage ratings), and 3) personnel shock hazard (safety). The plasma sheath surrounding the Earth interacts with voltages on the spacecraft (particularly the solar array) to cause leakage currents, corona, and (ultimately) arc discharges as voltages increase. The energy lost to leakage is ±1% for 400 Vdc. The semiconductor voltage rating must be about 2.5 times the working voltage. This restricts system voltage to below 400 V for impending availability of 1000 V components. Personnel safety for electrical shock is adequate below 400 Vdc. These factors indicate that the choice of system voltage should be more than 120 V and less than 270 V for ac or dc.

Most spacecraft distribution systems have been dc. However, study is under way to investigate the use of ac distribution or some hybrid of ac and dc. This effort includes the development of components for ac systems. The choice will be determined by the lowest life cycle cost, flexibility, and by compatibility with user requirements.

Station or platform loads will consist of many small loads ranging from 0.1 to 2 kilowatt each; power source buses of very large kW capacity are not needed. For reasons of redundancy, fault protection, load control, and the practicalities of large current switching, several smaller buses are indicated. Several power source channels, each with one distribution bus and one energy storage element (such as a battery), are optimum from a cost and reliability standpoint. Alternate connecting options of the loads to the buses can be provided, and therefore, direct paralleling of the channels may not be required. This avoids the complication of parallel operation of batteries or RFCs.

The number of channels are defined by the battery (or RFC) capacity. Because a large number of channels would be used (at least 12 for a 150-kW station), a natural capacity degradation occurs with channel failures, and therefore, extra redundancy is not required.

Good bus regulation could reduce or eliminate the regulation needs of each user. Practical considerations (i.e., on/off control, overload protection, and fault isolation) suggest that avoiding user load regulation is difficult to realize. Also, many loads do not require regulation (heaters, motors, etc.). A better system design will allow the distribution bus voltage to vary as provided by the energy source output characteristics and regulated (as needed) for each user.

Other Subsystems

Temperature could be controlled by a large centralized radiator using one- or two-phase fluid transport. Attitude could be controlled by control moment gyros with magnetic torquers used for desaturation and propulsive sources used for docking distrubances, reboost, and control backup.

Commonality

Since optimum commonality has been identified as essential for the space station/space platform programs, the degree of modularity (the smallest element to be replicated) to be implemented must be addressed. This is especially important, since the optimum size of the space platform may be too small to meet space station utility requirements. If so, weight and size of the station, as well as its cost and complexity, would increase. Another consideration is the minimum quantity needed to provide designed redundancy and safety (see Fig. 72). The energy storage module (channel) sizing is shown as a function of number of modules and size (kW) of modules. For platforms in the 14-to-18-kW range, only module sizes of 5 to 9 kW are possible. For resource modules of a more likely size (37.5 or 75 kW), this dictates large numbers of modules (from 5 to 8 and from 9 to 15). Furthermore, smaller numbers of larger modules (which may be more efficient for the station) would be unsuitable to use on the space platform.

NUMBER OF MODULES	MODULE SIZE (kW)															
	5	6	7	8	9	10	11	12	13	14	15	16	17	18	19	20
2	10	12	14	16	18	20	22	24	26	28	30	32	34	36	38	40
3	15	18	21	24	27	30	33	36	39	42	45	48	51	54	57	60
4	20	24	28	32	36	40	44	48	52	56	60	64	68	72	76	80
5	25	30	35	40	45	50	55	60	65	70	75	80	85	90		
6	30	36	42	48	54	60	66	72	78	84						
7	35	42	49	56	63	70	77	84								
8	40	48	56	64	72	80										
9	45	54	63	72	81											
10	50	60	70	80												
11	55	66	77													
12	60	72														
13	65	78														
14	70															
15	75															

SPACE PLATFORM
37.5 kW RESOURCE MODULE
75 kW RESOURCE MODULE

Fig. 72 Energy storage modular sizing.

One solution to the module sizing question would be to have a space platform design that would comprise one resource module when used on the station. Several such resource modules would be used for the initial space station, and twice as many could be used for the later version. Another solution would be to use a modular structure that consists of variable numbers of bays. If subsystems were configured as modules that could plug into holes in the structure in these bays, size could be kept small for the platforms and made larger for a 37.5- or 75-kW resource module. The initial station could use four 37.5-kW modules, but would only be equipped as required. Additional modules would be plugged in for station growth; possible expansion options are shown in Fig. 73.

Similarly, solar arrays and radiators could be modularized into subwings and/or panels so that a suitable number of either could be assembled. Thus, an 18.75-kW platform, or four 18.75-kW resource modules (for the 75-kW station), or four 37.5-kW resource modules (for the 150-kW station) could be created by using different quantities of the identical building blocks. Figure 74 shows some projected commonality features that may be required. Additionally, commonality could use the same method of berthing/docking for both the station and the platform so that orbiter

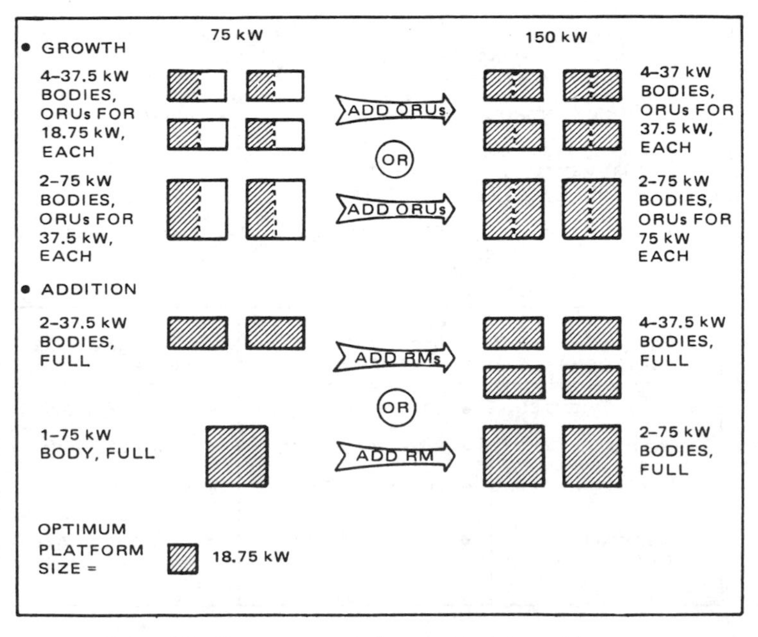

Fig. 73 Resource module expansion options.

changes would be not required. Using the same approach for ground control and ground stations would further reduce cost. Similar logistics and ground handling approaches should enable payloads to be used on either the platform or the station without change. Although reusable/permanent orbital platforms are a relatively new concept, a number of different platform types have been conceived: 1) reusable platforms, 2) co-orbiting platform (near the manned space station), 3) polar platform (in Sun-synchronous orbit), and 4) GEO platforms (35,800 km above the equator).

Reusable and co-orbiting platforms are primarily mission-dedicated, because special subsystems are required

- SOLAR ARRAY
  - WING
  - SUBWING
  - PANEL
  - ARTICULATION
- ENERGY STORAGE CHANNELS
  - MODULE SIZING
  - DISTRIBUTION (VOLTAGE, ETC)
- STRUCTURE
  - BODY SIZE/SHAPE
  - "PAYLOAD" INTERFACES
  - ORBITER INTERFACE
- ATTITUDE CONTROL
  - ORIENTATION
  - MOMENTUM MANAGEMENT
  - POINTING ACCURACY
- DATA
  - BUS
  - STORAGE
- COMMUNICATIONS
  - ANTENNAS
- THERMAL CONTROL
  - RADIATORS
  - INTERFACES
- REBOOST PROPULSION
- ORU
  - SIZING
  - INTERFACES

Fig. 74 Required resource module/space platform commonality.

ASTRO PLATFORMS	EARTH OBSERVATION PLATFORMS	MATERIAL PROC PLATFORMS
• STAR ORIENTED	• EARTH ORIENTED	• SUN ORIENTED
• HIGH ACCURACY	• HIGH ACCURACY	• LOW ACCURACY
• LOW POWER	• LOW/HIGH POWER (SAR)	• HIGH POWER
• MEDIUM DATA RATE	• HIGH DATA RATE	• LOW DATA RATE

Fig. 75 Co-orbiting and reusable platform types.

due to the payload requirements (see Fig. 75). Permanent orbital platforms are facilities that stay in orbit and only exchange payload (facilities). Regular maintenance operations, therefore, are required from a specially equipped servicing vehicle. A preferred mission for a permanent platform would be the permanent polar platform (PPP) in Sun-synchronous orbit for all types of regular global monitoring.

Acknowledgments

The authors wish to acknowledge the following persons and organizations for providing the material and assistance to prepare this chapter:

I. Skoog, Dornier Systems; W. Wienss, ERNO; H. Kraiman, General Electric; C. Poythress, Hamilton Standard; D. MacPherson, Hughes Aircraft; D. Koelle, MBB Space Division; D. Wenesley, McDonnell Douglas Astronautic West; C. Covington, NASA/JSC; C. Gregg, NASA/MSFC; G. Hanley, Rockwell International; and A. Sorensen, TRW.

Chapter VI. The Long-Range Future

The Long-Range Future

Jesco von Puttkamer
NASA Headquarters, Washington, D.C.

Introduction

When President John F. Kennedy, in 1961, reviewed potential goals for a major new national initiative in order to demonstrate to the world that the United States remained the leading nation in technology and social vitality even after the Soviet Union's surprising advances in space--which led to the first manned flight by Yuri Gagarin--his options in space flight were generally limited to an Earth-orbiting space station and a manned mission to the Moon because, as NASA Administrator James Webb, Defense Secretary Robert McNamara, and Vice-President Lyndon B. Johnson argued, "It is man, not merely machines, in space that captures the imagination of the world."(1)

By the year 2000, with a fully grown low-Earth orbit (LEO) space station and its orbital transportation, habitation, and operations infrastructure in place, the number of options for longer-range goals in space will be considerably larger. In addition, because the rationale for undertaking major space projects will have to consider numerous other arguments, such as scientific, industrial, commercial, international, etc., besides the quasisymbolic boost to national prestige that motivated Apollo, the job of selecting and coordinating between them will be considerably more complex.

This is because the existence of the space station will not only enhance Earth-orbital and deeper-space ventures, but also--as a unique space R&D facility, operations base as well as transportation node--enable entirely new initiatives for human advancement in space not possible before. Unlike the era of the Kennedy decision where there was little choice, the "name of the game" in the post-space station era is clearly "options" and "multiple objectives" because of

---

*Program Manager, Long-Range Studies, Office of Space Flight.

the presence, permanently, of humans in orbit in a position analogous to a beachhead on a new continent, with all the manifold possibilities of further progress inherent in such a staging point.

It is these possibilities that suggest the space station as a logical next step in mankind's sociotechnological evolution,(2) which is extending man's niche in the universe. If space is regarded as a frontier, the space station is required as a focal point for exploring and developing the new frontier, as an operations base for the continued exploration of the unknown, and as a major ingredient for expanding human knowledge.

To look at the long-range future beyond the space station, thus, is equivalent to asking, what are these manifold possibilities? And it also becomes immediately clear that the constraints of such an undertaking are set not so much by the limits imposed on a written paper but more by the limits of our imagination. In order to avoid focusing our access to these possibilities too narrowly, the long-range view, to be productive, must strain to glimpse the full "option bank" of the future and, thus, include a weighted measure of vision.

Human Evolution and Roles in Space

In the author's long-range view (Fig. 1), our increasing capability in space can be seen to advance in three major successive phases: 1) Easy access to and return from space; 2) permanent presence in LEO; and 3) limited self-sufficiency of man in space.

The development of the Space Shuttle/Space Transportation System (STS) for transportation and of the initial space station for orbital habitation and experimentation provides the main elements of the infrastructure of phase 1, to be accomplished by the end of this decade. But permanent manned presence requires considerably more than this: an orbital operations capability of a scale large enough to respond adequately to the projected socioeconomic needs of the 1990's. In particular, phase 2 will add the capability of manned access (sorties) to geostationary orbit (GEO) and the operational deployment of large space structures. To become more autonomous in space, man must proceed to develop closed-cycle life support systems and larger-scale industrial applications in space which, in phase 3, should lead to closed ecological systems (including space-grown food), space construction, space industrialization, and access to extraterrestrial materials. These developments will be discussed in turn below.

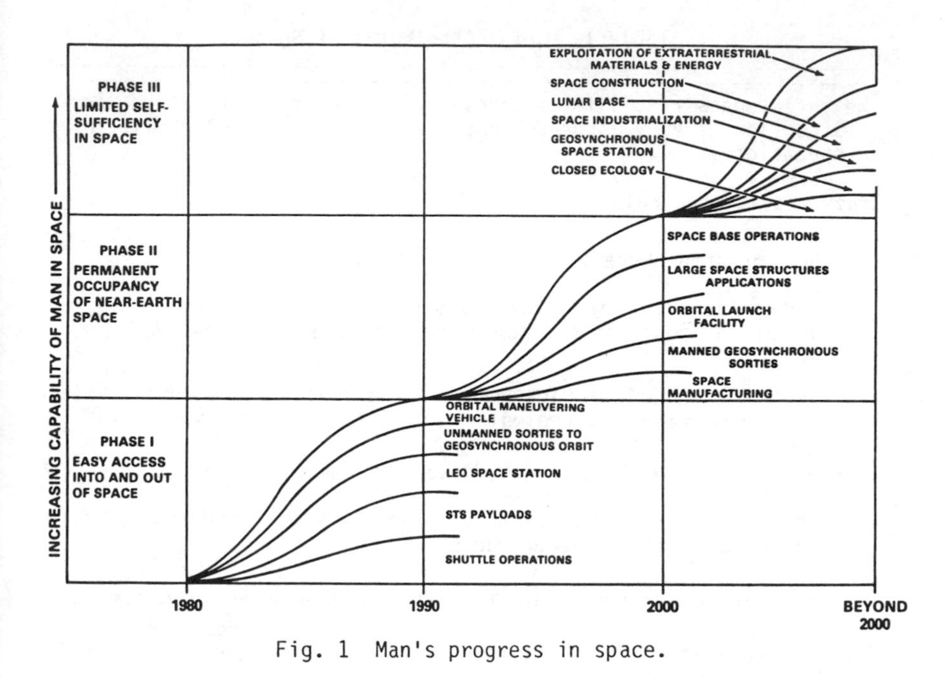

Fig. 1 Man's progress in space.

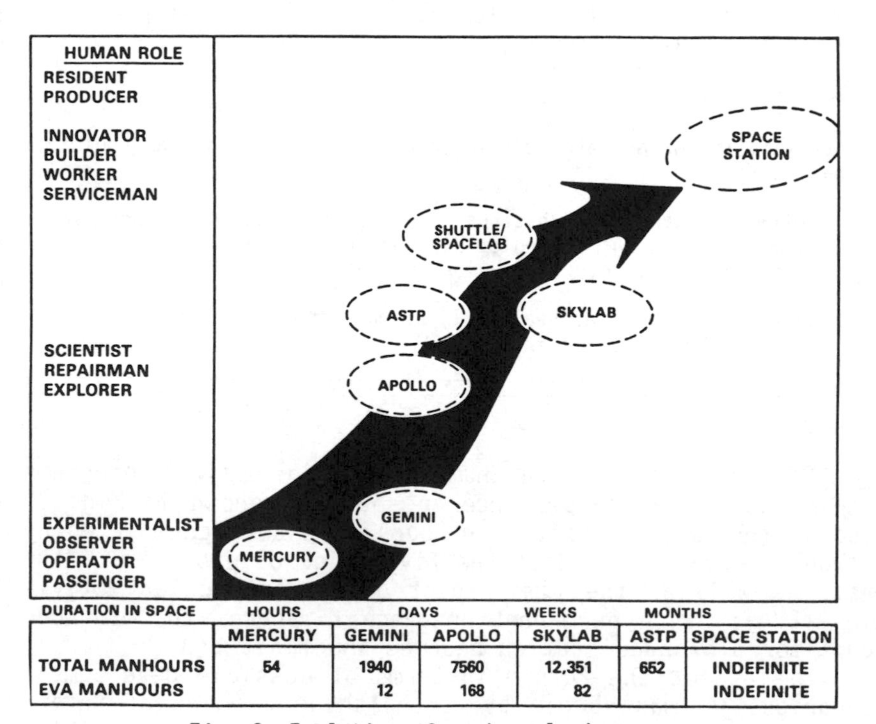

	MERCURY	GEMINI	APOLLO	SKYLAB	ASTP	SPACE STATION
TOTAL MANHOURS	54	1940	7560	12,351	652	INDEFINITE
EVA MANHOURS		12	168	82		INDEFINITE

Fig. 2 Evolution of man's role in space.

Table 1 Useful Attributes of Space

Weightlessness (facilitates special manufacturing activities, construction of very large delicate structures, and reliability of operations).
Easy gravity control.
Absence of atmosphere (unlimited high vacuum).
Comprehensive overview of Earth's surface and atmosphere, for communication, observation, power transmission, and other applications.
Isolation from Earth's biosphere, for hazardous processes: little or no environmental, ecological or "localism" issues.
Readily available light, heat, power (10 times rate on Earth),
Infinite natural reservoir for disposal of waste products and safe storage of radioactive products.
Super-cold temperatures (infinite heat sink near absolute zero).
Large, three-dimensional volumes (storage, structures).
Variety of nondiffuse (directed) radiation (ultraviolet, x-rays, gamma rays, etc.).
Magnetic field.
Availability of extraterrestrial raw materials on Moon and possibly on asteroids.
Avoidance of many Earth hazards (storms, earthquakes, floods, volcanoes, lightning, unpredictable temperatures and humidity, intruders, accidents, corrosion, pollution, etc.).
Potentially enjoyable, healthful, stimulating or otherwise desirable for human well-being.

The progress of the human role in orbital programs leading to permanent presence in space is shown in Fig. 2. Also listed are the total man-hours in space accumulated by astronaut crews in each of the five major US programs of the past, as well as the times spent on extravehicular activities (EVA). With permanent presence in space, the man-hour count for STS/space station becomes indefinite.(3)

To inspect the future in terms of possible development scenarios, we must begin by establishing a point of reference, an understanding of the plateau of "givens" from which to make our projections. What will our situation in space be by the year 2000, that is, when man, after reaching our current goal of achieving easy access into and out of space, has also accomplished phase 2 as discussed above?

Plateau of Departure: The Space Infrastructure at Year 2000

First, it behooves us to ask, "What are the most significant resources of space that provide the main motivation behind the Space Station, its infrastructure, and its subsequent further growth?" A listing of the major attributes of space that make it attractive for various types of industrialization, is given in Table 1 (Refs.4,5).

Because future progress in space is intrinsically connected with continued progress in many key technologies,(5) particularly for the manned operations in space now being envisioned (such as space laboratories, platforms, bases, industries, and, over the long term, space settlements), it appears appropriate to proceed with an assessment of space technology itself, based on past experience and projected out to year 2000. This is attempted in Table 2 (Refs.5,6).

The rapid growth in commercially significant technologies such as communication channels, viewing resolution, and cost of launching clearly suggests that space in the years around and beyond 2000 is likely to become a burgeoning frontier for the development of an increasing number of industrial and commercial ventures in LEO and GEO. Assuming that the world expenditure on all space developments was roughly $20 billion in 1983, an annual growth of only 2% of those investments would amount to investments of $75 billion annually in 2050 and $202 billion in 2100 (1983 dollars). For a somewhat more optimistic scenario of 8% growth, however, the annual investment in space would reach an astounding $3.5 trillion.(Ref.5)

Given the above projections, at year 2000, from the viewpoint of transportation, habitation, and operation, we will be able to operate with flight crews routinely--that is, safely, successfully, on schedule, and economically--not only between Earth's surface and LEO as the Shuttle will allow us to do soon, but also at geostationary orbits (GEO), between orbits, and eventually at lunar and planetary distances.

Taking this "year-2000 capability" as a mandate for today and as a starting position for our long-range view, the basic space station infrastructure will require certain augmentations by the end of this century which will avail us at least with the capability to 1) support the establishment of large permanent facilities for science, R&D, commercial,

Table 2 Progress in Space Technology, past and projected

System	1960	1975	1990	2000
1. Launch vehicle payload capability, lb	25	250,000	80,000 reusable	10^6 reusable
2. Communication channels	15 (LEO)	15,000 (GEO)	100,000 (GEO)	10^7 (GEO)
3. Communication bit rate (Mars-Earth)	8	10^5	10^7	10^9
4. Mission length, mandays	0.1	250	10^5	5×10^5
5. Resolution, km	5 (LEO)	0.1 (GEO)	0.05 (GEO)	0.02 (GEO)
6. Data storage onboard	15 pages (0.5 Mbits)	2000 books (20 Gbits)	1/2 Library of Congress (8000 Gbits)	10 times Library of Congress
7. Energy storage, kW-h/lb	0.02	40	800	1200
8. Active circuits, per $in.^3$	4	120,000	5×10^8	10^{10}-10^{12}
9. Onboard computer speed, operations/s	0.002×10^6	0.5×10^6	30×10^6	$100\text{-}1000 \times 10^6$
10. Launch costs, 84$/lb to LEO	20,000	3000	1000	100-300
11. Position error, m	1000	50	0.1	0.02
12. Failure rate, No./h/Mbits	10^{-2}	10^{-4}	10^{-6}	10^{-7}-10^{-8}

Note: Adapted largely from NASA Technology Forecast Report, Ref. 6.

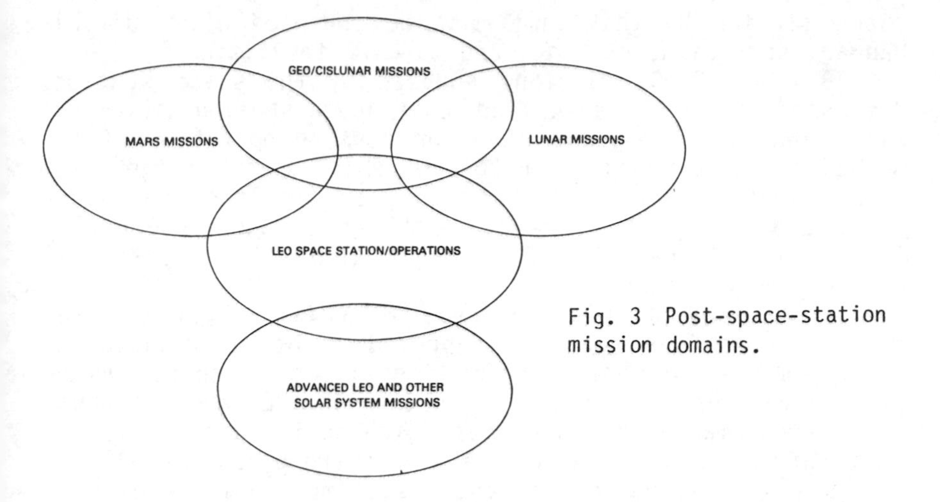

Fig. 3 Post-space-station mission domains.

and operations support in LEO and GEO; 2) provide routine, economical, and flexible access to all orbits by men and robotic systems; 3) institute routine checkout, refueling, repairing, and upgrading of space facilities, as well as debris removal in all orbits to GEO; and 4) devise and implement innovative STS uses and missions such as large space tether applications for power generation, nonpropulsive transportation, and satellite constellations.

By year 2000, an evolutionary cryogenic concept of the orbital transfer vehicle (OTV) family will provide reusable manned sortie flight capability to at least GEO. However, this system should also be capable of providing the basis for extended transportation for a lunar base, and for such planetary projects as a robotic Mars Sample Return Mission which are seen among the leading optional goals for the period around the turn of the century. Other planetary exploration missions have been identified by the Solar System Exploration Committee (SSEC) of the NASA Advisory Council. They include a recommended Core Program(7) comprising the Venus Radar Mapper, Mars Geoscience/Climatology Orbiter, Comet Rendezvous, and Titan Probe missions, as well as such augmentation missions as Mars Surface Science, Mars Sample Return, and Comet Sample Return that are not included in the Core Program.

The interorbital transportation system OTV, supplemented at times by a remotely controlled orbital maneuvering vehicle (OMV) with special kits for remote servicing, tanking, and debris capture, will be able to fly with or without crews with minimum change. Because servicing, maintaining, and refueling these reusable vehicles appears to be more advantageous if done in orbit, rather than on the ground, the space station by that time must be equipped with suitable hangar, servicing, storage, and tanking facilities.

By year 2000, missions enabled by the space station, discussed below as so-called post-space-station missions, will clearly introduce a need for routine operation of unmanned cargo-carrying Earth-to-LEO vehicles with larger payload diameter, greater lift weight, and/or lower cost-per-pound capability than the Space Shuttle. Such a Shuttle-derived launch vehicle (SDLV) might also satisfy potential military requirements.

By the end of this century, commercialization of materials processed in space is expected to be at an advanced level leading to manufacturing facilities in space. Among those areas that currently appear to hold the greatest promise are pharmaceutical products, high-purity advanced semiconductors, and unique glass materials and glassy metals.

Besides the various advanced systems, tools, and techniques necessary for routine servicing operations throughout the LEO/GEO/Cislunar regimes utilized by satellites and spacecraft, for improved permanent manned presence by year 2000, we must also have developed important augmentations to crew and life support systems in order to provide--for the long range beyond 2000--closed-cycle, regenerative operation (water and air loops) of onboard environment control and life support (EC/LS); regenerable space-maintained extravehicular EC/LS backpack systems; advanced high-productivity mobility systems; Earth-norm food, hygiene and interior habitability standards (the latter flexible to allow the growth of new "space-culture" standards over time); and advanced on-board automation capable of autonomous handling of tasks that do not necessarily require continued manned attendance, while being flexible enough to allow easy upgrading as electronics, avionics, and robotics technologies advance at their rapid pace.

The View Beyond Year 2000

Mission Spectrum: General Trends

As we look beyond year 2000 to those long-range future scenarios whose emergence and evolution are made possible by the space station as both a plateau in human progress and a major nodal point in time, we can see that their goals will include, at the least, exploration, commercialization, earth applications, and expanding human presence, as well as sustained space R&D for the conception and development of innovative systems and techniques currently not available, not understood, or even not yet known. Table 3 examines these potential mission goals with respect to their relevance to mankind's materialistic, intellectual, and humanistic needs. Also included is an attempt to identify the most likely space domains involved to meet these needs.

It is apparent that major manned and automated space missions that would answer to these goals and that would provide both unprecedented scientific/technical benefits and a vivid example of the dynamicism of the United States and its partners throughout the world, can be envisioned in all regimes of space accessible to us by year 2000: LEO, GEO, and other high-energy orbits, lunar orbit and lunar surface, and at the distances of the inner and, later, outer planets.

Thus, our visions of the era beyond year 2000 must include a mission spectrum with programs such as a complex of advanced scientific and industrial facilities in LEO, permanently manned scientific and communications facilities in GEO, a permanent US base on the Moon, a manned expedition to Mars, etc., all of these being programs that offer great opportunities for international participation up to and including joint missions. In each case, the LEO space station and its infrastructure would serve as the R&D facility and staging base with a maximum degree of commonality and modularity; but since this infrastructure and its future capabilities are evolutionary, we will require continued assessment and "parallel" (instead of "sequential") study of the requirements imposed on these novel capabilities by the future programs. It is of eminent importance to realize the fundamental reasons for this need of a broad-based, rather nontraditional approach at planning after the initiation of the space station as a "beachhead": The various space domains of interest cannot be isolated from each other; they "overlap" and are mutually depend on each other (Fig. 3). No longer will the future be as simple and straightforward as it seemed at the time of the Apollo decision. The complex mission spectrum that now becomes possible and the costs of its implementation will be affected deeply by the network relationships and synergisms between program and mission systems, design elements, and technologies.

Increasingly in the long range, we will also see an era of growing robotics capability, of automation systems working in space along with humans to help improve our life on Earth and pursue further challenges in space. There can be no doubt that with the establishment of permanent manned presence in orbit in our future there will be--more than ever before--a place for both men and "intelligent" machines in space and, through technology transfer, on Earth. Machines will certainly not replace humans in space, but they will free us for more productive undertakings. While artifi-

Table 3 Post space station mission goals vs. domains

Goals	Benefits	Type[a]	Domain involved
Science and exploration	Determine solar system origin/early history	I	LEO and other solar system science
	Understand galactic structure/dynamics	I	LEO and other solar system science
	Understand cosmology	I	LEO and other solar system science
	Verify physical laws	I	LEO and other solar system science
	New resources from space	M	Lunar surface/Mars moons/asteroids
Commercialization	Optimize industrial activity	M	LEO and GEO space
	Hazard removal	M	LEO, GEO and cis-lunar space
	Energy from space	M	GEO/cislunar space
	New resources from space	M	Lunar surface/Mars moons/asteroids
Earth applications	Agriculture, forestry, fishery management	M	LEO and GEO space
	Protect environment	H	LEO and GEO space
	Aid individuals in peril	H	LEO and GEO space
	Aid crime control	H	LEO and GEO space
	Aid internal security	H	LEO and GEO space
	Improve government	H	LEO and GEO space
	Energy from space	M	GEO/cislunar space
	New resources from space	M	Lunar surface/Mars moons/asteroids
	International cooperation	H	LEO and GEO space, lunar base, Mars missions
Expanding human presence	Space habitation	H	LEO and GEO space, lunar base, Mars missions
	New technological capabilities	M	LEO and GEO space, lunar base, Mars missions
	New resources from space	M	Lunar surface/Mars moons/asteroids
	Hazard removal	H	LEO, GEO and cis-lunar space
	International cooperation	H	LEO and GEO space, lunar base, Mars missions
	Promote international peace	H and M	LEO and GEO space, lunar base, Mars missions

[a]I = Intellectual, M = Materialistic/Utilitarian, H = Humanistic

cial intelligence systems will not have a major impact on the initial space station design for the early 1990's, an evolutionary station should be designed with the future availability of highly autonomous intelligent machines in mind.(8) People and machines in space will demonstrate new types of interactions and will thrive, not just survive. From the first stages of planning and design, they must be viewed as an integral system.

It should be obvious that the number of people who could be involved, along with machines, in space industrialization and other space ventures may depend to a considerable degree on transportation costs. Indeed, as transportation costs decline, the optimum division of labor between humans and various kinds of robots should shift toward an increasing demand for humans. Thus, it is clear that the industrialization of space will not only create new markets and more jobs here on Earth, but many of the space industries of the decades ahead will also require highly qualified workers in habitable orbital facilities.

LEO Regime Beyond 2000

If we start our preview of post-space-station programs beyond year 2000 at the LEO regime, the optimum balance between the numbers of technicians taken into space and the amount of automation will be determined by the space science, space exploration, and space industrialization ventures under-way. But, as Brown has pointed out,(5) there is at least one aspect of the expected developments in space which seems almost ready to get started and which inherently would attempt to maximize the presence of people: space tourism.(9) Current estimates by one commercial company with experience in "adventure tourism" of the price of a seat on a Shuttle Orbiter modified for passenger transportation run between $2 and $3 million, and it is believed that there will be a worldwide market for 300 to 350 passengers a year at this price during the second half of the 1990's. This market will grow explosively as transportation costs decline as indicated in Fig.4,(5) and as average personal income continues to rise about 1% to 2% annually, as it has in the past.

Brown has identified a number of critical factors that are likely to affect the pace of future investments in space in a major way.(5) These key elements are:

1) Successful technology (declining transportation costs to LEO, efficient OTV engines, orbiting space bases, and profitable space industries, increasingly reliable and

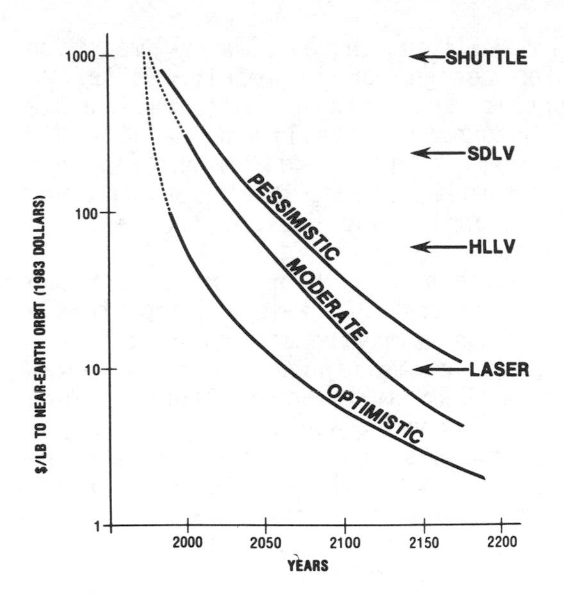

Fig. 4 Transportation costs to near-Earth orbit.

"intelligent" robots, lunar base, planetary space stations and settlements).

2) Tourism ("inexpensive" safe transport, space hotels, hospitals, convention centers, exciting journeys).

3) Favorable health expectations (Physical/mental/emotional, longevity).

4) Attitudes on Earth (increasing private investment, national enthusiasm, international cooperation, and peaceful competition).

5) Quality of "frontier life" (favorable social dynamics, benign politics, rapid economic growth).

Of the factors listed above, the fourth is related to some of the social and/or political driving forces on Earth.(5) These may be considered on three levels: First, commercial space activities must show at least a reasonable potential for becoming profitable. Whether financed privately, by quasipublic institutions, or by governments, they will need to appear to be profitable investments if they are to receive the required capital infusions for prolonged growth.

As we saw above, even as little as a 2% annual growth can provide huge space-based investments in the 21st Century. In addition, as shown on Fig. 4, by the late 21st Century or early in the 22nd Century, transportation costs alone may well be lower by a factor between 20 and 50 from those of today. When that factor is coupled with the expected miniaturization of electronics, advances in artificial intelligence, automation, versatile robots, the availability of materials from lunar and other extraterrestrial sources, relatively maintenance-free designs, etc., as suggested by extrapolating from the pre-2000 growth of technologies discussed on Table 2, it becomes obvious how incredibly difficult it is today to visualize what array of projects will be commonplace 100 years or more in the future that would require annual investments of a trillion dollars or more.

The second driving force for near-term space development is related to national enthusiasm. Whether or not the commercialization of space is profitable and rapid, many aspects of space exploration and development are known to be extremely important for basic scientific knowledge. In addition, the many valuable spin-offs from those technological developments that have occurred and can be anticipated in the future will provide valuable innovations and render important services to society. National pride in such accomplishments is a significant factor in many countries. Prolonged national enthusiasm appears to be an important prerequisite for a relatively optimistic long-range space scenario.(5)

The third driving force, and one of potentially great importance over the long term, is the requirements for both international cooperation and peaceful competition in space ventures. For example, many space developments will be very expensive if they are to be accomplished over the next several decades. The costs are so large that unless several nations participate over a relatively long period of time, the project could be postponed almost indefinitely. Thus, it is relatively easy to envision that cooperation would become essential for some of the more expensive projects. On the other hand, competition may often be desirable as a way to foster innovation and enthusiasm for space ventures, just as it is for many enterprises on Earth.

Routine space station operation involving assembly and launch, retrieval, maintenance, and repair will clearly enable the growth of very large science projects in LEO in such disciplines as Earth and life sciences, astronomy, and astrophysics. Of particular significance for the latter two appear to be the station-supported assembly and occasional tending of a 10-m optical telescope as a follow-on to the current Space Telescope, a large x-ray facility as the next step beyond the current Advanced X-Ray Astrophysics Facility (AXAF), a 100-m infrared/submillimeter telescope as next step in Large Deployable Reflector (LDR) evolution, and a high-energy astrophysics laboratory as follow-on to the High-Energy Astronomy Observatorys (HEAO).

Thus, in the years of the early 21st Century, very useful large-scale science activities and applications services will be accomplished in LEO space.

The ultimate location for the most effective utilization of a manned space system, however, appears to be at GEO.

GEO/Cislunar Regime Beyond 2000

A geosynchronous site has extremely attractive features such as continuous direct line-of-sight to large areas on Earth, nearly constant viewing to study celestial objects and the sun, an inherently stable orbit, and maximum avoidance of Earth albedo and occultation.

For these reasons, there would be a science and applications complex at GEO containing facilities for communications, Earth-pointing observation, navigation, tracking, astrophysics, and solar-terrestrial observations. The communications facility would include large antennas with flexibility for switching between beams/antennas, permitting efficient use of the frequency spectrum, and smaller, less complex ground terminals than today's. Because of the new capabilities of assembly, repair, maintenance, and checkout activities in space, it will be possible to make satellites much larger and more complex; at the same time, in a reversal of present practices, the user equipment on the ground can be made tiny, simple, highly portable, and inexpensive. This principle of complexity inversion, which is the key to making space flight directly beneficial and affordable to users, including the "man in the street," was put forward in a study for NASA in 1976.(10)

Because of the vantage point of geosynchronous altitude, millions of Earth-based users can be serviced by one or only a few satellite platforms. Because of their increased size and complexity, these satellites will be able to perform functions not possible with simpler and smaller satellites. Due to the weightlessness of space, it will be practical to assemble much larger antennas, reflector structures, and science platforms than those we now have in orbit. Typical multibeam antennas for radio frequency and microwave output could be from 200 to 600 ft in diameter; and passive space reflectors for beaming light to Earth or relaying energy to space-based users could reach diameters of 1000 to 3000 ft.

The new, almost unlimited opportunities offered by space communications using these advanced GEO systems in the beyond-2000 era, have the potential to answer many serious needs of mankind through numerous personal, civic, governmental, industrial, and international applications. For example, improved educational opportunities would become available with a large, high-power, and direct-broadcast satellite bringing televised programs to mountainous, rural, and remote areas of the globe. Multibeam satellites could reduce crime rates by providing the police with jam-proof communications from any location.(10)

Improved public and governmental services will be obtained by using multibeam satellites for direct and immediate communications to disaster areas. These satellites would also provide a way to hold direct and instantaneous votes and polls. They can improve national security and international air/sea traffic control; large arrays in space can detect ships, people, and goods that illegally cross borders and coasts; low-frequency loop antennas can provide global communications with ocean vessels; and 24-hour, all-weather monitoring of global air and ocean traffic can be accomplished with a large microwave antenna satellite. Public service satellite systems will also be used to provide services such such as wide-area health care, teleconferencing, electronic commuting, nuclear fuel tracking, news services, and search and rescue.(10)

High-resolution Earth observation devices could continuously monitor resources and the environment of our planet. The GEO science observatory would be capable of simultaneous full-spectrum observations at all wavelengths from 10^5 Hz radio to Gamma rays. Celestial observations would benefit from long exposure times of a given source, permitting smaller instruments and continuous monitoring of variable and transient sources. Other advanced GEO missions include public service and commercial functions such as hazardous waste management, environmental modification programs, trash management, and tourism, as well as military operations.

Man's presence will be required at GEO to provide space operations services to satellites as well as to large, long-life platforms and antennas. The appropriate mix of man, remote control, and automation will have been determined during the development of service operations know-how in LEO, but it is expected that GEO will introduce additional R&D requirements for communications and service operations. Because GEO operations will become increasingly complex and because they are sensitive to transportation costs, this human involvement will tend to grow from intermittent to permanent presence.

For accommodation and support of space workers at geostationary altitude, adequate GEO habitation will likely be required beyond the year 2000. While not a necessary part of a subsequent lunar transportation system, such a manned GEO space station, perhaps preceded by a temporarily mannable "shack" (circa 2004), would probably be highly beneficial also to a subsequent lunar program concerned with generating lunar material resources of value such as oxygen in gaseous and liquid form. Because of its high orbital energy, a GEO facility would generally tend to become an important transportation node for missions to and from cislunar and heliocentric space, and the availability of a nonterrestrially supplied resource such as liquid oxygen might well enhance the advantage of this facility by a large degree.

In addition to GEO, a number of other peculiar orbits in cislunar space appear to hold out promise for useful application to the post-space-station mission goals assumed above (Table 3). They include the family of pseudogeosynchronous orbits (for high-inclination coverage of Earth), long-lifetime orbits for waste storage, high-sunlight orbits outside Earth's shadow, and orbits passing through various portions of the plasma sphere. A typical cislunar science project could be an extra-long baseline interferometer (XLBI) with radio telescopes of 30-100 m diameter stationed at the Earth-Moon or Earth-Sun Lagrange points.

Lunar Regime Beyond 2000

As another one of the potentialities introduced by the space station in its "beachhead" role, the establishment of a manned lunar base during the early years of the decade after year 2000, earliest approximately in 2006, could well become a major goal of this nation and its partners-in-space for a number of persuasive economic, sociopolitical, cultural, and philosophical reasons.

Such a base would be the first permanent human activity on another natural body in space, and a landmark in human history. The Moon is a large, stable platform on which important scientific activities can be pursued, particularly scientific studies of a unique terrestrial planet other than Earth, with access to a unique record of solar and cosmic history preserved in the lunar surface materials as lucidly--to the expert--as in the pages of a book, and astronomical observations, to advance our understanding of the solar system, the universe, and their/our origins.

A lunar base would also be a potential source of nonterrestrial materials for use in space, and a major task of it would be to assess such uses and to develop techniques for them toward an eventual mining operation. If these initial steps are successful, they will clearly not remain limited to assaying, surveying, mining, excavation, beneficiation, gaseous oxygen production, and oxygen liquifaction on the Moon, but they will extend to metal extraction and manufacturing of construction materials.

There is a plethora of other resources: The lunar environment may have important scientific and engineering uses for research and technological development because of its low gravity, remoteness (isolation), vacuum, sterility, low background noise, Earth visibility, large temperature variation, low magnetic field, long day/night cycle, unique geological features, and stable station/platform character for the emplacement of large instruments and facilities.

With the successive evolution of lunar humans from Earth-dependent food and other resources to increasingly space-derived subsistence and growing self-sufficiency (Fig. 1), a lunar base would expand our understanding of man's ecology, i.e., our interrelationships with the environment, and of interhuman interactions and psychological processes, while at the same time becoming more economic to sustain.

A lunar base would be a highly visible symbol of human exploration and spirit, and of national/international commitment to a permanent presence in space, with the potential to stimulate creativity in all forms. As an initial effort in the long-term settlement of space, it would allow the evaluation and extension of man's capabilities in space, provide for the opportunity to develop new forms of social, political, and economic structures, and, in summary, enrich our lives.

The technological and management experience gained by this nation during the Apollo Program will clearly help immensely in turning this visionary scenario of the future into reality, should the United States one day decide to go ahead with this project. But no less important to its establishment and subsequent operation than Apollo's legacy will be today's STS and tomorrow's LEO space station. Both are essential stepping stones. (For some past studies see Refs. 11-14).

The Domain of Mars and the Planetary Regime Beyond 2000

Another larger goal of our space activities and a major national objective in the post-2000 era that requires the space station as a springboard will be a manned Mars landing mission. Whereas such a program would be unlikely to provide nonterrestrial materials in the foreseeable future as a lunar base or an asteroid mining program might do, its principal objectives would be humanistic and intellectual in kind: exploration and advancement of mankind in space.

In a manned planetary exploration program, Mars will undoubtedly be given priority before any of the other planets of the solar system because it offers the least severe environment for man: Due to its atmosphere, its accessible surface, its probable availability of water, and its relatively moderate temperatures (-120°C to +20°C), it is the most habitable of all planets other than Earth (and it has therefore traditionally appealed to the general public). Its resources include materials that could be adapted to support human life: air, food, fuels, fertilizers, and building materials. In addition, Mars could be an excellent gateway to the asteroid belt, facilitating the scientific exploration of asteroidal bodies by man in the next century and the eventual use of asteroidal resources such as water, oxygen, iron, nickel, and carbon.

Mars is a complex planet and a highly desirable target for scientific research. Geologic mapping of the Martian surface from Viking and Mariner data suggests almost 100 different geological units; thus, there should be a large number of exploration sites and considerable flexibility in moving and changing sites. Therefore, the main emphasis of a Mars science program will be on manned surface activities.

A manned Mars landing program would be technologically exceedingly demanding; thus, it could play a major role as a new "cutting edge" and stimulant to progress and economy.

Even more important in its long-range benefits, with a mission as large and challenging as a manned Mars expedition, international participation on a large scale would be encouraged. Space may indeed now become a potential surrogate for warfare, as some have seen it in the past.(15) While in today's world of negotiations aiming at limited arms reductions it may still appear to be a quasiutopian dream to achieve the ultimately desirable option of a total nuclear disarmament, this paradigm could become more realistic if it can be shown that the defense industrial potential of the nations can, in due time, be gainfully redirected and reutilized in peaceful programs of immensely challenging "cutting-edge" technologies and resources. If so, it is not entirely inconceivable that at the "green table" of some advanced SALT or START negotiations, a future US President could propose a joint international manned expedition to Mars.(16) Consequently, its scientific/technical/sociological benefits would undoubtedly be of an unprecedented scope and magnitude as well as, ultimately, value returned.

During the decade of the 1960's, NASA has supported considerable study activities of manned Mars mission alternatives. Over 40 different in-house and contracted studies were undertaken, with over $4 million in "then"-dollars spent on contracts (for some examples see Refs. 17-26). In August 1969, Dr. Wernher von Braun testified before Congress that a manned Mars landing could be accomplished in the

1980's as part of an integrated total NASA program, with funding requirements of $4 billion in 1971, increasing to a peak of about $8 billion in the late 1970's. The integrated program included a space station, a Space Shuttle, unmanned planetary exploration including the "Grand Tour" mission as well as a "balanced" program of applications, science, lunar exploration, etc. As time frame for the mission, a departure date of November 1981 and a return date of August 1983 were assumed.

A NASA study team reviewed the 1969 von Braun baseline in 1978 and presented an updated summary to NASA management in August 1978. Among the conclusions was the finding that a heavy-lift version of the Shuttle would be required to accommodate both the size and weight of the manned modules used in the 1969 study. Other comments were: While application of current reliability concepts (fault tolerance) would probably result in lighter and less complex systems, it is doubful whether long-duration orbital testing would be required to achieve required reliability confidence. Experience in orbital assembly of modules and booster systems should be obtained in actual flight demonstrations prior to mission start, and fluid transfer and long-term fluid storage technology must be developed in orbital precursor missions. The 1969 crew size requirements (six per ship) were considered reasonable unless scientific requirements dictate a change upward, involving the addition of mission specialists; crew accommodations would need reevaluation for improved human factors knowledge as well as male/female mixed crews (which were not considered in 1969).

Some special science conclusions of the 1978 NASA in-house study were:

1) There should be emphasis on sample return (rocks, etc.) in pristine condition.

2) Samples are desired from volcanic and temperate zones as well as from the polar cap(s).

3) Subsurface samples are highly desired.

4) En-route bioscience emphasis should be on humans (male and female, group dynamics).

5) The mission will be enhanced by man's presence, through his personal observations and his selection capability.

6) In addition to sample return, Apollo Lunar Surface Experiment Package (ALSEP)-type experiments would be considered, to be left on Mars' surface.

7) Back-contamination and quarantine requirements will impose severe constraints (different from Apollo requirements).

Also in 1978, a manned mission to the Mars moons Phobos and Deimos was proposed by Dr. S. Fred Singer to NASA. The mission scenario involved a single vehicle with six crew members, using a chemical OTV for Earth (GEO) departure, Deimos capture/escape, and Earth return (GEO insertion), and a solar-electric propulsion system (SEPS) of 2 MW array power for en-route propulsion. There would be a 2 to 4 months stay on Deimos, with an excursion to Phobos by two crewmen. The mission would require about 40 Shuttle flights for assembly in GEO.

As seen today, a manned Mars landing scenario would typically utilize the STS as well as the SDLV envisioned for the year-2000 era, to assemble an all-chemical multistage mission vehicle in Earth orbit. Total mass in Earth orbit of the expeditionary ship at departure would depend on the particular Earth-Mars constellation chosen, i.e., on the year of departure. For some of the oppositions after year 2010, these mass requirements are on the order of only 2 million pounds for a chemically propelled spacecraft delivering six crewmen to Mars and landing three of them on the planet's surface, with subsequent return of the entire crew--after a 60-day stopover--to Earth orbit for rendezvous with the LEO space station. Total mission duration would be 600 to 700 days. Much like the first Apollo lunar landing, this scenario assumes a one-flight exploration mission, leaving no usable infrastructure in Mars orbit or on the surface, but it does not exclude subsequent visits in the further course of time, with the eventual establishment of a synodic base camp on Mars.

With regard to technology requirements, one important point is that the environments involved are considerably better understood today than in 1969. The original allowances for uncertainties in meteorite shielding and Mars excursion module (MEM) weight can probably be reduced. The addition of aerobraking to an all-chemical system may deserve renewed evaluation. Small improvements in many or all systems will probably add up to significant mass reduction and improved reliability compared with 1969. This aspect should be addressed with the use of the Shuttle and the LEO space station as orbital R&D test beds. Some innovative concepts toward long-term planetary developments should be examined today, such as in situ propellant production on Mars for return missions. Manned planetary missions such as a Mars landing program are likely to reinforce the pursuit of advanced space capabilities on a growth space station in such areas as advanced space power and propulsion, in-space construction and assembly of large space systems, and long-term habitation and operations in space.

Planetary and solar system exploration missions beyond year 2000 other than to Mars have been suggested, including: 1) multiple main asteroid belt rendezvous missions with sample return, 2) Mercury surface rover, 3) large Venus balloons, 4) unmanned Venus surface station, 5) Jupiter aircraft and atmosphere balloon, and 6) advanced orbiters, landers, and rovers on the moons of Jupiter and Saturn, etc.

Other Programs Beyond 2000

Other candidate future large space initiatives that can be envisioned and that have been studied to varying degrees over the past years are: 1) in-space power generation systems for providing energy to space-based users and, perhaps, also to Earth,(27) 2) nuclear waste disposal in space,(28) 3) large automated deep-space/planetary missions such as an automated Titan Probe and an Interstellar Space Explorer,(29) 4) large defensive systems, 5) space settlement, 6) asteroidal mining, and others.

"Human" industries in space (such as medical, clinical, and biogenetic research), space science and space-based educational centers, space hospitals and sanitariums, as well as activities in areas such as entertainment and the arts are long-range possibilities that will eventually be brought within our grasp through the step-by-step development of permanent manned presence in LEO. Orbital vacation centers may also be somewhere in the future, adding a whole new dimension to space tourism.(4)

Future space scenarios were studied by NASA the last time in 1976;(30,31) the concept of Space Industrialization was evolved in 1976/77.(32) A "tree of relevance", first developed by the author at that time, showing a possible post-space-station program scenario with its synergistic interrelationships and dependencies between subscenarios, without particular temporal alignment or time scale, is suggested in Fig. 5.

Long-Range Technological Prerequisites

Some key technology advances that long-range post-space-station missions would require, are listed in Table 4.

In a more fundamental, generic assessment of the major technological prerequisites necessary for the three phases discussed in the section entitled Human Evolution and Roles in Space and in Fig. 1, none but the Space Shuttle is currently in a sufficiently advanced state to enable us to operate routinely even on the first level, phase I. Considerable progress is required in all areas for phase II capability (permanent occupancy of LEO space). This goes even more for phase III (limited self-sufficiency of man in space),

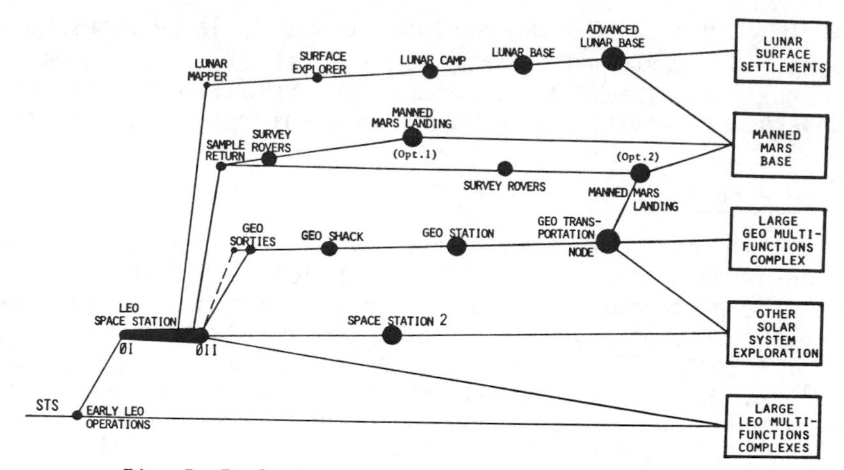

Fig. 5 Evolution of post-space-station programs.

where in some cases, to permit further expansion, considerable innovation and even breakthroughs appear necessary.(31)

Typical transportation requirements for larger-scale space endeavors of phase-III-type dimensions include a second-generation Shuttle (Shuttle 2) for low-cost personnel transport between Earth and LEO. A heavy-lift vehicle (HLV), which would utilize Shuttle-derived hardware elements, will be essential for manned lunar and planetary programs. A reusable space-based orbital transfer vehicle must become available for the advanced transportation requirements to GEO and beyond foreseen for the post-space-station period, with flexibility and growth potential for maximum-commonality use in both single-stage and multistage configurations. Further in the future is the introduction of low-thrust (electric) propulsion, both solar and nuclear powered, as well as high-thrust nuclear propulsion utilizing NERVA (Nuclear Engine for Rocket Vehicle Application)-type and advanced solid-core reactor (SCR) engines, and, later, perhaps also more remote gas-core reactor (GCR) propulsion. An advanced reusable cryogenic lunar lander would become necessary for support of a larger lunar facility, preceded by an expendable lander with storable propellants in the earlier phase of lunar encampment. Manned planetary programs will require development of an interplanetary mission module, perhaps a derivative of LEO space station modules, as well as a planetary landing system.

Space habitation requirements of the long-range future include facilities extending the early LEO space station toward becoming a more capable orbiting base. Part of these will be the equipment of a propellant depot and service sta-

Table 4 Technology requirements

Propulsion
- Advanced cryogenic (Lox/LH_2) engine for SB-OTV, MOTV-2
- Advanced storable engine for lunar and Mars landing and ascent
- Advanced electric propulsion
- Lox/hydrocarbon engine for HLLV and Shuttle 2
- Aeroassist for SB-OTV
- Propellant storage and transfer in space

Materials and Structures
- Extraterrestrial mining, processing, and manufacturing
- Space-based construction and assembly

Life Support
- Closed (regenerative) water and oxygen loops
- Advanced (regenerative) EVA and manned mobility equipment
- Advanced radiation protection
- Partially self-sufficient farms

Information transfer (communications)
- Microwave--GEO to Earth
- Laser--GEO to LEO, GEO to GEO

Data processing
- High speed, low power, fault tolerant
- User friendly interfaces
- Use of artificial intelligence

Robotics/automation
- Maintenance, checkout, servicing applications
- Unmanned rovers for lunar and Mars exploration

Power
- Multi-megawatt capability (10 MW for lunar resources development)

tion required for refueling and servicing space-based OTVs for high-rate traffic between Earth orbit and outlying outposts. Such a facility, acting as a demand buffer or accumulator, enhances the operational flexibility of space-based vehicles. A 100-plus-man space base is considered a necessary prerequisite in the first century of the next millenium for some of the far-term options. It would serve also as life sciences simulator and basic construction camp for large orbiting colonies of the more distant future. Other technology requirements of space habitation involve the geosynchronous space station, the lunar orbiting station, the lunar surface base, Mars base elements, a standard crew module (six-man) for an advanced manned orbital transfer vehicle (MOTV), and, eventually, probably a high-volume interorbital crew module (circa 500 passengers).

Among the more important engineering developments required for the long range, particularly for phase III, will be technology for space construction as opposed to space assembly. This involves on-site fabrication and finishing of construction elements as much as possible, minimizing prefab requirements on Earth and thus, presumably, both industrial effects on Earth's biosphere and transportation requirements. The use of extraterrestrial resources would also partially eliminate the need for Earth resources.

In the area of future propulsion, advanced chemical propulsion will be required for new or improved Shuttle transportation between Earth and LEO. Propulsion systems for this application may include high-performance/high-density propellant systems, high-performance hydrogen/oxygen systems beyond the advanced Shuttle main engine (ASME), dual-fuel engines on mixed-mode vehicles, composite systems (airbreathers/rocket), and advanced concepts such as fluoridized oxygen (Flox)/hydrocarbon and Flox/Monomethylhydrazine (MMH) systems, tripropellant engines (e.g., Li-F_2-H_2), and concepts using ozone, activated oxygen, atomic hydrogen, metallic hydrogen, activated helium, etc. The chief constraint on advanced chemical propulsion concepts for Earth-to-orbit flights will be environmental considerations concerned with prevention of damage or pollution of the atmosphere.

In nuclear propulsion, basic SCR technology on NERVA has essentially demonstrated flight readiness before development work was terminated in 1973. Nuclear propulsion beyond the basic SCR graphite engine of the NERVA type will need major development work (12 to 15 years from go-ahead) toward advanced solid-core concepts such as the fast-neutron-spectrum metall-clad engine with long operating life, for application to high-density transportation requirements in geolunar and heliocentric space. Core lifetimes of up to ten years will have to be demonstrated. For long-distance trips of phase III type, GCR systems such as the twin-vortex and "light bulb" concepts are indicated for higher-performance efficiency and economy. Substantial technology programs would be required to demonstrate containment of the fissionable material to maintain core criticality, radiation shielding concepts, and ability to handle the extreme pressures and temperatures in the engine.

For remote-future scenarios, one may also envision some "exotic" drives that are largely speculative. They include such ideas as detonation propulsion (chemical/solid, chemical/liquid, and nuclear, "Orion"), solar sail, laser/fusion, laser/photon, pure photon, matter/antimatter annihilation, interstellar ramjet, and others. Little or nothing is known regarding the feasibility of most of them.

New technology developments in the area of life sciences will be essential to the introduction of permanent human existence in space. While there are certainly no fundamental barriers to the development of semiclosed and fully-closed environmental control and life support (EC/LS) systems, the technology for them clearly is not ready at present but must be developed and tested in Earth orbit for long-duration missions and large groups of people. For large bases on the Moon and other celestial bodies, agricultural systems will replace stored food or onboard food regeneration. Much research is necessary, however, particularly on the effects of gravity and zero gravity on agricultural systems.

Major questions of life sciences in space include: How does the body adapt to weightlessness after very long periods in space? Does it stop losing calcium from the skeleton, or, if not, are there measures that prevent this deterioration? Will we require artificial gravity? What are the long-term effects on man's vestibular system? What are the cumulative effects on long-term exposure to cosmic radiation? How can one readapt the cardiovascular system prior to entry into gravitational fields after long stays in space? What exactly are the neuro-endocrine (hormonal) processes before, during, and after adaptation to weightlessness? Will the body lose its immunity to infectious diseases and how can we prevent it? Is the formation of antibodies and the transplantation-rejection mechanism affected by gravity or its absence? How can doctors practice medicine, surgery and dentistry in zero gravity? What does living in weightlessness do to the aging process? To procreation? To growth? To the psyche?

There are countless important questions like these, but nearly all of them can be attacked by a thorough-going program of biomedical research in LEO between now and 2000. We need a large statistical population for these studies; at least a hundred "average" men and women and eventually thousands must be exposed to weightlessness for many months before we can proceed to the megaprojects of the more distant future, such as space colonies. Work in ground laboratories will also be required to a large extent.

Of particular importance for long-duration flights--and a virtual blank area at present--are questions of group dynamics, psychology, and sociology of spacefarers other than highly trained, hand-picked professional astronauts. What are the optimal combinations of people for such missions, and how should the crews be structured? How should they be selected, trained, motivated, and indoctrinated? What will the group dynamics be during long confinement in a spaceship, in a closed colony?

In the area of operations technology, new approaches to space flight operations must replace the way we have been doing business in the past. For space-based vehicles and habitats, the data flow to and from Earth may become prohibitive if they are not provided with autonomous systems. Better data management, data compression concepts, and large-bandwidth communication systems will certainly not suffice. The degree of autonomy will have to be determined on a case-to-case basis, commensurate with new methods of communication and safety considerations. Techniques must be developed for routine countdown and launch of Earth-to-orbit shuttles and heavy-lift vehicles at high traffic rates, and to recover and refurbish returned elements of the transportation system at equally high rates. Earth-orbital operations will include new assembly and construction techniques, servicing of hardware not designed for return to Earth, and space-based tanking, countdown, and launching operations. The area of extraterrestrial mining and refining operations is a complete unknown now. It will probably require new engineering specialization to develop this technology to a point where it may become economically feasible.

Last but not least in this discussion of technological prerequisites are advanced management concepts that the megaprograms of the long-range future will require. Management has always been of importance when human beings have tried to achieve a common goal within a certain time span and with available resources, all the way back to the construction of the Egyptian pyramids or the Great Wall of China. But management systems have continually changed to keep pace with the needs of the program. What are some of the probable needs of the future?

Undoubtedly, current management concepts such as the "lead center" approach of the Space Shuttle and the space station program will have to change radically. Apollo- and Shuttle-type thinking will not necessarily apply to the megaprograms of the future, and even NASA's phased project planning approach may become inappropriate. Program managers will have to deal with very large undertakings requiring extensive commitments of resources, energy, and large portions of gross national and global products over long periods of time. Unlike the Apollo Program, a manned Mars program, in order to make sense, will not just focus on achieving first landing and return of some humans but must also include planning for subsequent sustaining operations.

Moreover, these megaprograms will involve a multitude of governments (which will change during the development period) and government agencies, industrial firms and combines, and other organizations on an international basis, funds in the multi-billion-dollar category, complex technologies that may reach beyond the contemporary ("then") state-of-the-art, large forces of manpower, and construction of extensive and highly specialized facilities on the Earth and in space.

Transcending the Horizon: The Humanization of Space

While space settlement as initial goal and prime objective is clearly not an answer to our current global problems(32,4), mankind's expansion into space will become unavoidable in the long run for sheer survival.(33) With the development of new worlds in space, humanity on Earth in the third millenium can win on all fronts: in the nearer term, the utilization of space and its manifold resources on Earth in the service of human needs; in the longer term, the progressive move of mankind into space in the course of natural evolution.

Thus, in our planning of the next steps in space, we must be responsive to mankind's near-term needs and wants, but we also have an important obligation to future generations to keep the growth options of space open at this time when we are but at the threshold of new frontiers. Our planning of a long-range space program beyond the space station, while based on essentially utilitarian aspects of the near-term, should not lose sight of the more humanistically significant long term.

In anticipating the long-range future beyond the space station, as attempted in this chapter, it is important to emphasize again that any limits that people can now set are most likely too "linear" and, thus, naive. The opportunity for the growth of new worlds in space, with all of the advantages that humans have gained from fresh starts in creating new societies, appears to be among the most intriguing potentials of space.(34)

A space program that has the potential, like the space station in its "beachhead" role, of exhibiting that there are no visible limits to humanity's future in the universe, could be a most important help in reviving faith in the idea of progress. Reconciliation of the two seemingly opposed views of solving pressing problems on Earth in the near term and of providing for man's humanistic growth in space in the longer term should therefore be foremost among our concerns. With few exceptions, it is difficult to imagine anything more relevant to our current problems.

Acknowledgment

William M. Brown of Hudson Institute in Indianapolis, Ind., has contributed to this paper by graciously supplying certain reference material, attributed as Ref. 5. His valuable assistance is gratefully acknowledged. (Dr. Brown's highly seminal report, Ref. 35, an update of an earlier Hudson Institute study of space futures, appeared after this chapter was written.)

Opinions contained in this chapter are those of the author and do not necessarily represent official opinion or policy of the National Aeronautics and Space Administration.

References

[1]Logsdon, J.M., "The Apollo Decision in Historical Perspective," in "Apollo--Ten Years Since Tranquillity Base," edited by Hallion, R.P. and Crouch, T.D.,National Air and Space Museum, Smithsonian Institution, Washington, D.C., 1979.

[2]von Puttkamer, J., "Space Shuttle--And the Future in Space," Futurics, Vol. 5, No. 3, 1981, pp. 271-277.

[3]von Puttkamer, J., "Roles and Needs of Man in Space," Journal of the British Interplanetary Society, Vol. 30, No. 8, 1983, pp. 351-356.

[4]von Puttkamer, J., "The Industrialization of Space: Transcending the Limits to Growth," in "Global Solutions--Innovative Approaches to Worlds Problems," World Future Society, Bethesda, Md., 1984.

[5]Brown, W.M., Private Communication (draft contribution).

[6]"A Forecast of Space Technology, 1980-2000," National Aeronautics and Space Administration, NASA SP-387, Jan. 1976.

[7]"Planetary Exploration through Year 2000, A Core Program," report by the Solar System Exploration Committee of the NASA Advisory Council, Washington, D.C., 1983.

[8]"Autonomy and the Human Element in Space," Executive Summary, Report of the 1983 NASA/ASEE Summer Faculty Workshop, Stanford University, Calif., 1983.

[9]Ehricke, K.A., "Space Tourism," AAS Preprint 67-127, American Astronautical Society 13th Annual Meeting, Dallas, Texas, May 1967.

[10]Bekey, I., Mayer, H.L., and Wolfe, M.G., "Advanced Space System Concepts and Their Orbital Support Needs (1980-2000)," Report ATR-76(7365)-1, Vols. 1-4. Contract NASW-2727, The Aerospace Corporation, El Segundo, Calif., April 1976.

[11]"Project Horizon. A US Army Study for the Establishment of a Lunar Military Outpost," Army Missile Command, Redstone Arsenal, Alabama, June 8, 1959.

[12]"Advanced Lunar Transportation Study," NASA-CR-50767. Lockheed Missiles and Space Company, Sunnyvale, Calif., Jan. 28, 1963.

[13]"Initial Concept for a Lunar Base," NASA-CR-56427. Boeing Company, Seattle, Wash., Sept. 15, 1963.

[14]"Lunar Base Synthesis Study," NASA Contract NAS8-26145, North American Rockwell, Downey, Calif., SD-71-477, 1971.

[15]von Puttkamer, J., "Spaceflight: A Force of Peace?" Futurics, Vol. 6, No. 1, 1982, pp. 35-39.

[16]von Puttkamer, J., "Spaceflight--Potential for Peace," Paper presented at World Future Society 4th General Assembly, Communications and the Future, July 18-22, 1982, Washington, D.C.

[17]Hornby, H., "Manned Planetary Missions--Exploration of Mars and Venus," in Planetary Studies, NASA Ames Research Center, published by NASA Headquarters, Washington, D.C., June 26, 1964.

[18]Callies, G.M. et al., "Alternative Mission Modes Study," final report, NASA-CR-89534, TRW Systems Group, Sept. 1967.

[19]Brown, H. and Coates, G.L., "Study of Low-Acceleration Space Transportation Systems," NASA-CR-89494, General Electric Co., Philadelphia, Penn., July 15, 1966.

[20]Jones, A.L., "Study of Manned Planetary Flyby Missions based on Saturn/Apollo Systems," NASA-CR-88293, North American Aviation, Inc., Downey, Calif., Aug. 1967.

[21]"Spacecraft Propulsion Systems for Manned Mars and Venus Missions," NASA-CR-83888, Martin Company, Denver, Colo. July 7, 1965.

[22]"Early Manned Interplanetary Missions Study," NASA-CR-67281, Lockheed Missiles and Space Company, Sunnyvale, Calif., Oct. 2, 1963.

[23]Jones, A.L.and McRae, W.V., "Manned Mars Landing and Return Mission Study," NASA-CR-59227, North American Aviation, Inc., Downey, Calif., April 1964.

[24]Ragsac, R.V., "Manned Interplanetary Missions," NASA-CR-56762, Lockheed Missiles and Space Co., Sunnyvale, Calif. Jan. 28, 1964.

[25]Ehricke, K.A., "A Study of Manned Interplanetary Missions-EMPIRE Follow-On," NASA-CR-56090, General Dynamics Company, San Diego, Calif., Jan. 31, 1964.

[26]Steinhoff, E.A., "Phobos Space Station--As A Possible Approach to Scientific Exploration of the Planet Mars," in From Peenemunde to Outer Space--A Volume of Papers Commemorating the Fiftieth Birthday of Wernher von Braun. NASA Marshall Space Flight Center, Huntsville, Ala., March 23, 1962.

[27]"Solar Power Satellites," report of the Office of Technology Assessment, US Congress, Aug. 1981.

[28]Priest, C.C., Nixon, R.F., and Rice, E.E., "Space Disposal of Nuclear Wastes," AIAA Journal of Astronautics and Aeronautics, Vol. 18, April 1980, pp. 26-35.

[29]"Advanced Automation for Space Missions," Proceedings of the 1980 NASA/ASEE Summer Study, University of Santa Clara, Calif., 1980.

[30]"Outlook for Space". NASA-SP-386, 1976.

[31]von Puttkamer, J., "Developing Space Occupancy: Perspectives on NASA Future Space Program Planning," Journal of the British Interplanetary Society, Vol. 29, March 1976, pp. 147-173, 1976.

[32]von Puttkamer,J., "The Next 25 Years: Industrialization of Space--Rationale for Planning," Journal of the British Interplanetary Society, Vol. 30, July 1977, pp. 257-264.

[33]von Puttkamer, J., "A Matter of Survival," Space World, Vol. U-7-247, National Space Institute, Washington, D.C., July 1984.

[34]von Puttkamer, J., "On Humanity's Role in Space," Futurics, Vol. 2, No. 3, 1978, pp. 76-81.

[35]Brown, William M., "Space Ventures and Society; Long-Term Perspectives," Report HI-3731/2-RR. NASA-Contract NASW-3724, Hudson Institute, Indianapolis, Indiana, May 31, 1985.

Author Index for Volume 99

PROGRESS IN ASTRONAUTICS AND AERONAUTICS SERIES VOLUMES

VOLUME TITLE/EDITORS

*1. **Solid Propellant Rocket Research** (1960)
Martin Summerfield
Princeton University

*2. **Liquid Rockets and Propellants** (1960)
Loren E. Bollinger
The Ohio State University
Martin Goldsmith
The Rand Corporation
Alexis W. Lemmon Jr.
Battelle Memorial Institute

*3. **Energy Conversion for Space Power** (1961)
Nathan W. Snyder
Institute for Defense Analyses

*4. **Space Power Systems** (1961)
Nathan W. Snyder
Institute for Defense Analyses

*5. **Electrostatic Propulsion** (1961)
David B. Langmuir
Space Technology Laboratories, Inc.
Ernst Stuhlinger
NASA George C. Marshall Space Flight Center
J.M. Sellen Jr.
Space Technology Laboratories, Inc.

*6. **Detonation and Two-Phase Flow** (1962)
S.S. Penner
California Institute of Technology
F.A. Williams
Harvard University

*7. **Hypersonic Flow Research** (1962)
Frederick R. Riddell
AVCO Corporation

*8. **Guidance and Control** (1962)
Robert E. Roberson
Consultant
James S. Farrior
Lockheed Missiles and Space Company

*9. **Electric Propulsion Development** (1963)
Ernst Stuhlinger
NASA George C. Marshall Space Flight Center

*10. **Technology of Lunar Exploration** (1963)
Clifford I. Cummings and Harold R. Lawrence
Jet Propulsion Laboratory

*11. **Power Systems for Space Flight** (1963)
Morris A. Zipkin and Russell N. Edwards
General Electric Company

*12. **Ionization in High-Temperature Gases** (1963)
Kurt E. Shuler, Editor
National Bureau of Standards
John B. Fenn, Associate Editor
Princeton University

*13. **Guidance and Control—II** (1964)
Robert C. Langford
General Precision Inc.
Charles J. Mundo
Institute of Naval Studies

*14. **Celestial Mechanics and Astrodynamics** (1964)
Victor G. Szebehely
Yale University Observatory

*15. **Heterogeneous Combustion** (1964)
Hans G. Wolfhard
Institute for Defense Analyses
Irvin Glassman
Princeton University
Leon Green Jr.
Air Force Systems Command

*16. **Space Power Systems Engineering** (1966)
George C. Szego
Institute for Defense Analyses
J. Edward Taylor
TRW Inc.

*17. **Methods in Astrodynamics and Celestial Mechanics** (1966)
Raynor L. Duncombe
U.S. Naval Observatory
Victor G. Szebehely
Yale University Observatory

*18. **Thermophysics and Temperature Control of Spacecraft and Entry Vehicles** (1966)
Gerhard B. Heller
NASA George C. Marshall Space Flight Center

*19. **Communication Satellite Systems Technology** (1966)
Richard B. Marsten
Radio Corporation of America

*Out of print.

***20. Thermophysics of Spacecraft and Planetary Bodies: Radiation Properties of Solids and the Electromagnetic Radiation Environment in Space** (1967)
Gerhard B. Heller
NASA George C. Marshall Space Flight Center

***21. Thermal Design Principles of Spacecraft and Entry Bodies** (1969)
Jerry T. Bevans
TRW Systems

***22. Stratospheric Circulation** (1969)
Willis L. Webb
Atmospheric Sciences Laboratory, White Sands, and University of Texas at El Paso

***23. Thermophysics: Applications to Thermal Design of Spacecraft** (1970)
Jerry T. Bevans
TRW Systems

24. Heat Transfer and Spacecraft Thermal Control (1971)
John W. Lucas
Jet Propulsion Laboratory

25. Communication Satellites for the 70's: Technology (1971)
Nathaniel E. Feldman
The Rand Corporation
Charles M. Kelly
The Aerospace Corporation

26. Communication Satellites for the 70's: Systems (1971)
Nathaniel E. Feldman
The Rand Corporation
Charles M. Kelly
The Aerospace Corporation

27. Thermospheric Circulation (1972)
Willis L. Webb
Atmospheric Sciences Laboratory, White Sands, and University of Texas at El Paso

28. Thermal Characteristics of the Moon (1972)
John W. Lucas
Jet Propulsion Laboratory

29. Fundamentals of Spacecraft Thermal Design (1972)
John W. Lucas
Jet Propulsion Laboratory

30. Solar Activity Observations and Predictions (1972)
Patrick S. McIntosh and Murray Dryer
Environmental Research Laboratories, National Oceanic and Atmospheric Administration

31. Thermal Control and Radiation (1973)
Chang-Lin Tien
University of California at Berkeley

32. Communications Satellite Systems (1974)
P.L. Bargellini
COMSAT Laboratories

33. Communications Satellite Technology (1974)
P.L. Bargellini
COMSAT Laboratories

34. Instrumentation for Airbreathing Propulsion (1974)
Allen E. Fuhs
Naval Postgraduate School
Marshall Kingery
Arnold Engineering Development Center

35. Thermophysics and Spacecraft Thermal Control (1974)
Robert G. Hering
University of Iowa

36. Thermal Pollution Analysis (1975)
Joseph A. Schetz
Virginia Polytechnic Institute

37. Aeroacoustics: Jet and Combustion Noise; Duct Acoustics (1975)
Henry T. Nagamatsu, Editor
General Electric Research and Development Center
Jack V. O'Keefe, Associate Editor
The Boeing Company
Ira R. Schwartz, Associate Editor
NASA Ames Research Center

38. Aeroacoustics: Fan, STOL, and Boundary Layer Noise; Sonic Boom; Aeroacoustic Instrumentation (1975)
Henry T. Nagamatsu, Editor
General Electric Research and Development Center
Jack V. O'Keefe, Associate Editor
The Boeing Company
Ira R. Schwartz, Associate Editor
NASA Ames Research Center

39. Heat Transfer with Thermal Control Applications (1975)
M. Michael Yovanovich
University of Waterloo

40. Aerodynamics of Base Combustion (1976)
S.N.B. Murthy, Editor
Purdue University
J.R. Osborn, Associate Editor
Purdue University
A.W. Barrows and J.R. Ward, Associate Editors
Ballistics Research Laboratories

41. Communications Satellite Developments: Systems (1976)
Gilbert E. LaVean
Defense Communications Agency
William G. Schmidt
CML Satellite Corporation

42. Communications Satellite Developments: Technology (1976)
William G. Schmidt
CML Satellite Corporation
Gilbert E. LaVean
Defense Communications Agency

43. Aeroacoustics: Jet Noise, Combustion and Core Engine Noise (1976)
Ira R. Schwartz, Editor
NASA Ames Research Center
Henry T. Nagamatsu, Associate Editor
General Electric Research and Development Center
Warren C. Strahle, Associate Editor
Georgia Institute of Technology

44. Aeroacoustics: Fan Noise and Control; Duct Acoustics; Rotor Noise (1976)
Ira R. Schwartz, Editor
NASA Ames Research Center
Henry T. Nagamatsu, Associate Editor
General Electric Research and Development Center
Warren C. Strahle, Associate Editor
Georgia Institute of Technology

45. Aeroacoustics: STOL Noise; Airframe and Airfoil Noise (1976)
Ira R. Schwartz, Editor
NASA Ames Research Center
Henry T. Nagamatsu, Associate Editor
General Electric Research and Development Center
Warren C. Strahle, Associate Editor
Georgia Institute of Technology

46. Aeroacoustics: Acoustic Wave Propagation; Aircraft Noise Prediction; Aeroacoustic Instrumentation (1976)
Ira R. Schwartz, Editor
NASA Ames Research Center
Henry T. Nagamatsu, Associate Editor
General Electric Research and Development Center
Warren C. Strahle, Associate Editor
Georgia Institute of Technology

47. Spacecraft Charging by Magnetospheric Plasmas (1976)
Alan Rosen
TRW Inc.

48. Scientific Investigations on the Skylab Satellite (1976)
Marion I. Kent and Ernst Stuhlinger
NASA George C. Marshall Space Flight Center
Shi-Tsan Wu
The University of Alabama

49. Radiative Transfer and Thermal Control (1976)
Allie M. Smith
ARO Inc.

50. Exploration of the Outer Solar System (1976)
Eugene W. Greenstadt
TRW Inc.
Murray Dryer
National Oceanic and Atmospheric Administration
Devrie S. Intriligator
University of Southern California

51. Rarefied Gas Dynamics, Parts I and II (two volumes) (1977)
J. Leith Potter
ARO Inc.

52. Materials Sciences in Space with Application to Space Processing (1977)
Leo Steg
General Electric Company

53. Experimental Diagnostics in Gas Phase Combustion Systems (1977)
Ben T. Zinn, Editor
Georgia Institute of Technology
Craig T. Bowman, Associate Editor
Stanford University
Daniel L. Hartley, Associate Editor
Sandia Laboratories
Edward W. Price, Associate Editor
Georgia Institute of Technology
James G. Skifstad, Associate Editor
Purdue University

54. Satellite Communications: Future Systems (1977)
David Jarett *TRW Inc.*

55. Satellite Communications: Advanced Technologies (1977)
David Jarett
TRW Inc.

56. Thermophysics of Spacecraft and Outer Planet Entry Probes (1977)
Allie M. Smith *ARO Inc.*

57. Space-Based Manufacturing from Nonterrestrial Materials (1977)
Gerard K. O'Neill, Editor
Princeton University
Brian O'Leary, Assistant Editor
Princeton University

58. Turbulent Combustion (1978)
Lawrence A. Kennedy
State University of New York at Buffalo

59. Aerodynamic Heating and Thermal Protection Systems (1978)
Leroy S. Fletcher
University of Virginia

60. Heat Transfer and Thermal Control Systems (1978)
Leroy S. Fletcher
University of Virginia

61. Radiation Energy Conversion in Space (1978)
Kenneth W. Billman
NASA Ames Research Center

62. Alternative Hydrocarbon Fuels: Combustion and Chemical Kinetics (1978)
Craig T. Bowman
Stanford University
Jorgen Birkeland
Department of Energy

63. Experimental Diagnostics in Combustion of Solids (1978)
Thomas L. Boggs
Naval Weapons Center
Ben T. Zinn
Georgia Institute of Technology

64. Outer Planet Entry Heating and Thermal Protection (1979)
Raymond Viskanta
Purdue University

65. Thermophysics and Thermal Control (1979)
Raymond Viskanta
Purdue University

66. Interior Ballistics of Guns (1979)
Herman Krier
University of Illinois at Urbana-Champaign
Martin Summerfield
New York University

67. Remote Sensing of Earth from Space: Role of "Smart Sensors" (1979)
Roger A. Breckenridge
NASA Langley Research Center

68. Injection and Mixing in Turbulent Flow (1980)
Joseph A. Schetz
Virginia Polytechnic Institute and State University

69. Entry Heating and Thermal Protection (1980)
Walter B. Olstad
NASA Headquarters

70. Heat Transfer, Thermal Control, and Heat Pipes (1980)
Walter B. Olstad
NASA Headquarters

71. Space Systems and Their Interactions with Earth's Space Environment (1980)
Henry B. Garrett and Charles P. Pike
Hanscom Air Force Base

72. Viscous Flow Drag Reduction (1980)
Gary R. Hough
Vought Advanced Technology Center

73. Combustion Experiments in a Zero-Gravity Laboratory (1981)
Thomas H. Cochran
NASA Lewis Research Center

74. Rarefied Gas Dynamics, Parts I and II (two volumes) (1981)
Sam S. Fisher
University of Virginia at Charlottesville

75. Gasdynamics of Detonations and Explosions (1981)
J.R. Bowen
University of Wisconsin at Madison
N. Manson
Université de Poitiers
A.K. Oppenheim
University of California at Berkeley
R.I. Soloukhin
Institute of Heat and Mass Transfer, BSSR Academy of Sciences

76. Combustion in Reactive Systems (1981)
J.R. Bowen
University of Wisconsin at Madison
N. Manson
Université de Poitiers
A.K. Oppenheim
University of California at Berkeley
R.I. Soloukhin
Institute of Heat and Mass Transfer, BSSR Academy of Sciences

77. Aerothermodynamics and Planetary Entry (1981)
A.L. Crosbie
University of Missouri-Rolla

78. Heat Transfer and Thermal Control (1981)
A.L. Crosbie
University of Missouri-Rolla

79. Electric Propulsion and Its Applications to Space Missions (1981)
Robert C. Finke
NASA Lewis Research Center

80. Aero-Optical Phenomena (1982)
Keith G. Gilbert and
Leonard J. Otten
Air Force Weapons Laboratory

81. Transonic Aerodynamics (1982)
David Nixon
Nielsen Engineering & Research, Inc.

82. Thermophysics of Atmospheric Entry (1982)
T.E. Horton
The University of Mississippi

83. Spacecraft Radiative Transfer and Temperature Control (1982)
T.E. Horton
The University of Mississippi

84. Liquid-Metal Flows and Magnetohydrodynamics (1983)
H. Branover
Ben-Gurion University of the Negev
P.S. Lykoudis
Purdue University
A. Yakhot
Ben-Gurion University of the Negev

85. Entry Vehicle Heating and Thermal Protection Systems: Space Shuttle, Solar Starprobe, Jupiter Galileo Probe (1983)
Paul E. Bauer
McDonnell Douglas Astronautics Company
Howard E. Collicott
The Boeing Company

86. Spacecraft Thermal Control, Design, and Operation (1983)
Howard E. Collicott
The Boeing Company
Paul E. Bauer
McDonnell Douglas Astronautics Company

87. Shock Waves, Explosions, and Detonations (1983)
J.R. Bowen
University of Washington
N. Manson
Université de Poitiers
A.K. Oppenheim
University of California at Berkeley
R.I. Soloukhin
Institute of Heat and Mass Transfer, BSSR Academy of Sciences

88. Flames, Lasers, and Reactive Systems (1983)
J.R. Bowen
University of Washington
N. Manson
Université de Poitiers
A.K. Oppenheim
University of California at Berkeley
R.I. Soloukhin
Institute of Heat and Mass Transfer, BSSR Academy of Sciences

89. Orbit-Raising and Maneuvering Propulsion: Research Status and Needs (1984)
Leonard H. Caveny
Air Force Office of Scientific Research

90. Fundamentals of Solid-Propellant Combustion (1984)
Kenneth K. Kuo
The Pennsylvania State University
Martin Summerfield
Princeton Combustion Research Laboratories, Inc.

91. Spacecraft Contamination: Sources and Prevention (1984)
J.A. Roux
The University of Mississippi
T.D. McCay
NASA Marshall Space Flight Center

92. Combustion Diagnostics by Nonintrusive Methods (1984)
T.D. McCay
NASA Marshall Space Flight Center
J.A. Roux
The University of Mississippi

93. The INTELSAT Global Satellite System (1984)
Joel Alper
COMSAT Corporation
Joseph Pelton
INTELSAT

94. Dynamics of Shock Waves, Explosions, and Detonations (1984)
J.R. Bowen
University of Washington
N. Manson
Universite de Poitiers
A.K. Oppenheim
University of California
R.I. Soloukhin
Institute of Heat and Mass Transfer, BSSR Academy of Sciences

95. Dynamics of Flames and Reactive Systems (1984)
J.R. Bowen
University of Washington
N. Manson
Universite de Poitiers
A.K. Oppenheim
University of California
R.I. Soloukhin
Institute of Heat and Mass Transfer, BSSR Academy of Sciences

96. Thermal Design of Aeroassisted Orbital Transfer Vehicles (1985)
H.F. Nelson
University of Missouri-Rolla

97. Monitoring Earth's Ocean, Land, and Atmosphere from Space – Sensors, Systems, and Applications (1985)
Abraham Schnapf
Aerospace Systems Engineering

98. Thrust and Drag: Its Prediction and Verification (1985)
Eugene E. Covert
Massachusetts Institute of Technology
C.R. James
Vought Corporation
William F. Kimzey
Sverdrup Technology AEDC Group
George K. Richey
U.S. Air Force
Eugene C. Rooney
U.S. Navy Department of Defense

99. Space Stations and Space Platforms — Concepts, Design, Infrastructure, and Uses (1985)
Ivan Bekey
Daniel Herman
NASA Headquarters

(Other Volumes are planned.

TL 507 .P75 1985
Space stations and spa
platforms
ce
MERCYHURST COLLEGE
HAMMERMILL LIBRARY
ERIE, PA 16546
MAY 4 1987